Dating Buildings and Landscapes with Tree-Ring Analysis

This book presents guidance, theory, methodologies, and case studies for analyzing tree rings to accurately date and interpret historic buildings and landscapes.

Written by two long-time practitioners in the field of dendrochronology, the research is grounded in the fieldwork data of approximately 200 structures and landscapes. By scientifically analyzing the tree rings of historic timbers, preservationists can obtain valuable information about construction dates, interpret the evolution of landscapes and buildings over time, identify species and provenance, and gain insight into the species matrix of local forests. Authors Darrin L. Rubino and Christopher Baas demonstrate, through full-color illustrated case studies and methodologies, how this information can be used to interpret the history of buildings and landscapes and assist preservation decision-making.

Over 1,000 samples obtained from more than 40 buildings, including high style houses, vernacular log houses, and timber frame barns, are reported. This book will be particularly relevant for students, instructors, and professional readers interested in historic preservation, cultural landscapes, museum studies, archaeology, and dendrochronology globally.

Darrin L. Rubino is a Professor of Biology at Hanover College, USA. He joined the Hanover faculty in 2002. He is a graduate of Thiel College (1995) and earned his Masters from Clarion University of Pennsylvania (1997) and his PhD from Ohio University (2002). He has been a member of the Indiana Academy of Science since 2002. In addition to serving as the President of the Academy, Darrin served as Chair and Vice Chair of the Academy's Botany Section on several occasions. He became a Fellow of the Academy in 2013. Darrin is a forest ecologist and studies wood decay, vascular plant community composition, non-native plants, and woody debris (dead and down trees). His main research interest is dendrochronology, the study of tree rings. He uses patterns of tree rings to understand the influence of historic human impacts and natural processes on forest structure, function, and composition. In 2007, 2012, and 2018 Darrin was awarded the Arthur and Ilene Baynham Award for Outstanding Teaching at Hanover College, and he won the College's Daryl R. Karns Award for Scholarly and Creative Activity in 2014.

Christopher Baas is an Associate Professor in the Department of Landscape Architecture at Ball State University, USA, and a Registered Landscape Architect. Following a 20-year career as a practicing Landscape Architect, he joined the Ball State University faculty in 2008. He is a graduate of Ball State University (1988) and earned his Masters from the University of Wisconsin–Madison (1996). He teaches planning and design studios, as well as historic preservation and landscape restoration courses. Christopher's scholarly focus is documenting and interpreting historic cultural landscapes. He is interested in the work of early 20th-century landscape architect George Kessler's park and boulevard design of Hoosier cities, and Danish landscape architect Jen Jensen's conservation of the Indiana dunes. Baas is also active in the documentation and interpretation of southeast Indiana's 19th-century culture of growing, pressing, and shipping hay.

Dating Buildings and Landscapes with Tree-Ring Analysis

An Introduction with Case Studies

Darrin L. Rubino and Christopher Baas

Routledge
Taylor & Francis Group
LONDON AND NEW YORK

First published 2019
by Routledge
2 Park Square, Milton Park, Abingdon, Oxon OX14 4RN

and by Routledge
52 Vanderbilt Avenue, New York, NY 10017

Routledge is an imprint of the Taylor & Francis Group, an informa business

British Library Cataloguing-in-Publication Data
A catalogue record for this book is available from the British Library

Library of Congress Cataloging-in-Publication Data
Names: Rubino, Darrin L., author. | Baas, Christopher, author.
Title: Dating buildings and landscapes with tree-ring analysis : an introduction with case studies / Darrin L. Rubino and Christopher Baas.
Description: New York, NY : Routledge, 2019. | Includes bibliographical references and index.
Identifiers: LCCN 2018061196 | ISBN 9781138503953 (hbk) | ISBN 9781138503960 (pbk) | ISBN 9781315145679 (ebk)
Subjects: LCSH: Dendrochronology. | Landscape archaeology. | Archaeological dating. | Historic buildings—Dating.
Classification: LCC CC78.3 .R83 2019 | DDC 930.1/0285—dc23
LC record available at https://lccn.loc.gov/2018061196

ISBN: 978-1-138-50395-3 (hbk)
ISBN: 978-1-138-50396-0 (pbk)
ISBN: 978-1-315-14567-9 (ebk)

Typeset in Charter and FS Albert
by Apex CoVantage, LLC

Visit the eResource: www.routledge.com/9781138503960

Contents

Figures

Tables

Acknowledgments

We acknowledge our friends and colleagues who share their time and interest in our work. For our analysis of sites in southeastern Indiana: Martha Bladen, Director, Switzerland County Historical Society; and Herschel Brichto, ERJO Restoration. In New Harmony, Indiana: Amanda Bryden, State Historic Sites Collections Manager; and Meagan Patterson, Collections Assistant. In Fort Wayne, Indiana: Jill McDevitt, ARCH Executive Director; and Fort Wayne, Indiana, preservationist Jill Downs. For our work at the *Musée de Venoge*: Donna Weaver and Pat Cunningham. For Jefferson County, Indiana: Stan Totten and Bill Jackson. For our chapter on teaching dendrochronology, we thank Jonathan Spodek, Professor of Architecture, Director of the Graduate Program in Historic Preservation; and preservationist and homeowner Jeff Harmon for sharing documents from his family history. We thank Amy Johnson, State Archaeologist, Indiana Division of Historic Preservation and Archaeology, who has been supportive of promoting and disseminating our work.

We acknowledge the institutions granting permissions to use amazing images from their collections: Allen County Public Library; Fort Wayne History Center; Old Economy Village, Pennsylvania Historical and Museum Commission; Indiana State Museum and Historic Sites; and Gordon Weddle, Professor of Biology, Campbellsville University, Director of Clay Hill Memorial Forest. We thank Don Ruch, co-Editor of *Proceedings of the Indiana Academy Science*, for use of our journal article in the text. We also thank Iain Banks and Tony Pollard (Editors of *Journal of Conflict Archaeology*) and Taylor and Francis (https://www.tandfonline.com/) for granting us permission to include content from the article "The fort in the doctor's house: using tree-ring growth patterns to discover historic Fort Wayne, Indiana" by C. Baas, T.N. Davis, and D.L. Rubino (*Journal of Conflict Archaeology* 13(2): 97–116). We thank co-author Taylor N. Davis for allowing us to include portions of this article in the book.

We acknowledge our students, colleagues, and friends at Hanover College. Students Gina Wyn, Cassie (Morris) Lothery, M. Ross Alexander, Anna (Selby) Glowinski, Jon Ward, Lindsay Beasley, Rachel Turner, Taylor Davis, Brian Lawrence, Nate Minyard, Ben Lye, Kyra Graham, Paul Killian, Kim Fong, Sarah Biles, and Evie Sehr. DLR thanks the Hanover Faculty Development Committee and the Patricia Walne Student Research Fund. DLR thanks Kelly Joyce, Director of the Duggan Library at Hanover College, and Dr. Sarah McNair Vosmeier for tracking down hard to find literature. Debbie Quinn and Celeste Sutter provided constant logistical support.

We acknowledge our students, colleagues, and friends at Ball State University. CB thanks students who have assisted with fieldwork and graphic presentation of our projects: Kevin Henn, Pei Li, Daniel Liggett, Weilun Xia, Bixia Zhou, and Yan Zhou. Thank you to colleagues from the Department of Landscape Architecture and the College of Architecture and Planning, especially Joe Blalock for the artwork shown in Figures 3.5 and 4.3. Many thanks to Angela Gibson, Geographic Information Systems Specialist, and Melissa Gentry, Map Collections Assistant, Ball State University Libraries. Thank you Kevin Nolan, Director and Senior Archaeologist, Applied Anthropology Laboratories. Thank you Pan Zhang, Department of Landscape Architecture, Harbin Institute of Technology, Harbin, China.

We are thankful to all of the property managers and owners that allowed us to sample their historic buildings. This book would not have been possible without your assistance and time.

We are, most of all, thankful for the love, patience, and support of our families, especially Michelle, Kathryn, Russell, Henry, and Paul.

To our readers, it is our sincerest hope that this book opens up new perspectives in both science and history.

Conventions

Throughout the text we present the size of structures in feet, the unit of length in which the builders most likely constructed the buildings. Tree-ring measurements are reported in millimeters, since metric units only are used in scientific investigations.

All of the data collected from historical sources that we utilize in interpreting historic buildings and landscapes are reported in non-metric units. We have converted these values in the text and provide the following for your reference:

1 inch = 2.54 centimeters = 25.4 millimeters
1 foot = 0.3 meters
1 acre = 0.4 hectares
1 pound = 0.45 kilograms

Plant taxonomy follows H.A. Gleason and A. Cronquist (1991). The scientific name for nearly all of the different species discussed in the text can be found in Table 3.1. The scientific name for all others will be found immediately after they are first mentioned in the text.

r = correlation coefficient or Pearson product-moment correlation coefficient.

Literature cited

Gleason, H.A. and A. Cronquist. 1991. *Manual of Vascular Plants of Northeastern United States and Adjacent Canada* (2nd edition). The New York Botanical Garden, Bronx, New York, 910 pages.

Chapter 1

Can you date my building

An introduction to tree-ring analysis for dating buildings and landscapes

1.1 Introduction

Trees are reliable timekeepers. Undoubtedly you have learned that trees can be aged by counting their rings – count the number of rings, and you know the age of the tree. Yes, trees can be aged by counting rings, but tree rings can tell more than just age (Figure 1.1). Trees are autobiographers. A tree records in its growth rings the annual layers of wood formed each year, a story written in one-year-long chapters. The size of a tree ring in a given year is determined by a variety of factors experienced by the tree in that year (amount of precipitation, temperature, light, etc.). Reading a tree's "story" requires a researcher to translate unique patterns in growth (patterns of large and small tree rings) to gain a better understanding of what a tree has experienced over its lifetime. One can determine when a tree began growing, when it died, and how it reacted to changes in its environment. The science of tree rings is called dendrochronology (which roughly translates from its Greek roots to the study of time using trees). Using dendrochronological techniques (acquiring samples, prepping them for "reading," assigning accurate dates to individual rings, etc.) researchers have been able to ask and answer a multitude of questions. For example, tree rings have been used to investigate and/or date insect outbreaks and natural disasters (volcanoes, windstorms, and fires) and to study long- and short-term climatic variability.

The wide application of tree rings to investigate the natural world is a testament to the versatility of tree-ring science. Many subdisciplines of dendrochronology have evolved, such as dendroecology (using tree rings to explore ecological questions and phenomena) and dendroclimatology (tree-ring analysis of climate). The subdiscipline of dendroarchaeology relates to the use of tree-ring methodology to date historical wooden objects. The wooden object could be a piece of furniture, a panel used in a painting, a ship, a musical instrument, or a building (Baillie 1982, 1995; Speer 2010). This text focuses on how tree rings and tree-ring science can be used to date historically constructed buildings and the landscapes in which they are found. Tree rings offer a unique ability to establish construction dates because they offer annual resolution. In other words, dendrochronological techniques allow for actual calendar dates to be assigned to individual tree rings – tree rings are precise! Dating historic buildings and landscapes has been done since the 1920s and is one of many methods used to establish construction dates (Stokes and Smiley 1968; Dean 1997; Nash 1999; Taylor 1999).

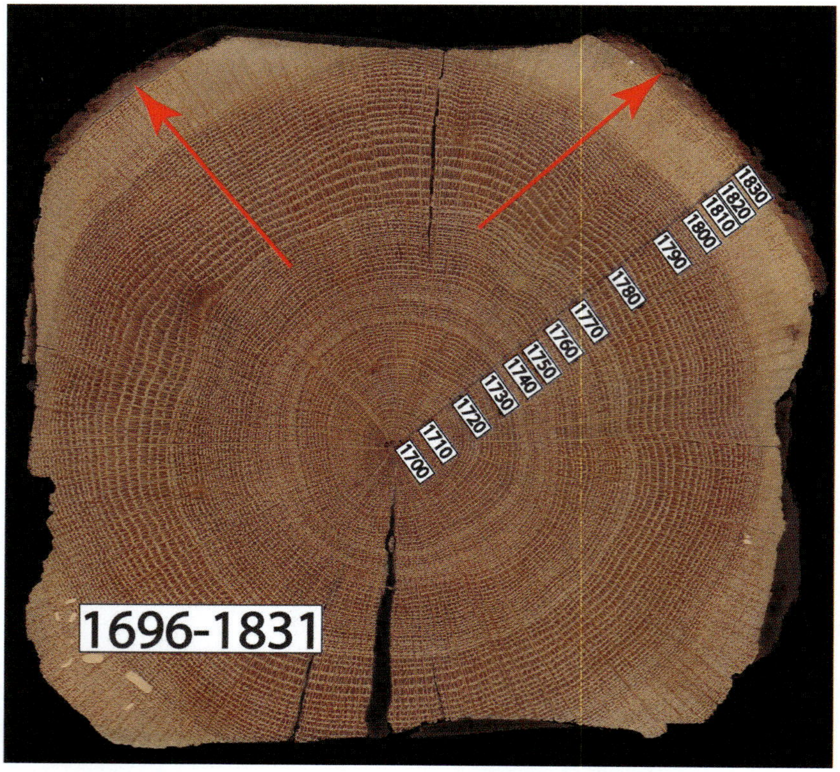

Figure 1.1
Cross section of a white oak beam from a southern Indiana log house. The tree rings in this timber were dated from 1696 to 1831; tree rings can be identified by the alternating light and dark woody tissue.

Note that not all of the rings have the same width. The variation in ring size is related to the growth conditions experienced by a tree in a given year. Depending on the amount of resources available for growth, such as light or water, trees will produce larger or smaller rings. Tree rings, therefore, record what a tree experienced while it was growing. Since the last ring in this timber is next to the bark (arrow), we can determine that the tree was felled in 1831 and subsequently was used for construction.

Source: Authors.

1.2 An overview: can you date my building?

To introduce the reader to this text, we thought we would begin with a commonly asked question: "Can you date my building?" We are often asked by landowners, museum directors, and private citizens if we can provide construction dates for their buildings using tree-ring analysis. They are often initially perplexed by our answer of "maybe." We have found that many stewards of historic buildings have heard of using tree-ring science to determine construction dates, but very rarely do they understand what is necessary for date determination. Once, while performing fieldwork on a barn, the owner came out after approximately two hours of work, asked us for the construction date, and wondered why we were still there! This experience was a bit of a revelation for us, and we realized that we needed to do a better job of explaining our methods and the potential and limitations of tree-ring analysis. What follows is an explanation of how we approach dating a structure and how we communicate with building owners before, during, and after an analysis is performed. We think it serves as a useful introduction to this text and gives the reader a basic summary of how we approach assigning dates to historical structures and landscapes.

1.2.1 Buildings and landscapes

The first step of dating a building's construction requires a thorough examination of the timbers used in its construction. Even though they may be quite old and were not constructed with the benefits of modern technology, historic buildings and landscapes are complex. Trees were harvested, and timbers were fashioned using time-tested techniques. How a building is constructed and its role in a landscape must be understood to fully interpret it. In Chapter 2, we provide a framework, a sort of set of guidelines, of how to interpret buildings and landscapes.

1.2.2 Understanding how trees grow

One needs to have a basic knowledge of the biological processes that lead to the formation of tree rings to truly appreciate the results obtained when dating the timbers of historically erected buildings. The production of annual growth increments by trees is a complex process that is influenced by a wide variety of factors. In Chapter 3, we introduce the reader to the basic botanical concepts needed to understand tree-ring formation, tree growth, and how wood is formed.

1.2.3 Basics of tree-ring science

Like the understanding of basic botany, an understanding of the basics of tree-ring science is necessary to interpret and appreciate the results of any investigation focusing on providing a construction date for a building. In Chapter 4, we discuss the basic principles of dendrochronology. Dendrochronological examination of a building can provide information about a building that is unavailable from any historic source. Without a thorough understanding of how dates are assigned to tree rings, a simple construction date will surely be a less powerful piece of information when one is trying to interpret or restore a historical structure.

1.2.4 Obtaining a construction date

Most building owners and managers are quite interested in how we arrive at a construction date for a building or landscape. This interest likely stems from both curiosity and also a need to know how their buildings will be impacted by the sampling process. A thorough understanding and description of how we sample structures, prepare samples for study, and analyze data is imperative; Chapter 5 provides a description of field and laboratory work required for determining a building's construction date. In order to date a building's construction, many factors must be taken into consideration. In other words, obtaining the "right" samples is paramount for dating a building's construction. Given how important these considerations are in the dating process of historic buildings and landscapes, we provide them here early in the text so that the reader can better appreciate what is necessary to answer the "Can you date my building?" question and to appreciate the important factors that must be kept in mind while reading subsequent chapters:

- *Presence of bark or wane*. In order to give an accurate construction date to a building, it is necessary to determine when the trees used to erect it died. Determining a death or felling date of a tree requires assigning a calendar date to the last year it formed a ring. This is the ring adjacent to the bark (Figure 1.1) or a ring associated with wane (the outermost ring if the bark has been removed or sloughed off. Wane is identified by noting a smooth, rounded outer surface on a log that is free of tool marks (Figure 1.2). When multiple timbers from a structure have the same or similar death dates we can determine the construction date of a building. If none of the accessible timbers of a building have bark or wane, calendar dates can still be assigned to the tree rings in the timbers, but an exact construction date cannot be determined (Figure 1.3).

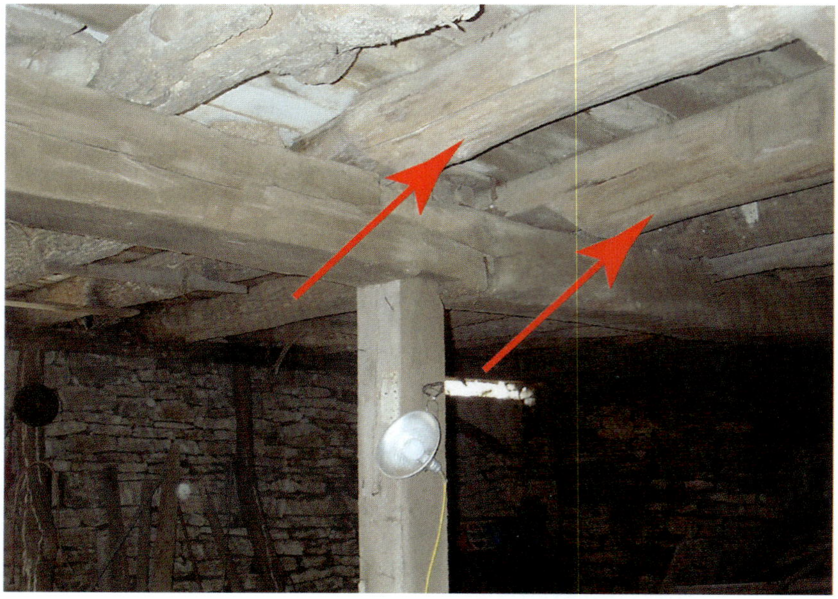

Figure 1.2
Floor joists from a southeastern Indiana barn. These ash floor joists (arrows) exhibit wane, a smooth outer surface with no bark present. If the outermost ring in these joists can be dated, the felling or death date of the tree can be determined. When several timbers from throughout a building have the same or comparable death date, one can determine a construction date for a building.

Source: Authors.

Figure 1.3
Cross section of a white oak post from a log house. This timber was made by removing woody tissue from a fallen tree so that a square shape was achieved. This timber has no bark or wane, and it is not possible to determine how many rings were removed when it was squared. Therefore, the exact year of felling cannot be determined. The outermost ring in this sample is 1813 so we know this timber was put into service at some undeterminable year after 1813. The structure from which it was taken was built sometime after this date.

Source: Authors.

- *Timber accessibility*. Dating historically erected buildings requires extraction of wood samples from timbers from throughout a structure. In some structures, such as barns, sampling is fairly straightforward, because the timbers are readily accessible. Buildings such as houses and churches pose sampling problems because construction timbers are covered with plaster or other wall finishes, and access to structural timbers is not possible without causing major damage. In these buildings, we rely on sampling in crawlspaces, cellars, and attics (Figure 1.4). During renovations we have had great success in accessing timbers for dating (Figure 1.4; see also the Huxford House in Chapter 10, the Wyneken House in Chapter 8, and *Musée de Venoge* in Chapter 9).

Figure 1.4
To provide accurate construction dates for a building, samples from timbers must be obtained for analysis. Top: Sampling in buildings with finished walls is often done in cellars or crawlspaces, since the floor joists of the ground floor are usually accessible. Bottom: During renovation, samples are easily extracted, since wall coverings have been removed. During the renovation of Eleutherian College (Chapter 9), all of the baseboards (white boards in the foreground) were removed from the plastered walls. We were able to access and sample a large number of timbers without causing damage to finished walls. The vertical wooden timbers (center) are wall studs.

Source: Authors.

- *Multiple samples are needed.* Determination of a building's construction date requires more than dating a single timber. Historical buildings have stood for extended periods of time (approximately two-century maximum where we work, but much, much longer in other parts of the world such as Europe and Asia). During this time, the fabric of buildings is routinely altered, and renovations and additions are made (see the Fauntleroy House, Chapter 11). New timbers are added to repair damaged timbers due to rot and insect damage (see Delphi's Canal Park, Chapter 8). Additionally, buildings are often erected fully or in part with recycled timbers (see the Thiebaud and Posey Township Hay Press Barns, Chapter 12). To accurately date the construction of a building, many samples are needed, and these samples need to be taken from throughout the structure – different elevations, floors, rooms, cellar/crawlspace, and attic. Through thorough sampling, one is able to increase the odds that a building, not an addition or repair, was dated.
- *Reference chronologies.* A chronology is a series of accurately dated (and, for our work, measured) tree rings. A chronology will contain the long- and short-term growth patterns of many trees from throughout a region (Chapter 4). A regional chronology informs us as to what we could "expect" in regards to ring size in a given year. Dating the timbers from a building of unknown age requires matching the pattern of small and large growth rings in the timbers to those found in a regional chronology. This matching procedure is called crossdating (Chapters 4 and 5), and it permits dating of historically constructed buildings and landscapes. Regional growth signals are impacted by many factors, with climate being of great importance in determining the size of a ring produced in a given year. For example, a small ring (in relation to its neighboring rings) is produced during a droughty year. Climate, therefore, imparts an expected pattern of tree-ring widths throughout an area; this pattern is contained in a region's chronology. Additionally, chronologies used for crossdating should not be too far removed geographically from the building or landscape being dated; the closer the reference chronology to the structure, the better the odds of successfully dating a building.
- *Timber species.* Different tree species differ in the way they grow and in how they form rings (Chapter 3). Certain species crossdate well, while others do not. How well a species crossdates is dependent on many factors. From our experience, we have found that a subset of species used to construct buildings crossdates very well (white oaks, tulip poplar, ash, and American beech) and can be used for dating construction while others do not (maples and certain elms; Figure 1.5). We have also found that crossdating is more likely successful when we crossdate timbers with regional chronologies of the same species or taxon (closely related subset of species). Consequently, successful crossdating often relies on having reference chronologies that are both local and from a particular type of timber.
- *Number of rings in a sample.* In order to crossdate reliably, a sample taken from a timber must contain a sufficient number of tree rings to confidently match growth patterns in regional chronologies or the growth patterns found in other timbers obtained from a structure. A several-decade series of tree rings will not suffice (Figure 1.6).
- *Overlap with existing chronologies.* The tree-ring patterns in a building's timbers must overlap in time with those found in existing chronologies. We have unsuccessfully attempted to date several early- to mid-20th-century structures because our reference chronologies simply do not overlap. Crossdating requires matching patterns in samples of unknown age with

those having verified dates. If samples don't overlap with a chronology, their growth patterns cannot be matched.

- *Sensitive trees*. In order for crossdating to be successful, timbers must show variation in their ring widths from year to year. This variation is called sensitivity. A sensitive tree-ring series can be crossdated, whereas a complacent series, one that shows little variation in ring width, cannot (Figure 1.7).

Figure 1.5
Not all samples are datable. Note the irregular growth in this red elm. In many of the individual tree rings (a single year of growth), there is much variation in ring width. Depending on which radius is chosen for tree-ring analysis, different ring-width patterns would be noted. Dating requires matching the ring-width patterns in samples of unknown age with the patterns observed in a set of dated samples. Given the great variation in width of individual rings, dating this specimen would likely be impossible. Compare to the bull's-eye-like growth (resulting from fairly consistent ring width in a given year) in the tree rings of Figure 1.1.

Source: Authors.

Figure 1.6
In order to accurately date (crossdate) the tree rings in a timber, it is necessary to have many rings in the sample. These two ash tongue and groove floorboards were salvaged from the Wyneken House in northeastern Indiana (Chapter 8). The top floorboard contains 87 tree rings, enough for reliable dating (1735 to 1821), while the bottom floorboard contains only 28 rings, far too few to be dated with any reliability. One floorboard from this structure contained 126 rings.

Source: Authors

Figure 1.7
Dating the timbers from buildings necessitates tree-ring growth patterns that show annual variation. The unique pattern of small and large rings makes crossdating and assigning calendar dates to tree rings possible (Chapters 4 and 5). These ash samples were obtained by removing a core of wood from individual building timbers. The top sample shows variability in ring width over time and would be considered sensitive; it dated from 1725 to 1858 (Braytown North Hay Press Barn; Chapter 13). The bottom sample would be considered complacent – there is little variation in ring width from year to year, and accurately dating it is unlikely.
Source: Authors.

So, "can you date my building?" As can be deduced from the preceding discussion, tree-ring dating of historic buildings requires that several conditions be met. Not all buildings can be dated with tree rings. There are limitations, and construction date determination using tree rings may not be possible for all buildings.

What does a construction date tell you? When bark and wane dates are determined on timbers, we can infer a construction date for a building. A death date represents the last year a tree was alive and the first year it could be incorporated into a building as a timber(s). To fully understand and interpret a landscape or a building, more than tree rings is needed. We use archival and secondary sources to support construction dates established through tree-ring analysis. In Chapter 6, Archival and Scholarly Sources for Interpreting Tree-Ring Analysis, we present a wide range of archival sources we use to interpret the results of dendrochronology. Archives, such as historical societies, libraries, universities, and governmental institutions, are repositories of historical documents and data. The sources come in a variety of forms. For instance, in the case studies included in this text, we use historic maps and drawings, census data, letters, deeds, and photographs. Secondary sources are studies and conclusions drawn by other researches. Many of these sources (prepared by historians, archaeologists, preservationists, etc.) provide excellent histories for historic sites, but without dendrochronology they are often only able to provide approximate construction dates. For example, seven of the sites in this text are listed on the National Register of Historic Places, and four are National Historic Landmarks. Each of these sites have detailed histories prepared by architectural historians and archaeologists. By combining tree-ring and historical evidence we are able to better understand a building or landscape and to provide a historic framework or context for a site.

1.2.5 Reporting findings

Once a date of construction (and, often, modification) and historical documentation are interpreted, results must be disseminated. Reporting can be done in a variety of ways, depending on the audience (professional peer-reviewed journals, magazines read by the general public, museum newsletters, public presentations, etc.). In Chapter 7 we report how we share our interpretations from our analyses by using data tables, photos, and measured drawings.

Over the past 15 years, we have presented our research at many different venues ranging from professional academic conferences focusing explicitly on tree-ring science or historic preservation, to civic groups, to conferences aimed at local citizens who privately own historic buildings (mostly barns), and to elementary students. Over this time period, we have had the pleasure of interacting with scientists and nonscientists who are truly interested in the dating of historic buildings with tree rings. We have found that explaining our research entails not only a tidy listing of dates and pictures of old timbers but also a fairly thorough explanation of how we do our work and the basic principles of tree-ring science and botany (general tree physiology and anatomy).

When presenting to non-dendrochronologists, we have found that spending at least half of our allotted speaking time to explaining tree-ring dating makes the talk relevant and that many audience members enjoy learning the basics of the science. Again, without a knowledge of the process and what must be done to date a building, results, regardless of how interesting or thought-provoking they may be, mean little if the methodologies are not understood. We have found that the audience enjoys the "basics" as much as they do the findings (building construction techniques, history of the landscape, construction dates, and the ages of the trees used to make a building's timbers). We have chosen to construct this book in much the same manner. Chapters 2 through 7 provide the reader with the background and methodologies of tree-ring dating, and Chapters 8 through 13 provide case studies of what tree-ring analysis has taught us about various buildings and landscapes. The case studies, which help to make this text unique, are categorized into topics, and we present a brief history, a review of archival sources, photos, and illustrations for each site along with dendrochronological results.

- Chapter 8 provides results obtained by analyzing tree rings and historic resources for a number of museums, including open air and individual house museums.
- Chapter 9 explores not only dates of unique buildings and landscapes but also the importance of understanding the people who made and used the buildings. Among the buildings and landscapes included in this chapter are a central-Kentucky slave cabin and a purported slave cabin, Eleutherian College (a mid-19th-century educational institution that accepted women and African-Americans), and the *Musée de Venoge*, a timber frame house built in the French colonial tradition but located in a region dominated by the descendants of German and English immigrants with their own building traditions.
- Chapter 10 reports the transformations of landscapes over time and how these transformations can be documented using tree-ring analysis. In this chapter, we explore an evolving southern Indiana farmstead and Fort Wayne, Indiana, where a military garrison was erected in the late 18th century.
- Chapter 11 provides a historical and dendrochronological account of a unique communal group, the Harmonists, who carved a thriving community out of the wilderness while awaiting the second coming of Christ.
- Chapter 12 investigates how a unique agricultural system (pressing hay into dense bales for export) developed and changed the way the landscape was used and how barns were built.
- Chapter 13 documents the use of tree-ring analysis to not only provide construction dates of buildings and landscapes but also to learn about unique construction aspects of buildings, including determining the geographic source of timbers and how builders used the vast timber resources available to them when erecting structures. There is more to the study of tree rings than the widths of individual tree rings!

In the final chapter of the text (Chapter 14), we provide an opportunity for the reader to apply tree-ring dating techniques and historic document research to interpret the Daniel L. Harmon House. This "lesson" utilizes historical documents (e.g., plat maps, census data, and newspaper articles), oral histories, and tree rings. The exercise can be used in the classroom, laboratory, and studio to demonstrate how tree-ring analysis is combined with other historical resources to aid interpretation. The historical sources and tree-ring material are real; we wanted to provide a genuine experience for the reader. Performing the exercise does not require expensive equipment, and it can be custom tailored to accommodate time constraints, pedagogical objectives, and student maturity. Our hope is that upon completion, the reader better understands that by simultaneously using several types of evidence, a fuller, more complete, picture of what happened in a landscape is possible.

1.3 Who we are

The authors of this text come from two very different backgrounds and have received very different training. Christopher Baas is a landscape architect who teaches landscape history, planning, and design at Ball State University (Muncie, Indiana, USA). Darrin L. Rubino was trained as a plant/forest ecologist and is a botany professor at Hanover College (Hanover, Indiana, USA). The two authors have collaborated on the dating of historically constructed buildings and landscapes for a decade. Our partnership has allowed us to meld our unique training, experiences, and expertise into providing an understanding of dating historic structures. Many of our students have joined us in this endeavor, and many of those students are pursuing advanced degrees in the sciences or have become successful landscape planners and designers.

1.4 Our goals

Our major goal for this book is to create a text that introduces dendrochronology to students, teachers, practitioners, technicians, property managers, and those with a general interest in both tree-ring science and historical preservation. Our hope is to show how dendrochronological techniques can be used, along with archival sources (such as letters, deeds, maps, etc.), to best interpret, preserve, present, and understand historic structures and landscapes. The text is not meant to be an exhaustive treatise on the use of tree rings in archaeological contexts or one on tree-ring science (e.g., history, application, etc.). For a more thorough review of these topics, we strongly recommend the reader consult the sources found throughout the text. An excellent introduction to the history and application of dendrochronology can be found in texts by Frits (1976), Nash (1999), and Speer (2010).

The satisfaction of dendroarchaeological research comes from the involvement with truly remarkable historic sites and the people that manage them. This book provides case studies of how we have used tree rings to date, understand, and interpret historic buildings and landscapes. This book, therefore, will serve both as an example of how tree rings inform and enhance interpretation and as a report, or record, of the work we have done at various historic sites. We also created this text to demonstrate how important historically erected buildings are, both culturally and scientifically. Historic structures help us to better understand how landscapes were transformed by culture over time. They serve as a treasure trove of information that can help in explaining how certain basic needs, such as providing a place to live, a place to store and raise crops and livestock, or a place to worship, were built. However, historic structures are more than a part of a region's cultural legacy and fabric. The tree rings in the timbers of these buildings tell us how trees and forests grew over several centuries. Sadly, most old-growth forests have vanished. In many regions, our only repository of information concerning past forest composition and tree growth lies within the timbers of historic buildings (Rubino 2014; Trouet et al. 2017).

In *Log Buildings of Southern Indiana*, a classic text written and updated in the latter decades of the 20th century (and a source we heavily rely upon to interpret vernacular structures), folklorist Warren Roberts lists a variety of sources for dating the log buildings in his study (written records, oral tradition, features of construction, etc.). However, he laments how the "exact dates of construction" for the historic timber buildings in his study are "unknown" (Roberts 1996). Similar statements have been made by other authors. Stotz (1966) bemoans the inability to date the earliest log houses of western Pennsylvania except in the "rare instances where valid records are available." Similarly, Rempel (1980) states that log houses cannot be dated unless historic records are available. Conversely, in 2017, the journal *Dendrochronologia* published an editorial and special volume proclaiming the revival of dendrochronology in eastern North America (Nash and Copenheaver 2017). As active participants in this revival, and in response to Roberts' and other's desire for precise construction dates to interpret material culture, we want this text to make dendrochronology more accessible to those preserving and interpreting historic buildings and landscapes.

Although the case studies in this text are regional (mid-Ohio River Valley, North America), similar investigations can be and have been performed in diverse locations throughout the world. For example, tree-ring dating of historic buildings and landscapes has been performed extensively throughout Europe. In our study region and beyond, we strongly advocate for locating, documenting, dating, and interpreting historically constructed buildings before they are lost to history. Regionally, we have witnessed the loss of many historic buildings, especially barns (Figure 1.8). Once these buildings are destroyed, due to neglect or, more recently, for the selling of timbers for architectural accents, their history is gone. Historic buildings represent irreplaceable pieces of a region's cultural heritage, and their study and interpretation by scientists, historians, and preservationists is important.

Figure 1.8
Regional historic buildings are disappearing from the landscape. Once these buildings are gone, an important piece of cultural heritage is lost. Scientifically, their disappearance is problematic, because these buildings serve as a repository of vast amounts of irreplaceable tree-ring data.

Source: Authors.

1.5 Literature cited

Baillie, M.G.L. 1982. *Tree-Ring Dating and Archaeology*. The University of Chicago Press, Chicago, IL. 274 p.

Baillie, M.G.L. 1995. *A Slice through Time: Dendrochronology and Precision Dating*. Routledge, London, UK. 176 p.

Dean, J.S. 1997. Dendrochronology. Pp. 31–64. In: *Chronometric Dating in Archaeology*. (R.E. Taylor and M.J. Aitken, eds.). Plenum Press, New York, NY.

Frits, H.C. 1976. *Tree Rings and Climate*. Academic Press, New York, NY. 567 p.

Nash, S.E. 1999. *Time, Trees, and Prehistory: Tree-Ring Dating and the Development of North American Archaeology*. The University of Utah Press, Salt Lake City, UT. 294 p.

Nash, S.E., and C.A. Copenheaver. 2017. Editorial: Revival of dendroarchaeology in the Eastern United States. *Dendrochronologia* 43: 1–3.

Rempel, J.I. 1980. *Building with Wood and Other Aspects of Nineteenth-Century Building in Central Canada*. University of Toronto Press, Toronto, Canada. 454 p.
Roberts, W. 1996. *Log Buildings of Southern Indiana*. Trickster Press, Bloomington, IN. 228 p.
Rubino, D. 2014. Tree-ring analysis in Indiana with special attention to dendroarchaeology. *Proceedings of the Indiana Academy of Science* 123: 7–13.
Speer, J.H. 2010. *Fundamentals of Tree-Ring Research*. The University of Arizona Press, Tucson, AZ. 333 p.
Stokes, M.A., and T.L. Smiley. 1968. *An Introduction to Tree-Ring Dating*. University of Chicago Press, Chicago, IL. 73 p.
Stotz, C.M. 1966. *The Architectural Heritage of Early Western Pennsylvania*. University of Pittsburgh Press, Pittsburgh, PA. 293 p.
Taylor, R.E. 1999. Science-based dating methods in historic preservation. *Archives and Museum Informatics* 13: 227–247.
Trouet, V., M. Domínguez-Delmás, C. Pearson, N. Pederson, and D. Rubino. 2017. "Dendro-archeo-ecology" in North America and Europe: Re-purposing historical materials to study ancient human-environment interactions. Pp. 365–394. In: *Dendroecology: Tree-Ring Analyses Applied to Ecological Studies*. (M. Amoroso, L. Daniels, P. Baker, and J.J. Camarero, eds.). Springer, Cham, Switzerland.

Chapter 2

Understanding cultural landscapes and historic buildings

Frameworks for interpreting and communicating tree-ring analysis

2.1 Introduction

The audience for this book represents various academic backgrounds with varying theoretical frameworks. They use specific vocabularies to communicate their work (e.g., archaeology, history, museum studies, and historic preservation). In this chapter we explain how we approach our study of and interpretation of historic sites within the fields of cultural landscape studies and cultural geography. Additionally, we provide a list of terms and definitions we routinely use to communicate our findings.

2.2 Cultural landscape studies

Material culture is the objects "man has learned to make." It is the tangible product of a culture that "embraces those segments of human learning which provide a person with plans, methods, and reasons for producing things that can be seen or touched" (Glassie 1968). A cultural landscape is the "combined works of nature and man" (UNESCO 2012). Cultural landscapes have material culture. When we use tree-ring data to interpret cultural landscapes and buildings, we are, in essence, asking why and how material culture is at a particular place at a specific time. And, we often ask why it is not in other places at the same time. Simply stated: Why is this landscape or building "here"? And, when and how did it get "here"?

Dendrochronology, used with archival and secondary sources, can provide the temporal, or time, element to answering these questions. How these types of questions are answered depends upon a researcher's discipline. For instance, while establishing a construction date for a landscape or building is one goal of tree-ring analysis, the results might be interpreted within a theoretical framework distinct to archaeology, history, historic preservation, museum studies, and cultural landscape scholars.

Our approach to interpreting the results of our work is grounded in frameworks developed within the cultural landscape studies and cultural geography fields of study. These theoretical frameworks seek to explain the form as well as the geographic and time-based (temporal) distribution of material culture. Although we ask questions and approach interpretation with frameworks typical to cultural landscape studies and cultural geography, the methods for determining the death, harvest, or construction date of a timber sample are the same regardless of how the results are interpreted by various disciplines.

We do not want to belabor our readers with theoretical concepts of material culture, but a basic understanding of how we interpret our work will help in understanding the case studies in the ensuing chapters. Richard Schein, whose theoretical framework for cultural landscapes is described later, states "no matter how abstract, how theoretical, the formulation of the idea of cultural landscape, what really matters is investigating particular landscapes to understand how they have come to be, to wonder how they are received and lived in and through, to ask how and why they matter, and to figure out how they work" (Schein 2009). Within a context of tree-ring analysis identifying construction dates for landscapes and buildings, we appreciate Schein's assertion that "details are important, and so is the specificity of those details in a system of [local, regional, and national] scalar relations. The deep contours of historical geography generally are relevant to understanding contemporary landscapes" (Schein 2009). We believe dendrochronology is a viable and time-tested means for adding the specificity of time and place. Tree-ring analysis has been used extensively in Europe and North America to determine the construction and modification of historically constructed buildings when scant or dubious historical data is available.

2.3 Defining cultural landscapes and the cultural hearth theoretical framework

Scholars have described how landscape, as a term, has a myriad of meanings ranging from a scenic view to a place or location (Meinig 1979; Jackson 1984; Groth and Wilson 2003). We approach dendrochronology through a framework of cultural landscape studies. This framework was formed in the early 1920s by geographer Carl Sauer (1925). Sauer's straightforward description of the cultural landscape is often quoted: "Culture is the agent, the natural area is the medium, the cultural landscape is the result" (1925). The United Nations states it even more directly as the "combined works of nature and man" (UNESCO 2012). Therefore, when we interpret the results of tree-ring analysis, we do so with the understanding that cultural landscapes contain all types of material culture, including buildings. Furthermore, a building cannot be effectively interpreted without an understanding of the cultural landscape. Examples of cultural landscapes examined in this book include a military fort, communal village, regional agricultural systems, the Underground Railroad, and a 19th-century college campus.

Alanen and Melnick (2000) explain that Sauer's "landscape school" evaluated the landscape starting with the natural environment and "continuing through all the subsequent culture groups that inhabited an area or region." Therefore, interpreting material culture cannot be adequately analyzed without understanding the context of its natural and cultural history. Cultural landscapes, and the buildings contained within, have layers of history that dendrochronology can isolate and date. Additionally, as you will see in Chapter 13, we attempt to "reverse engineer" this process by exploring what a cultural landscape's built structures can tell us about the pre-European settlement natural environment.

The National Park Service (NPS) is the lead governmental agency for the preservation of cultural landscapes in the United States, and there are equivalent agencies in other countries. The NPS defines four types of cultural landscapes: historic designed landscapes, historic sites, historic vernacular landscapes, and ethnographic landscapes (National Park Service 2018). We want to highlight the term vernacular, which means common or ordinary, because it represents the majority of the sites we have analyzed. A vernacular landscape, as defined by the NPS is a:

> landscape whose use, construction, or physical layout reflects endemic traditions, customs, beliefs, or values; in which the expression of cultural values, social behavior, and individual actions over time is manifested in physical features and materials and their interrelationships, including patterns of spatial organization, land use, circulation, vegetation, structures, and objects; in which the physical, biological, and cultural features reflect the customs and everyday lives of people.
>
> (National Park Service 2018)

The case studies presented in this book mostly represent the historic vernacular landscape type and exhibit examples typical of North American vernacular architecture with European antecedents. Like the vernacular landscapes in which they are found, vernacular architecture represents cultural traditions of form, construction, and use (Glassie 1968). Immigrants brought these traditions to North American "cultural hearths," described by geographer Fred Kniffen as New England, Middle Atlantic, and Lower Chesapeake (Kniffen 1965). The traditions were then accepted, rejected, or hybridized in their new environment, and then diffused across the landscape over generations as settlement moved into the interior of the continent (Kniffen 1965; Stanton 1985). We have found that the results of published peer-reviewed dendrochronological studies are often lacking any in-depth interpretation of their historic resource. Misinformation, as we have seen, results in log buildings displayed without siding, and smokehouses and dogtrots presented as pioneer fortifications. We encourage readers to review classic texts and articles (for instance Sauer 1925, Kniffen 1965, Glassie 1968, Zelinsky 1973, and Noble and Seymour 1982 to name a few) as well as recent contributions to cultural landscape studies cited in the ensuing chapters. Students of the vernacular will recognize many of the traditional structural forms in the case studies chapters (e.g., I-house, English barn, and dogtrot) as well as construction methods (e.g., horizontal log, timber frame, and *fachwerk*).

2.4 Discourse materialized: a theoretical framework

Cultural hearth–based theories like Sauer's and Kniffen's do not effectively explain the postcolonial material culture that is distinct to North America. While there are multitudes of theoretical frameworks explaining the formation and evolution of the cultural landscape, we have come to rely on geographer Richard Schein's (1997, 2009) theoretical framework for evaluating the results of our analysis. Schein's "discourse materialized" framework offers a plausible means for why and how a landscape came to be. Schein (1997) views the landscape as ongoing and overlapping local, regional, and national discourse (i.e., politics, economies, religion, popular culture, etc.). Cultural landscapes are "embedded within a discourse" that result in "discourse materialized." We used Schein's framework to explain how Samuel Hewitt's Mormon beater hay press, a machine used to produce compressed bales of hay for easy transport, was a product of a discourse of economic entrepreneurship, regional cultural changes, improvements in transportation systems, the region's abundance of construction material, growing East Coast urban horse populations, the development of agricultural science, and a culture of invention (Chapter 12, especially Figure 12.2; Baas and Rubino 2013).

Regardless of the theoretical frameworks used to explain the cultural landscape, we have tended to focus on describing how common people interacted with their surroundings. When choosing particular buildings to analyze, we did not limit the scope of our work to only those structures directly associated with prominent individuals. Consider that there are far more common people than famous. Fortunately, many of their landscapes and buildings still exist – some still serving their initial purpose as houses, barns, churches, and commercial buildings. By analyzing building attributes, such as construction techniques or timber types selected for construction, we gain a clearer understanding of how the environment was viewed, used, modified, and adapted over time. The majority of these places will never attain landmark status but are historically relevant and have important stories to tell to a wide audience of scholars.

In contrast to vernacular landscapes, a designed landscape "was consciously designed" by a trained architect or landscape architect "working in a recognized style or tradition" (National Park Service 2018). The Grouseland and the Sullivan House museums are case study examples of designed landscapes described in Chapter 8. Additionally, case studies of vernacular ethnographic landscapes "containing a variety of natural and cultural resources that associated people define as heritage resources" are described in Chapter 9 (National Park Service 2018).

2.5 Terms and definitions

Tree-ring analysis requires a precise vocabulary to consistently communicate the timber-based material being sampled. This is important to the archival recording of sampling locations for both the current and future analysis of a site. Future researchers and scholars should be able to understand the nomenclature included in documentary drawings and field notes to find sample locations from past analysis (i.e., for the replication of your research by others). Investigators also need to communicate tree-ring analysis in reports and journal articles.

In writing this text we strived to limit the use of jargon by assembling a list of building terms we commonly use. The list is not exhaustive, but the terms will be found in the methodologies and case studies in Chapters 8 through 13. These terms are common to the architecture in our region, and may vary in other locations (for example, see Harris 1989 and Alcock 1996). We have included generic images of house and barn frames to illustrate many of the terms (Figures 2.1 and 2.2).

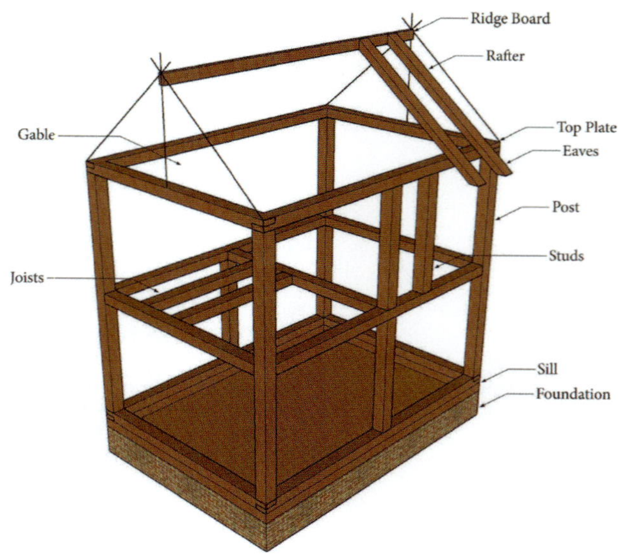

Figure 2.1
Illustration of a basic timber frame house showing the different terms defined in Chapter 2.
Source: Authors.

Beam. The horizontal member in post and beam construction typically found in timber frame buildings. The beam either sits on top of, or is fastened to two or more posts (i.e., through mortise and tenon joinery), and spans the distance between. It typically carries the dead load of a building and its contents, and the live load of the activities that take place within a building.

Bent (or barn bent). Timber posts, beams, and braces assembled into a frame (Figure 2.3). Barn bents are often designed to accommodate specific agricultural activities (e.g., housing animals or threshing wheat), and some bents are constructed to support heavy loads (e.g., storing hay or farm machinery; Chapter 12).

Brace (or angle brace, corner brace, knee brace). A diagonal timber connected to both a post and beam to provide structural stability.

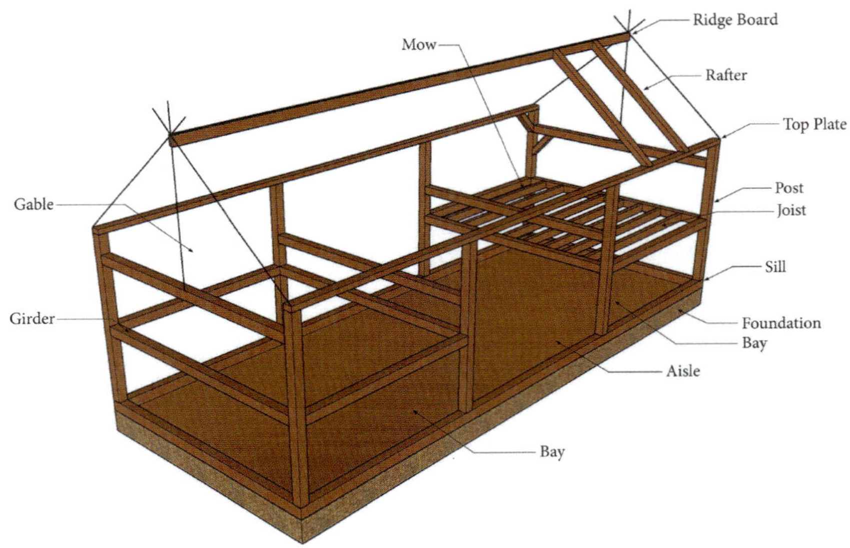

Figure 2.2
Illustration of a basic timber frame barn showing the different terms defined in Chapter 2.
Source: Authors.

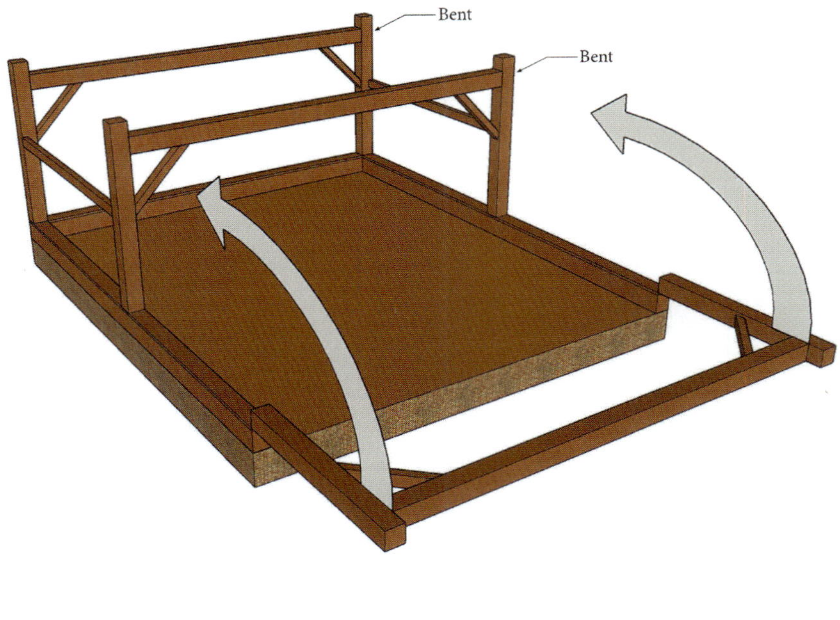

Figure 2.3
A barn bent is an assemblage of timber posts, beams, and braces assembled into a frame and raised into place.
Source: Authors.

Chinking. The material used to fill (and close) the gaps between horizontal timbers in a log building (Figure 2.4). It typically consists of pieces of wood or stone laid at a 45 degree angle and covered with clay (Roberts 1996; see the timber chinking sample from the Sanders House in Chapter 9).

Death date (or harvest date or build date). The year, and often the season, a tree died from being cut for use in a timber structure. Since green (recently cut) timber is easier to shape into building materials (e.g., logs, posts, beams, etc.), tree-ring analysis assumes that timber is placed into a building relatively soon after it was harvested, and the death date is representative of the construction date.

Eaves (as in roof eaves). The part of a roof that overhangs the walls of a building to shed water away from a building. Eaves often have rain gutters.

Fachwerk. An Old World building technique brought to America by German and other European immigrants (Blair 1964; Kniffen 1965; Glassie 1968; Stanton 1985; Tishler 1986). The walls of timber frame structures contain an infill of low-fired brick (see the Harmonist case studies in Chapter 11).

Frame construction (or balloon-frame construction). A building method where structures are assembled from wood members nailed together.

Gable. The triangular-shaped part of a wall that encloses a pitched roof (see the Wyneken House in Chapter 8 and the hay press barns in Chapter 12).

Girder. A horizontal structural timber in a wall that provides structural support.

Figure 2.4
Chinking is the material used to fill the gaps between horizontal timbers in a log building. It helps prevent wind and rain from entering the structure. This American chestnut sample is from the Sanders House reported in Chapter 9.

Source: Authors.

Green timber. A piece of hewn or milled building material that was recently cut. Green timber, as opposed to seasoned, was used in the construction of timber frame buildings because it was easier to shape.
Hewn timber. A harvested log prepared for use in a building by flattening its sides with an ax (Figure 2.5).
Jamb (as in doorjamb or window jamb). The side posts that frame a door or window opening.
Joist (as in floor joist or ceiling joist). The horizontal structural members of a building that are typically arranged in a series of parallel planks or timbers that support both a floor and ceiling. For reporting tree-ring analysis, we use the term floor joist to describe sampling locations. We only use the term ceiling joist when sampling between a building's top story and attic.
Kick plate (riser). The vertical surface of a stair step.
Lath (or plaster lath). The thin and narrow wooden strips nailed to joists and studs as a base for plaster on finished walls and ceilings (Figures 2.6 and 5.14). Accordion lath is made from a single plank that has been split, stretched, and fastened to joists. Lath that has been removed from a house typically leaves trace marks on studs and joists. These marks help to interpret the use of a room or explain the provenience of a timber.
Log or wall timber. The trunk or branch of a tree harvested for building materials.
Loose (as in loose timber or plank sample). "Loose" indicates that a specific timber was not fastened to any other timber in the structure. For some loose timbers, provenience can be determined by its shape, size, and tooling (tongue and groove floorboards, for example). However, provenience cannot always be determined. A timber may become "loose" for a variety of reasons. For instance, during renovation salvageable timbers are often removed from the structure and stored in stacks. Additionally, timbers may have extensive damage to one area, and the damaged portion is removed. Such cut off sections are often discarded, but we have found that, despite damage, they are

Figure 2.5
Hewn logs are prepared for use in a building by flattening their sides with an ax. These two logs are from the Loom House described in Chapter 8. The chop marks from the ax are visible. Modern chinking has been used to fill the spaces between the logs.
Source: Authors.

still valuable for tree-ring analysis, and readily collect them for study. Finally, some buildings or sections of buildings are in complete disrepair (caved in upper floor, collapsed roofs, missing exterior walls). Again, datable material is still present in such a structure, but exact location and provenience of a timber is often undeterminable.

Lumber. Timber or logs prepared for building materials.

Material culture. The objects “man has learned to make” (Glassie 1968).

Milled timber. A harvested log prepared for use in a building at a sawmill.

Mortise. One of the two components of a mortise and tenon joint used in carpentry and woodworking. The mortise is a hole in a timber that receives the tenon (the tongue or tab) from another timber. The two timbers are typically pinned together with wood pegs. An empty mortise indicates that a structure's framing member has been removed, or that timbers from one building have been recycled into a different one (Figure 2.7 and 5.13; see the Posey Township and Thiebaud hay press barns in Chapter 12).

Mow. A barn's mid or upper-story floor where hay and feed are stored (see the Robert Glenn Barn in Chapter 10 and the Bakes Road Barn Chapter 12).

Figure 2.6
Lath is the thin and narrow wooden strips nailed to joists and studs as a base for plaster on finished walls and ceilings. This example of accordion lath is from the upper story of the *Musée de Venoge* (Switzerland County, Indiana), and shows the inside of a frame wall. The marking of the lath that has been removed can be seen on the face of the wall studs.

Source: Authors.

Notching (or corner notching). The corner joints used in the construction of log buildings (i.e., V-notch, half-dovetail, and square; Figure 2.8). A carpenter's choice of a notch type often indicates a cultural bias (Kniffen and Glassie 1966; Roberts 1996; Chapters 8, 9, 10, and 13).

Peg (as in wood peg, tree nail, or trunnel). Hard wooden pin that fastens timber together (Figures 2.7 and 2.9). Pegs are typically made from durable wood timber species that accommodate the pounding (hammering) required of construction.

Pen (as in single-pen or double-pen architecture. Also, known as room or crib). A square or rectangular construction unit for log houses and barns. To enlarge a structure a new pen was added.

Plate (or top plate). The top horizontal member of a frame wall that is attached to the vertical members of the building (i.e., posts and wall studs), and on which the roof's rafters are fastened.

Poteaux-sur-sol (post-on-sill) Construction. A vernacular building method typical of French colonial architecture in North America where buildings are constructed with vertical posts inserted and pegged into horizontal timber sills (Kniffen and Glassie 1966). The spaces between the posts are often infilled with brick (see The *Musée de Venoge* in Chapter 9; Figure 2.10).

Figure 2.7
A mortise and tenon joint from the Robert Glenn English Barn (Jefferson County, Indiana). Although they are not visible, the post contains a mortise and the beam contains the tenon. Their connection is secured with the wood pegs.

Source: Authors.

Post. The vertical member in post and beam construction typically found in timber frame buildings. Posts support the beam, and transfer dead and live loads to a structure's foundation.
Purlin. The timber roof member running perpendicular to the rafters, and parallel to the roof ridge. Purlins brace rafters, and provide a timber to attach the roof sheathing.
Plank. A thin, rectangular piece of timber, typically used for flooring, siding, or roof sheathing. Planks are typically sawn lumber (i.e., milled) but can also be from logs that are riven (wedged and split).
Puncheon. A hewn timber used as a floorboard.
Rafter (or roof rafter). The angled wooden framework of a roof. Rafters typically run from the top of a building's wall (from the top plate) to the peak of the roof (Figure 2.11).
Ridge board. A timber or plank located at a roof's peak that supports the roof rafters (Figure 2.11). The rafters are pegged or nailed to the ridge board.
Riven. Wood that is split into building materials (rather than hewn or sawn) using a fro.

Figure 2.8
Notching is the corner joinery used in the construction of log buildings. The notching locks the timbers together. This example of the V-notch is from an American elm log house located in Markle, Indiana.

Source: Authors.

Figure 2.9
A prepared cross section that includes a peg. The log is maple and the peg is oak. Pegs are typically made of oak because the hardness of the species can take the pounding required to pin timbers together.

Source: Authors.

Saw marks. The markings left from the preparation of a timber using a saw (i.e., pit, sash, or circular). The pattern of saw marks can indicate the type of saw and the time period the timber was made.
Scab. A piece of lumber attached to a building's structural member (typically with nails) for added strength, or as framing for an addition to the structure (e.g., a new door or window).
Shingle (or shake). Thin, wood-plank roof and wall covering laid in courses (rows). Historically, shingles (or shakes) were split, or riven, from a short timber log called a bolt (Figure 2.12).

Figure 2.10
Poteaux-sur-sol (post-on-sill) construction at the *Musée de Venoge* reported in Chapter 9.

Source: Authors.

Figure 2.11
A photograph of the attic of a commercial building located in Vevay, Indiana. The attic displays rafters and a ridge board. The carpenter's Roman numeral marriage marks can be seen on two of the rafters (IIV and VI).
Source: Authors.

Sheathing (as in roof sheathing). Thin wooden planks that are attached to roof rafters and purlins, and to which roof shingles are attached.
Sill (or sill plate). The bottom horizontal member of a timber frame or frame wall that sits on a foundation or footings. The vertical members of the building (i.e., posts and wall studs) are attached to the sill.
Sleeper. A support timber laid on the ground (see the corncrib sample for the Thiebaud Farmstead in Chapter 12). Also, a term that is often used to describe the floor joists of a log house.
Stretcher. A wooden horizontal support beam.
Stringer. The two, step-shaped planks on each side of a flight of stairs onto which the treads and kick plates are attached.
Stud. The vertical wood framing members of a frame wall.
Tenon. One of the two components of a mortise and tenon joint used in carpentry and woodworking. The mortise is a hole in a timber that receives the tenon (the tongue or tab) from another timber. The two timbers are typically pinned together with wood pegs. When a timber has been removed from a structure (i.e., removed on purpose or through structural failure), the tenon is often left behind. The presence of an empty mortise suggests missing beams or braces, or that a timber is recycled (Figure 5.13).

Figure 2.12
Cross section of riven oak shingles from a slave dwelling reported in Chapter 9.
Source: Authors.

Timber. Wood prepared for constructing buildings. Typically hewn or milled.
Timber frame construction. A building method where structures are assembled from heavy timbers pinned together with mortise-and-tenon joints, and strengthened with diagonal bracing.
Tongue and groove. A method of wood joinery in which a plank's ridge (tongue) is inserted into a neighboring plank's groove (Figure 1.6). Tongue and groove joinery is typically found in plank flooring and siding (see the Frederick and Sophia Wyneken House in Chapter 8).
Tread. The horizontal surface of a stair step.
Wall timber. A mid-wall timber in a log house (i.e., not the top plate or bottom sill logs or timbers).

2.6 Conclusion

In Chapter 2, we explained how we approach our study of and interpretation of historic sites within the fields of cultural landscape studies and cultural geography. Additionally, we provided a list of terms and definitions that we routinely use to communicate our work. The frameworks and nomenclature will help you understand the presentation of tree-ring analysis and case studies in the ensuing chapters.

2.7 Literature cited

Alanen, A., and R.Z. Melnick. 2000. *Preserving Cultural Landscapes in America*. The John Hopkins University Press, Baltimore. 264 p.

Alcock, N.W. 1996. *Recording Timber Framed Buildings: An Illustrated Glossary*. Council for British Archaeology, London, UK, 29 p.

Baas, C., and D.L. Rubino. 2013. The most successful press in this or any other country: The material culture of 19th-century beater hay presses in the mid-Ohio valley. *Material Culture* 45: 1–20.

Blair, D. 1964. *Harmonist Construction*. Indiana Historical Society, Indianapolis, IN. 82 p.

Glassie, H. 1968. *Pattern in the Material Folk Culture of the Eastern United States*. University of Pennsylvania Press, Philadelphia, PA. 316 p.

Groth, P.E., and C. Wilson. 2003. The polyphony of cultural landscape study: An introduction. Pp. 1–22. In: *Everyday America: Cultural Landscape Studies After J.B. Jackson* (C. Wilson and P.E. Groth, eds.). University of California Press, Berkeley, CA. 385 p.

Harris, R. 1989. The grammar of carpentry. *Vernacular Architecture* 20: 1–8.

Kniffen, F.B. 1965. Folk housing: Key to diffusion. *Annals of the Association of American Geographers* 55: 549–577.

Kniffen, F.B., and H. Glassie. 1966. Building in wood in the eastern United States: A time-place perspective. *Geographical Review* 56: 40–66.

Jackson, J.B. 1984. *Discovering the Vernacular Landscape*. Yale University Press, New Haven. 165 p.

Meinig, D.W. 1979. The beholding eye: Ten versions of the same scene. Pp. 33–48. In: *The Interpretation of Ordinary Landscapes* (D.W. Meinig, ed.). Oxford University Press, Oxford. 255 p.

National Park Service. 2018. *Cultural Landscapes*. www.nps.gov/subjects/culturallandscapes/understand-cl.htm. Accessed 29 June 2018.

Noble, A.G., and G.A. Seymour. 1982. Distribution of barn types in Northeastern United States. *Geographical Review* 72: 155–170.

Roberts, W.E. 1996. *The Log Buildings of Southern Indiana*. Trickster Press, Bloomington, IN. 228 p.

Sauer, C. 1925 (1969). The morphology of the landscape. In: *Land and Life: A Selection from the Writings of Carl Ortwin Sauer* (J. Leighly, ed.). University of California Press, Berkeley, CA. 435 p.

Schein, R.H. 1997. The place of landscape: A conceptual framework for interpreting an American scene. *Annals of the Association of American Geographers* 87: 660–680.

Schein, R.H. 2009. A methodological framework for interpreting ordinary landscapes: Lexington, Kentucky's courthouse square. *The Geographical Review* 99: 377–402.

Stanton, G.W. 1985. *Brought, Borrowed, or Bought: Sources and Utilization Patterns of the Material Culture of German Immigrants in Southeastern Indiana, 1833–1860*. PhD Dissertation, Department of Folklore, Indiana University, Bloomington, IN. 444 p.

Tishler, W.H. 1986. Fachwerk construction in the German settlements of Wisconsin. *Winterthur Portfolio* 21: 275–287.

UNESCO. 2012. *Operational Guidelines for the Implementation of the World Heritage Convention*. United Nations Educational, Scientific, and Cultural Organization: Intergovernmental Committee for the Protection of the World Cultural and Natural Heritage, Paris. 165 p.

Zelinsky, Wilbur. 1973. *The Cultural Geography of the United States*. Prentice-Hall, Englewood Cliffs, NJ. 164 p.

Chapter 3

Botany for the dendrochronologist

3.1 Introduction to botany

Woody plants of temperate regions have a unique life history. They perennate or survive from year to year with periods of active growth alternating with dormant periods. During active growth periods, when conditions are favorable, they enlarge in length and girth to produce increasingly larger plant bodies. During the growth periods, they produce woody tissue around their trunks, limbs, and roots. This woody tissue is deposited in increments we recognize as individual tree rings.

The physiology (various mechanisms responsible for how an organism lives, survives, grows, etc.) of woody plants is very complex, and researchers are actively investigating the many intricacies of woody plant growth and function. We view this chapter as a primer in botany and not an exhaustive explanation of the complexities of the wood formation process. We hope to provide the reader with the basic botanical knowledge needed to understand tree-ring formation as it relates to successfully dating the construction of historic buildings and landscapes. Among woody plants, there is variation in the formation of tree rings and woody tissue. Consequently, the following discussion of tree growth and ring formation is presented in generalizations, but these generalizations will not hamper your understanding of tree-ring formation. For those interested in a deeper understanding of the topic, we recommend a general botany text such as Evert and Eichhorn (2013). For those wanting an even deeper comprehension of the wood formation process and the functioning of woody plants, texts such as Pallardy (2007) or Kozlowski et al. (1991) are very useful.

We remind the reader that providing precise construction dates for buildings is dependent on assigning calendar dates to individual tree rings. In a best-case scenario, we hope to assign a calendar date to the last ring formed by a tree to determine its felling or death date. We can identify this last year of growth if bark is present – the most recently formed ring in a woody stem is next to the bark. Additionally, we can date the last year of growth of a tree if a waney edge is present on a timber. Wane is characterized by a uniform, curved outer surface that is free of any tool marks (ax, adze, hatchet, plane, or saw). A waney surface can be seen if the bark of a tree is removed or has simply sloughed off over time (Figure 1.2). When multiple timbers from throughout a structure have similar or comparable death or cutting dates, one is able to identify a construction date for a building.

3.2 Anatomy and morphology of a woody stem

Generally speaking, trees produce a layer of wood, a tree ring, around their circumference each year in regions characterized by a temperate or seasonal climate. When performing tree-ring analysis of historic buildings, we examine and note the year-to-year variation in tree-ring width in a building's timbers. The pattern of ring widths over time permits accurate dating. Observation of series of tree rings is made possible by preparing cross sections from timbers (Figure 3.1). Cross sections are obtained by cutting a timber perpendicular to its long axis (the grain); you can think of a stump as a cross section of a tree. Also, a type of cross section can be obtained by removing a core from a timber with a borer. Cores are taken by drilling a hollow boring bit perpendicularly to the length of a timber (Figure 5.4; Chapter 5). A core, therefore, is simply a small portion of a cross section.

Through an examination of cross sections of stems of various ages, one can gain a better understanding of how tree rings are deposited and how annual layers of growth can be dated. First, an understanding of how the different tissues in a woody plant are distributed is essential. Figure 3.2 shows a cross section obtained from a one-year-old basswood stem. The center of the stem is called the pith. The pith is formed during the first year of growth of the stem; it is often distinguished from surrounding tissue by the type and size of cells present. Adjacent to and surrounding the pith is a layer of xylem. This layer represents the first tree ring produced in the stem. As a tree grows over time, additional layers of xylem or tree rings are deposited around the most recently formed ring of wood. Therefore, the oldest growth ring in a stem is found in the center of the stem, next to the pith (Figure 3.3). The youngest or most recently formed growth ring is toward the

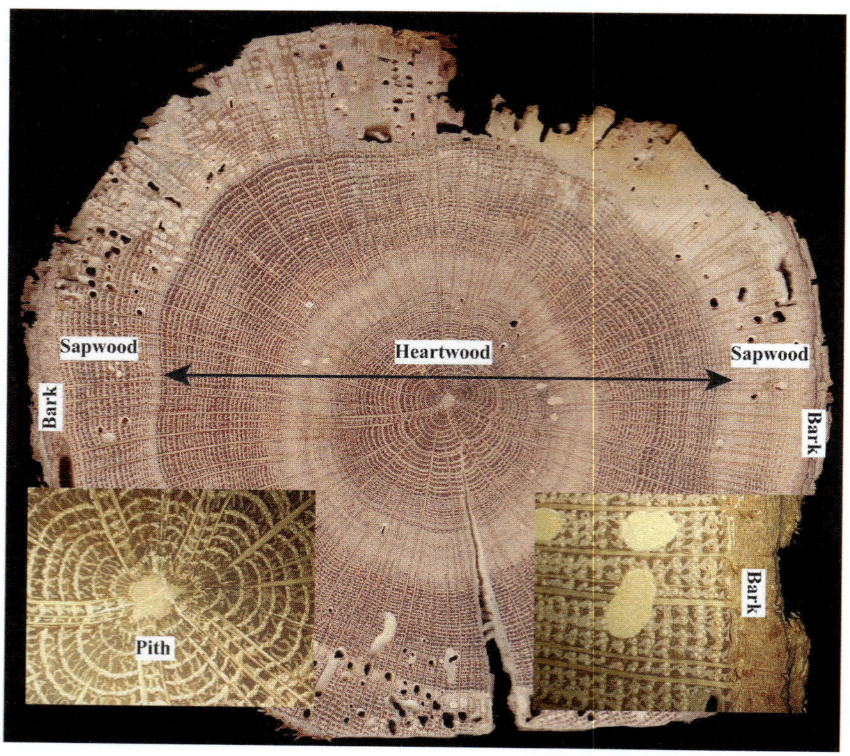

Figure 3.1
A cross section (1728–1853) of a white oak floor joist obtained from the Clay Hill Cabin (Chapter 9). Creating a cross section is achieved by cutting a timber perpendicular to the long axis of the stem. Cross sections of timbers are necessary for dating historic buildings because they allow for the identification of individual tree rings and the study of growth patterns in a tree. A magnified view of the pith and bark are shown in the lower left and lower right inserts, respectively. The black voids in the cross section and the large oval-shaped structures in the bark inset are insect damage.

Source: Authors.

outside of the stem, adjacent to the bark. Bark is a complex tissue that consists of many different cell types that serve a variety of functions in a woody stem (e.g., protection, gas exchange). Compare the cross section of Figure 3.1 (cross section taken from a mature tree that was harvested to make a floor joist of a log house) to Figure 3.2. The tree in Figure 3.1 is much older than the stems in Figure 3.2, but the distribution of the various tissues is still the same.

During each growing season, a tree ring is produced by a unique set of cells called the vascular cambium. Cells of the vascular cambium are found between the bark and the outermost cells of the tree ring formed in the previous year (Figure 3.2). When these cells divide, they produce unique tissue types to the inside (pith side) and toward the outside (bark side) of the stem. The vascular cambium is active during the growing season; it is reactivated each spring and is dormant during the winter. Tissue produced by the vascular cambium to the outside is called phloem. The phloem is considered to be part of the bark. Phloem serves several roles in a plant, including transport of sugars produced by photosynthetically active tissues of the plant. We often think of phloem as moving dissolved sugars from the leafy top of the plant down to the rest of the plant tissues that require energy; this is but one of its functions in a plant.

The tissue produced to the inside of the stem is xylem. Wood is xylem, and tree rings are layers of xylem that are produced by the vascular cambium each year. Xylem is a complex tissue and consists of several different types of cells (Figure 3.4) that serve different functions in a plant, such as a way to send chemical messages

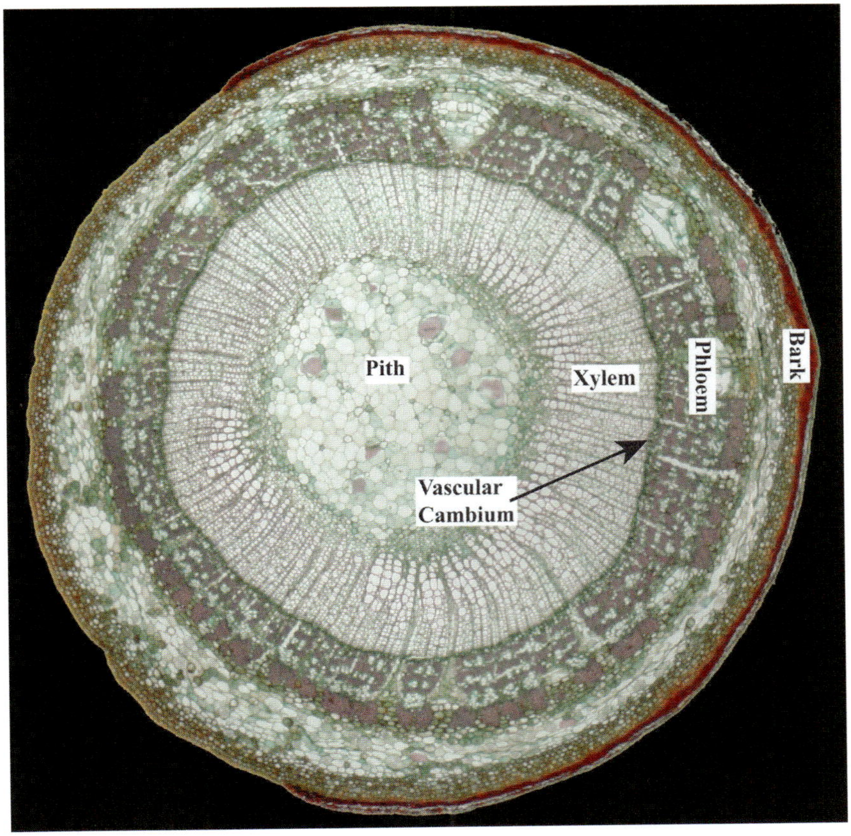

Figure 3.2
Cross section of a one-year-old basswood stem. The layer of xylem (wood) represents one year's growth. This layer is a tree ring. At this young age, bark may not have yet formed on the stem, but its position is marked.
Source: Authors.

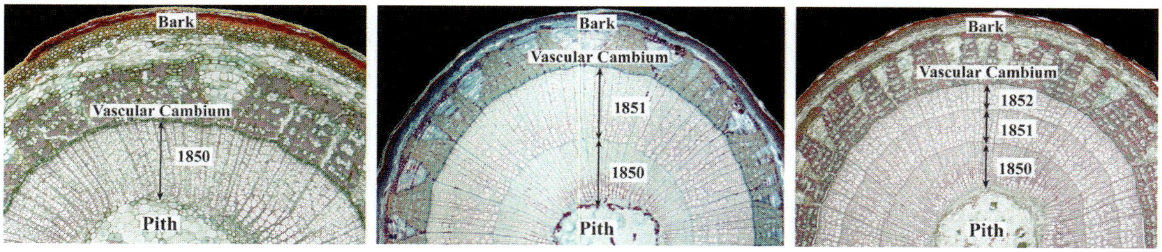

Figure 3.3
Cross sections of one-, two-, and three-year-old basswood stems. The oldest tree ring is the innermost layer of xylem cells (wood) surrounding the pith, and the most recently formed ring is next to where the bark will eventually form.
Source: Authors.

throughout a plant body, food storage, structural support, and conduction of water and dissolved minerals. We generally think of xylem as the transport system that caries water and nutrients from the roots to the rest of the plant body. At the beginning of each growing season, the vascular cambium creates new phloem and xylem cells; xylem toward the inside of the stem and phloem to the outside.

As a tree grows, it increases in both height and diameter. Height growth results from specialized tissue found in the buds at the tips of each branch. When these cells in the buds divide, elongation of the plant occurs. Elongation, therefore, occurs at the tips of the plant; this is why a sign nailed to a tree or initials carved in bark do not move higher and higher up the trunk as the tree grows. This leads to a very interesting property of a tree trunk, that is, it is not the same age from the base to the tip. Each year the vascular cambium produces a layer of cells around the circumference of the stem to produce a tree ring to the outside of the ring produced last year. In essence, a woody stem is produced by stacking or nesting layers of wood, like cones, on top of previously formed layers (Figure 3.5). The oldest part of the stem (with the most tree rings) will be toward the base, and the youngest part (with the fewest number of rings) will be toward the tip.

We stated previously that trees "generally" produce a single tree ring during each growing season. However, in some instances when trees are severely stressed or injured, they may not form a complete ring around the entire circumference of the stem. When a tree does not form a complete ring in a particular year, missing rings or locally absent rings occur (Figure 3.6). When dating wood of unknown age, it is essential to locate where rings are missing; otherwise, incorrect dates will be assigned to all of the rings that are formed after the missing ring. Crossdating, which is explained in Chapters 4 and 5, allows one to locate missing rings in a sample.

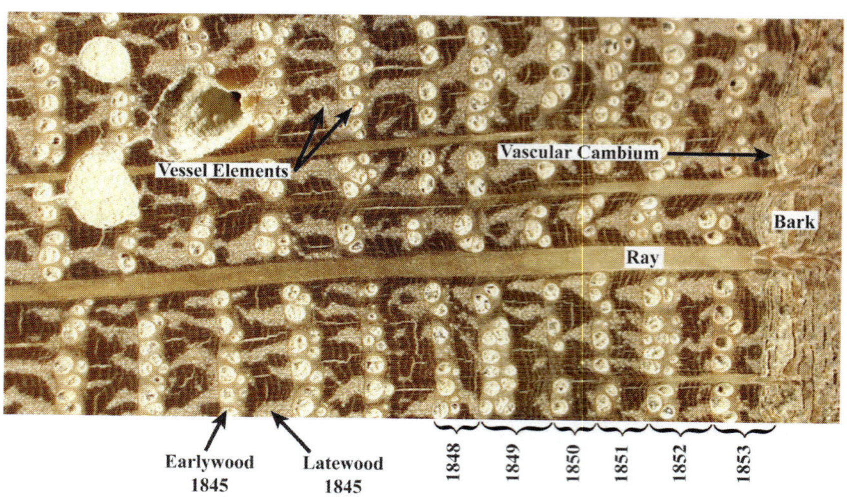

Figure 3.4
White oak wood in a highly magnified cross section. The distribution of the different cell types permits identification of wood type and the boundaries of individual tree rings. The tree from which this timber was made was harvested after the initiation of the 1853 growing season; 1853 is the ring adjacent to the bark. Note the large difference in vessel element size (white circular structures) between the earlywood and latewood. The earlywood and latewood formed in a single year form that year's tree ring. The structures running from left to right are rays (structures used to transport substances across a stem); the large void and circular areas (upper left) are insect damage.
Source: Authors.

Figure 3.5
A woody stem varies in age along its length. Each year a layer of wood is deposited on top of the previously formed increments of wood once the vascular cambium is activated during the growing season. The oldest part of the stem, is found at the base, and the youngest part near the growing (elongating) tip. The vertical line represents the pith of the tree, and the numbers represent the year of growth in the stem over four years.

Source: Authors.

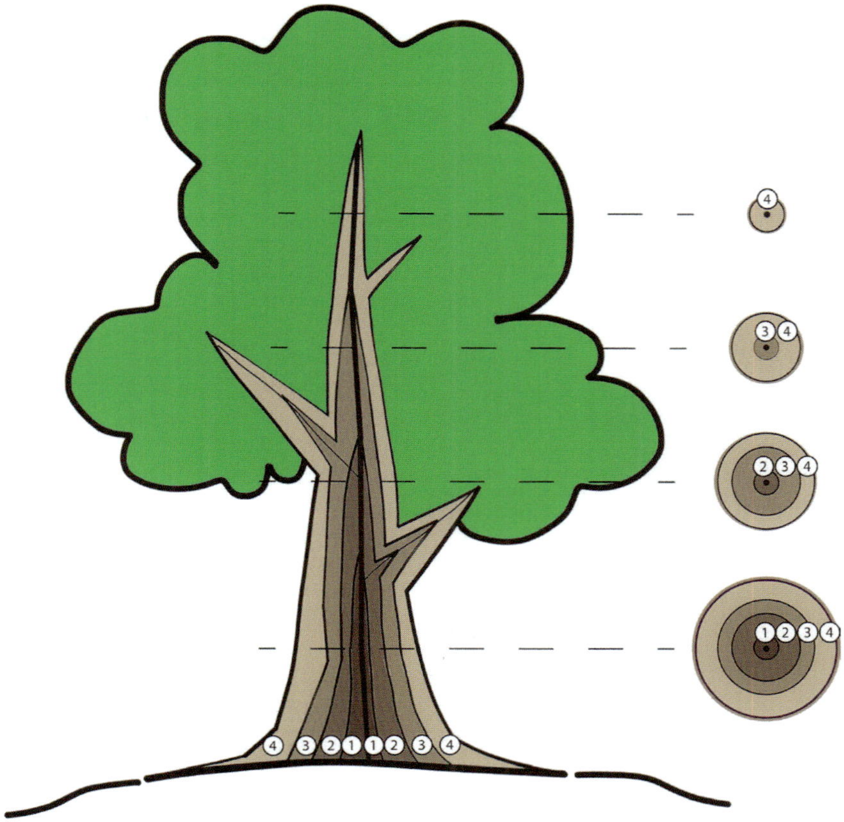

Figure 3.6
A locally absent or missing ring can occur when a tree is under stress or is injured. In this cross section of a tulip poplar, we placed pencil marks on ring 10 (left) and ring 20 (right; Chapter 4). Note that ring 15 is visible on one radius (bottom) but missing in another (top); the black arrow indicates where ring 15 would be in the upper radius. Missing rings can be identified through the process of crossdating tree-ring growth patterns (Chapters 4 and 5; scale bar equals 1 mm).

Source: Authors.

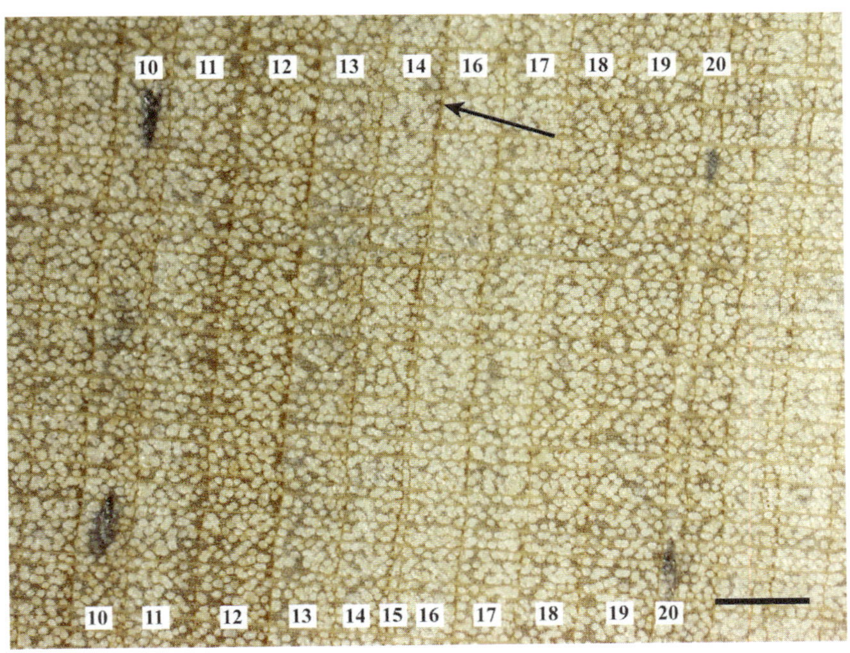

3.3 Wood words

There are many "wood" words, and they can be easily confused. A basic understanding of these words is necessary for understanding tree-ring analysis of historic buildings.

3.3.1 Hardwoods and softwoods

Technically, "hardwood" refers to any wood produced by a flowering plant (angiosperm). A flowering plant may not produce a large, showy flower like a tulip poplar, dogwood, or magnolia. For example, oaks, ashes, and maples are flowering plants but their floral structure is relatively inconspicuous; an oak is not planted for its floral beauty. Flowering plants produce fruits that enclose the seeds. "Softwood" refers to the wood produced by gymnosperms; they do not produce flowers or fruits. Common gymnosperms, often called conifers, include pines, spruces, firs, hemlocks, and Douglass fir. The use of the words "hard" and "soft" is a bit misleading. Not all hardwoods are harder or denser than softwoods. For example, hard pine (a group of anatomically similar pine species) is often used in outdoor construction (e.g., decks). It is harder than hardwood species such as aspen or basswood. Generally, hardwoods come from trees that lose their leaves each season (deciduous) while softwoods come from trees that maintain them (often needles) for longer periods than one growing season. However, exceptions do occur; larch, for example, is a softwood that keeps its leaves for only a single growing season.

Anatomically, the wood or xylem produced by hardwood and softwood trees is quite different, even though they both contain cells that are used to move water and materials up a woody stem, provide structural support, and store food. A major difference between the two wood types lies in the type of cells used to transport water up the stem. Hardwoods contain vessel elements, which are absent in softwoods. In contrast, softwoods use tracheids for conduction (Figure 3.7). Vessel elements and tracheids can be thought of as tiny straws; their long axis runs parallel with the length of a stem. These tiny, hollow straws are arranged in a stem so that each cell can move water up a stem. Movement can be accomplished by transferring water from adjacent cell to adjacent cell or, with vessel elements, from cells stacked one upon another to create a long, continuous conduit in the xylem tissue called a vessel. Hardwood (angiosperm) and softwood (gymnosperm) wood varies in several other key features, including the proportion of different types of cells and arrangement of cells. The differences in wood anatomy allows for wood identification (see Section 3.4).

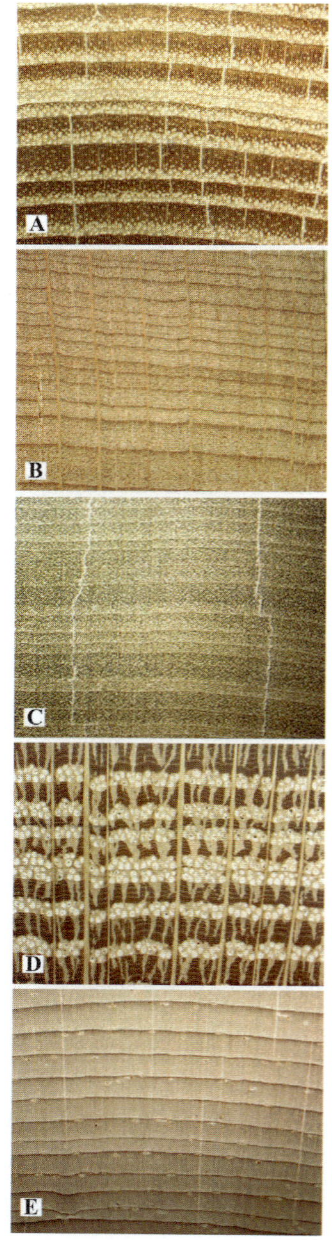

Figure 3.7
Cross sections of commonly found timber types in historically constructed buildings: (A) ash, (B) American beech, (C) tulip poplar, (D) white oak, and (E) white pine. For each cross section, the pith would be toward the bottom, and the bark would be toward the top, if the entire cross sections were pictured. Note the differences in size and distribution of the various cells in each of the different types of wood. These differences permit timber type identification and identification of individual tree-ring boundaries. Compare the diameter of the conducting elements in white pine (softwood) to the other timber types (hardwoods). The tracheids are so small in diameter that they appear as a mass of small white dots while the vessel elements of oak and ash are so large that individual cells can be identified. All of the cross sections were taken at the same magnification.
Source: Authors.

3.3.2 Earlywood and latewood

In stems, tree-ring growth and wood deposition does not occur year-round in temperate regions. We refer to the period during which wood is deposited around the circumference of a stem as the growing season. In the region where we perform our work (where the case studies in this text were performed), the growing season can be generally defined as the period between April and August. Trees will be in leaf for a longer time than these five months, but the current year's tree ring will not be increasing in size. The length, initiation, and cessation of the growth season is variable among individual years and different tree species (Pallardy 2007).

When the vascular cambium is active during the growing season, it produces new xylem and phloem cells. Depending on the time of the season, the xylem cells produced by the vascular cambium vary in size, type, and arrangement. In the beginning of the growing season, earlywood is formed, while later in the season latewood is formed. Conductive cells (tracheids and vessel elements) formed early in the season are often larger with thinner cell walls (the outermost layer of a cell that makes a cell an individual living unit) than cells formed later in the growing season (Pallardy 2007). Hence, in a single growing season, the wood (xylem tissue) that makes up a tree ring is composed of two distinct regions (Figure 3.4). An individual tree ring can be identified by noting the xylem found between two successive regions of earlywood (Figure 3.4). Simply put, an individual growth ring consists of the earlywood and latewood produced during one season of growth. This change in anatomy, along with other clues, permits identification of individual tree rings.

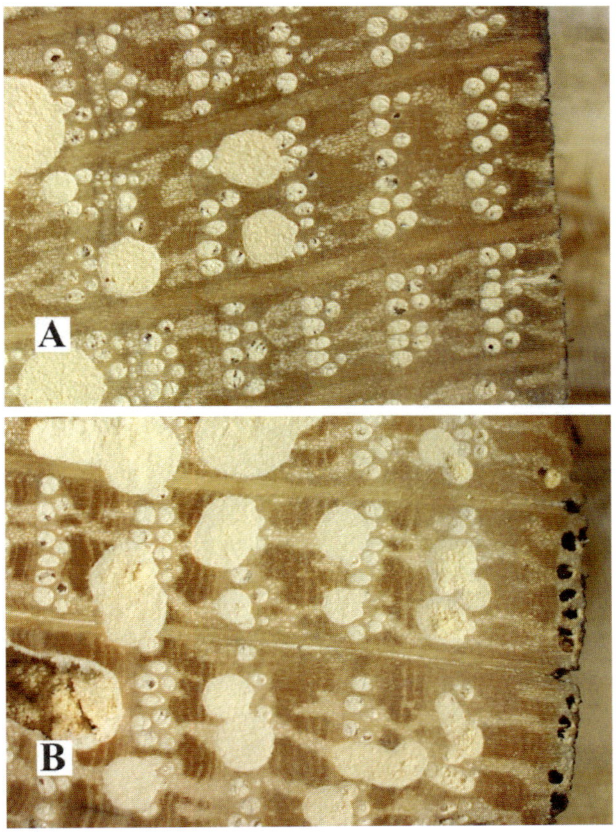

Figure 3.8
The season of felling can be determined by noting the anatomy of the tree ring that is adjacent to bark or taken from a waney edge of a timber. In the top sample (A), the outermost ring is dated 1844; it contains both earlywood and latewood that was formed in that year. In the bottom sample (B), the outermost ring consists of only earlywood formed in 1845. The outermost ring in each of these samples was obtained from white oak timbers with a waney edge. These samples were obtained from the Reed Case House at the Delphi Canal Park (Chapter 8).

Source: Authors.

As was stated earlier, when bark or wane is present on a timber, one can determine the year in which a tree was harvested by assigning a calendar date to the outermost ring. Furthermore, examination of the last tree ring formed can aid in determining when a tree was harvested (Baillie 1995). If the outermost ring (again, if it is adjacent to bark or from the wane) contains both earlywood and latewood, one can infer that the tree was cut sometime during the summer of the growth year but before the initiation of the next growing season (April). If only earlywood is present, the tree was likely cut during the early months of the year's growing season (April or May). For example, the samples obtained in Figure 3.8 were both obtained from the same structure. In Figure 3.8(A) we successfully dated the outermost tree ring in the timber to 1844. Since the outermost ring contains both earlywood and latewood, we can conclude that the tree was killed between June of 1844 but before April of 1845. In Figure 3.8(B) only earlywood is present in the outermost ring. The tree from which this timber was obtained died in the spring of 1845. From this information, we can gain a better idea of when wood was harvested to construct this building.

3.3.3 Heartwood and sapwood

Heartwood and sapwood refer to different regions in a stem (Figure 3.1). Heartwood is found toward the inside of a stem while sapwood is found toward the outside (near the bark). Though exceptions exist, generally, sapwood is the conductive portion of the stem (trunk) and the heartwood serves as structural support for the ever-growing and expanding stem. The two regions can often be identified by changes in wood color (Figure 3.1 and 3.9), with the sapwood usually being lighter than the heartwood. Sapwood width (number of rings) varies considerably among species and among individuals within the same species. Young stems may have no heartwood present. As a tree ages, its sapwood will be converted to heartwood. During this conversion, cells in the woody tissue die – changes in physiology and anatomy occur, food reserves are depleted, and chemicals are transferred across the stem and accumulate in the heartwood (Pallardy 2007).

Being able to identify the presence or absence of sapwood can be very beneficial when dating historic buildings. For instance, if a timber lacks bark or wane, an exact death or felling date for the tree used to make the timber is not determinable. In such a scenario, the outermost rings of a timber may have been removed during working of the timber (e.g., squaring a trunk to make a post; Figure 1.3) or lost to decay. However, if sapwood is present, one can reasonably assume that the outermost ring observable in the timber was near the bark, and an estimate of the cutting date of the tree can be made since the outer portion of the stem is available for dating (Baillie 1995; Haneca et al. 2009). Such estimates require that knowledge of the number of sapwood rings expected is available. Different species of trees have a differing number of rings in the sapwood. Also, tree age, growth rate, and environmental conditions can lead to variations in the number of sapwood rings within a single species (Hughes et al. 1981; Pallardy 2007; Sohar et al. 2012).

Figure 3.9
Tulip poplar window jamb from the Robert Glenn Apple House (Chapter 10). Note the change in color from the green heartwood to the yellowish-tan sapwood. Also note that the sapwood of this timber exhibits extensive insect damage, while the heartwood shows little.

Source: Authors.

Figure 3.9 is a cross section obtained from a window jamb made of tulip poplar. The building from which this sample was taken had no original timbers with wane or bark present so determination of an exact construction date was not possible. In this sample, there are many rings of sapwood present, but no bark or wane is present. The outermost date in this sample (1823) aids in determining the minimum age of the building. If the expected number of sapwood rings present in regional tulip poplar was available, we could better estimate when this building was constructed. Oral tradition states that it was built in 1825.

When no sapwood is present, it is not possible to determine how many rings were present between the outermost ring in the sample and the last ring formed by the tree (i.e., the amount of wood removed in fashioning a timber; Figure 1.3). When presented with only heartwood, one can only identify the earliest possible date of cutting of a tree to make a timber; the year of felling is some indeterminable time after the last year found in the sample. When dating historic buildings, we routinely focus sampling on timbers with wane or bark. If no timbers contain wane or bark, we must rely on those containing sapwood to provide an approximate construction date.

3.4 Wood identification

As noted earlier, xylem (wood) is a complex tissue made up of several types of cells. The types, sizes, and arrangement of cells in wood permit identification of the type of tree from which a timber was made. For example, hardwoods and softwoods can be distinguished by the presence or absence of vessel elements (described earlier). Hardwoods will contain large-diameter vessel elements whereas softwoods will not. Wood identification can be done at a much finer scale, to the species level for some timbers. For example, tulip poplar, a timber type found in many buildings we have analyzed, can be readily identified by noting the vessel arrangement and other anatomical characteristics of the wood (e.g., types and proportion of different cell types; Figure 3.7). Additionally, the common green color of the heartwood helps in initial identification of the wood (Figure 3.9). Likewise, using wood anatomy, we can positively identify American beech timber, another common species used in construction.

However, the exact species of tree from which a timber was obtained is not always possible since several sets of closely related species cannot be reliably separated using only wood anatomy (Panshin and de Zeeuw 1980). For example, timbers identified as "ash" are one of three species found in the region: white, green, or blue ash (Table 3.1). Wood anatomy, therefore only permits identification to the genus (*Fraxinus*) level and not the species level. Oaks (genus *Quercus*) can be divided into two subgenera, "white" oak (*Quercus* subgenus *Lepidobalanus*) and "red" oak (*Q.* subgenus *Erythrobalanus*). There are several different species present in both the "red" and "white" oak subgroups (hereafter, red oak and white oak, respectively). Using wood anatomy, red oaks can be separated from white oaks, but individual species within each group cannot be reliably identified. See Table 3.1 for a listing of the different regional species found in each group and other timber types encountered when sampling historically erected buildings in our study region. A summary of procedures used for identification of wood type is found in Chapter 5; Hoadley (1990) provides an excellent introduction to wood identification.

Table 3.1 Timber types encountered in historically constructed buildings and landscapes. The exact species of tree cannot be determined for certain timber types using only wood anatomy. Therefore, certain species are grouped into "timber type" (first column of the table). Timber type is what is reported throughout the text and in the tables providing sample information for each of the case studies presented in Chapters 8 through 13. In our analysis of regional buildings, we have encountered other timber types but have only listed those found in the case studies included in this text. Only those tree species that are found in the locations we sampled are listed (based on the county distributions prepared by Jackson (2004)). Taxonomy follows Gleason and Cronquist (1991).

Timber Type	*Common name*	*Species*
Ash	White ash	*Fraxinus americana* L.
	Green ash	*F. pennsylvanica* Marshall
	Blue ash	*F. quadrangulata* Michx.
Basswood	Basswood	*Tilia americana* L.
Beech	American beech	*Fagus grandifolia* Ehrh.
Black gum	Black gum; tupelo	*Nyssa sylvatica* Marshall
Black locust	Black locust	*Robinia pseudoacacia* L.
Butternut	Butternut	*Juglans cinerea* L.
Chestnut	Chestnut	*Castanea dentata* (Marshall) Borkh.
Cottonwood	Cottonwood	*Populus deltoides* Marshall
Elm		
Hard elm	Winged elm	*Ulmus alata* Michx.
	Rock or cork elm	*U. thomasii* Sarg.
Red elm	Red or slippery elm	*Ulmus rubra* Muhl.
White elm	White or American elm	*Ulmus americana* L.
Hickory	Pignut hickory	*Carya glabra* (Miller) Sweet
	Shellbark hickory	*C. laciniosa* (Michx. f.) Loudon
	Sweet pignut hickory	*C. ovalis* (Wangenh.) Sarg.
	Shagbark hickory	*C. ovata* (Miller) K. Koch
	Mockernut hickory	*C. tomentosa* (Poiret) Nutt.
Honey locust	Honey locust	*Gleditsia triacanthos* L.
Maple		
Hard maple	Black maple	*Acer nigrum* Michx. f.
	Sugar maple	*A. saccharum* Marshall
Soft maple	Boxelder	*Acer negundo* L.
	Red maple	*A. rubrum* L.
	Silver maple	*A. saccharinum* L.
Oak		
Red oak	Scarlet oak	*Quercus coccinea* Muenchh.
	Northern pin oak	*Q. ellipsoidalis* E.J. Hill
	Southern red oak	*Q. falcata* Michx.
	Shingle oak	*Q. imbricaria* Michx.
	Black-jack oak	*Q. marilandica* Muenchh.
	Cherrybark oak	*Q. pagoda* Raf.
	Pin oak	*Q. palustris* Muenchh.
	Northern red oak	*Q. rubra* L.
	Shumard oak	*Q. shumardii* Buckley
	Black oak	*Q. velutina* Lam.
White oak	White oak	*Quercus alba* L.
	Swamp white oak	*Q. bicolor* Willd.
	Overcup oak	*Q. lyrata* Walter
	Bur oak	*Q. macrocarpa* Michx.
	Swamp chestnut oak	*Q. michauxii* Nutt.

Timber Type	Common name	Species
	Chinkapin oak	*Q. muehlenbergii* Engelm.
	Rock chestnut oak	*Q. prinus* L.
	Post oak	*Q. stellata* Wangenh.
Pine		
Hard pine	Sand pine	*Pinus clausa* (Chapm. ex Englem.) Vasey ex Sarg.
	Yellow or shortleaf pine	*P. echinata* Miller
	Slash pine	*P. elliottii* Engelm.
	Spruce pine	*P. glabra* Walt.
	Longleaf pine	*P. palustris* Miller
	Table mountain pine	*P. pungens* Lambert
	Pitch pine	*P. rigida* Miller
	Pond pine	*P. serotina* Michx.
	Loblolly pine	*P. taeda* L.
	Scrub or Virginia pine	*P. virginiana* Miller
White pine	White pine	*Pinus strobus* L
Sweet gum	Sweet or red gum	*Liquidambar styraciflua* L.
Sycamore	Sycamore	*Platanus occidentalis* L.
Tulip	Tulip poplar	*Liriodendron tulipifera* L.
Walnut	Black walnut	*Juglans nigra* L.

Understanding what types of wood were used in the construction of historically erected buildings is important for various reasons. For example, do the species used in construction help us to better understand what tree species were found in the landscape when the forest was cleared for construction and/or agricultural purposes? Or, one could ask what were the impacts on the landscape of selecting a particular species or subset of species from local forests?

Additionally, determining wood type assists in the dating of a building. Growth patterns vary among species even when they are growing in close proximity (e.g., Graumlich 1993; Cook et al. 2001; Martin-Benito and Pederson 2015). From experience, we have found that dating timbers is most reliable when comparing samples of unknown age to dated samples of the same timber type. For example, we analyzed a log home built from tulip poplar and white oak. The growth patterns in the white oak samples matched (crossdated; Chapters 4 and 5) with each other very well – we found a common, unique growth pattern among them. Similarly, the tulip poplar samples matched each other very well. However, pattern matching between the two timber types was not reliable. Ultimately, to obtain calendar dates for the tulip poplar we had to use a regional tulip poplar chronology (a collection of previously dated timbers) instead of the white oaks that were obtained from the same structure. When we attempt to date a building or landscape we, therefore, identify the timber type of each timber prior to analysis and analyze each type separately.

Also, understanding what types of wood are used in construction helps in determining the geographic origin of timbers. When Europeans settled the region, they had at their disposal vast amounts of timber to construct their homes, smokehouses, barns, agricultural buildings, stores, churches, and schools. Most regional structures we have analyzed are constructed from a fairly consistent set of hardwoods: tulip poplar, white oak, American beech, and ash. The use of these timber types is expected since regional forests were dominated by hardwoods such as oak, maple, ash, basswood, tulip poplar, and American beech (Braun 1950). However, we do occasionally encounter softwoods such as hemlock and pine (white and hard) in regional buildings. The distribution of these softwoods is extremely spotty throughout the landscape and they only grow in very isolated, geographically removed stands in the region (Jackson 2004). Given the paucity of these species throughout the landscape, it is hard to imagine enough trees could be harvested for timber production. Therefore, when encountering these timber types, one must wonder about the origin of the timber. Tree-ring pattern matching and dating would not be possible since these timbers did not originate from the area, and local reference material is best for dating wood of unknown age. In Chapter 13, analysis and determination of provenance or origin of softwood timbers is discussed in detail.

3.5 Conclusion

An understanding of tree-ring formation and woody plant anatomy and morphology is needed to fully appreciate how tree rings can be utilized in the dating of historically constructed buildings and landscapes. Now that you understand the basics of how a tree grows and what a tree ring is, we will examine how dendrochronologists determine the exact year of formation for individual tree rings in Chapter 4.

3.6 Literature cited

Baillie, M.G.L. 1995. *A Slice Through Time: Dendrochronology and Precision Dating*. Routledge, London, UK. 176 p.

Braun, E.L. 1950. *Deciduous Forests of Eastern North America*. The Blackiston Co, Philadelphia, PA. 596 p.

Cook, E.R., J.S. Glitzenstein, P.J. Krusic, and P.A. Harcombe. 2001. Identifying functional groups of trees in west Gulf Coast forests (USA): A tree-ring approach. *Ecological Applications* 11: 883–903.

Evert, R.F., and S.E. Eichhorn. 2013. *Raven Biology of Plants*. 8th ed. W.H. Freeman and Company Publishers, New York, NY. 727 p.

Gleason, H.A., and A. Cronquist. 1991. *Manual of Vascular Plants of Northeastern United States and Adjacent Canada*. 2nd ed. New York Botanical Garden, New York, NY. 910 p.

Graumlich, L.J. 1993. Response of tree growth to climatic variation in the mixed conifer and deciduous forests of the upper Great Lakes region. *Canadian Journal of Forest Research* 23: 133–143.

Haneca, K., K. Čufar, and H. Beeckman. 2009. Oaks, tree-rings and wooden cultural heritage: A review of the main characteristics and applications of oak dendrochronology in Europe. *Journal of Archaeological Science* 36: 1–11.

Hoadley, R.B. 1990. *Identifying Wood: Accurate Results with Simple Tools*. The Taunton Press, Newtown, CT. 223 p.

Hughes, M.K., S.J. Milsom, and P.A. Leggett. 1981. Sapwood estimates in the interpretation of tree-ring dates. *Journal of Archaeological Science* 8: 381–390.

Jackson, M.T. 2004. *101 Trees of Indiana*. Indiana University Press, Bloomington, IN. 364 p.

Kozlowski, T.T., P.J. Kramer, and S.G. Pallardy. 1991. *The Physiological Ecology of Woody Plants*. Academic Press, San Diego, CA. 657 p.

Martin-Benito, D., and N. Pederson. 2015. Convergence in drought stress, but a divergence of climatic drivers across a latitudinal gradient in a temperate broadleaf forest. *Journal of Biogeography* 42: 925–937.

Pallardy, S. 2007. *Physiology of Woody Plants*. 3rd ed. Academic Press, Burlington, MA. 464 p.

Panshin, A.J., and C. de Zeeuw. 1980. *Textbook of Wood Technology*, volume 1. 3rd ed. McGraw-Hill Book Company. New York, NY. 705 p.

Sohar, K., A. Vitas, and A. Läänelaid. 2012. Sapwood estimates of pedunculated oak (*Quercus robur* L.) in Eastern Baltic. *Dendrochronologia* 30: 49–56.

Chapter 4

Tree-ring basics for the historian, archaeologist, and preservationist

4.1 Introduction to dendrochronology

We introduced this text with a simple question that is often asked by owners and managers of historically constructed buildings – "Can you date my building?" – and we gave an ambiguous answer, "maybe." In order to better explain this vague response, we will use this chapter (and the following chapter on methodology) to introduce you to some of the fundamentals of tree-ring science. By better understanding the basic tenets of dendrochronology we hope that the reader will better understand our seemingly flippant answer.

The goal of this chapter is to provide a basic introduction to tree-ring science and to demonstrate how dating historic buildings and landscapes is possible with tree rings. Without some concept of how dendrochronologists work and the basics of tree-ring science, dates of structures and landscapes, we feel, are not all that meaningful. When construction dates are provided for a structure, much focused work and analysis was performed by a scientist(s) who appreciates and understands the complexities and guiding principles of dendrochronology and their application. A thorough discourse on the rich and fascinating history of dendrochronology and all of the applications of tree-ring science (including the use of tree rings to date historic objects) is far beyond the scope of this text. Therefore, for accounts of the history of the development of this field of study, we strongly suggest that the reader consult Baillie (1982), Nash (1999), and Speer (2010).

Broadly speaking, dendrochronology utilizes accurately dated tree rings to address environmental and historical questions (Kaennel and Schweingruber 1995). Dendrochronology allows an investigator to study past events, if they can be dated by tree rings (Speer 2010). Researchers have successfully used tree rings to ask and answer a broad number of questions and to explore many natural phenomena. The science of dendrochronology can be broken down into a number of different subfields based on the type of question(s) a researcher is asking. For example, tree rings, and the annual precision they provide, are excellent for investigating ecological questions such as understanding the disturbance history of a forest. In such an investigation, a researcher can identify sudden and sustained increases in annual growth. Such a growth pattern could be attributed to a disturbance event (e.g., windstorm, tornado) that released a tree from competition for resources (light, for instance) from neighboring trees (Figure 4.1). An investigation such as this would fall under the subdiscipline of dendrochronology known as dendroecology.

Similarly, a researcher may be interested in knowing the years in which fires occurred in a forest. Such a study would fall into the subfield or subdiscipline of dendropyrochronology. Tree-ring research focusing on how climate affects tree growth or how climate has changed over long periods of time would be classified as dendroclimatology. Tree rings can be used to study volcanic eruptions, earthquakes, erosion, floods, insect outbreaks, plant community dynamics, and glacial movement (Dean 1997). An exhaustive list of the unique application of dendrochronological techniques and subfields of dendrochronology can be found in Speer (2010). The wide application of tree rings to investigate the natural world is a testament to the versatility of tree-ring science and the creativity and ingenuity of tree-ring researchers.

Application of dendrochronological techniques, however, is not limited to the study of natural events; tree-ring science is well suited for the study of human history and how humans interact with their natural world (Bannister 1963; Dean 1997). The subfield of dendroarchaeology utilizes tree rings to date wooden objects, which includes furniture, musical instruments, panels used in a painting, ships, or buildings (Baillie 1982, 1995; Speer 2010) Dendroarchaeological techniques can be used to determine when "timber has been felled, transported, processed, and used for construction" (Kaennel and Schweingruber 1995). In this text, we explore how tree rings are used to date the construction and modification of historical buildings and landscapes.

Figure 4.1
Cross section of a white oak floor joist. Note the variability in the tree-ring widths over the lifespan of the tree. The size of individual tree rings is an aggregate response to the environmental conditions experienced by the tree in a given year. During years of suppression (many-year-long period of little growth), very small rings are formed. Rings will appear to contain only earlywood during such periods (left inset). During releases, when resources are more plentiful, larger rings are formed (right inset). Both insets were taken at the same magnification.

Source: Authors.

Dendrochronology (and its numerous subdisciplines), like all scientific fields, is built on a set of basic principles and core concepts (Fritts 1976; Speer 2010; Grissino-Mayer 2017). These principles and concepts allow dendrochronologists to confidently date tree rings to the precise calendar year in which they were formed and to ask and answer tree-ring related questions. Later we discuss some of the basic principles that guide research focusing on dating the construction of historic buildings and landscapes: the aggregate tree growth model, limiting factors, replication, and crossdating.

4.2 Variability in tree-ring width

Tree-ring width varies from year to year. Examination of a series of tree rings in a cross section allows a dendrochronologist to note the variability in growth in a tree over time (Figure 4.1). Some years are marked by very little growth, while others show above average growth (ring width). Trees may exhibit extended periods during which they produce very small or quite large tree rings. Also, periods of average growth may be interrupted by a particularly small or a particularly large ring. Each tree, therefore, exhibits a unique growth pattern over its lifespan or over that portion of its lifespan that has been preserved in a timber.

What causes these growth patterns? In other words, why are not all tree rings produced in a tree the same size? A multitude of factors influence the amount of wood (the width of a tree ring) produced in any given year. These factors relate to the amount of various resources (water, light, etc.) that are available for growth in a particular year. According to the aggregate tree growth model, the ring width in a given year is a cumulative response to various factors experienced by a tree in a given year. These factors include, but aren't limited to, age, disturbance events, and climate (Cook 1992). For example, in droughty years, smaller than average tree rings will be deposited, because the tree is stressed (Figure 4.2). A severe drought may affect growth not only during the year in which it occurred but also in the subsequent year(s).

Disturbances in a forest also affect the width of a tree ring in a given year; these disturbances can come from within the forested stand in which a tree is growing or from outside the stand. For example, if a tree dies (windstorm, pathogens, etc.), neighboring trees will be released from resource competition with this tree (Rubino and McCarthy 2004; Trouet et al. 2017). An increase in light, a much needed resource, will result in an increase in photosynthesis and food (sugar) production and ultimately an increase in growth rate (Figure 4.1). Disturbances can also lead to a decrease in growth. For example, insect outbreaks that damage the tree's tissue (defoliation) will lead to the production of smaller annual increments.

The age (or size) of the tree will affect the amount of growth in a given year. When young, trees often produce larger rings. As trees age, individual growth increments are smaller, because the tree is depositing wood over an ever-increasingly larger plant body (Figure 3.5). According to the aggregate tree growth model, other unknown or not well-understood factors can also affect growth in a year. These factors may include phenomena such as masting, years in which a tree dedicates a large amount of energy to the production of fruits (Speer 2010).

So which factor is the most important in regard to how large or how small of a tree ring will be deposited in a given year. It can vary from year to year. A core concept of dendrochronology is the principle of limiting factors, which relates to Liebig's (1840) law of the minimum. This concept states that, in a given year, a plant is only able to grow as much as the most limiting factor allows; the environment does not

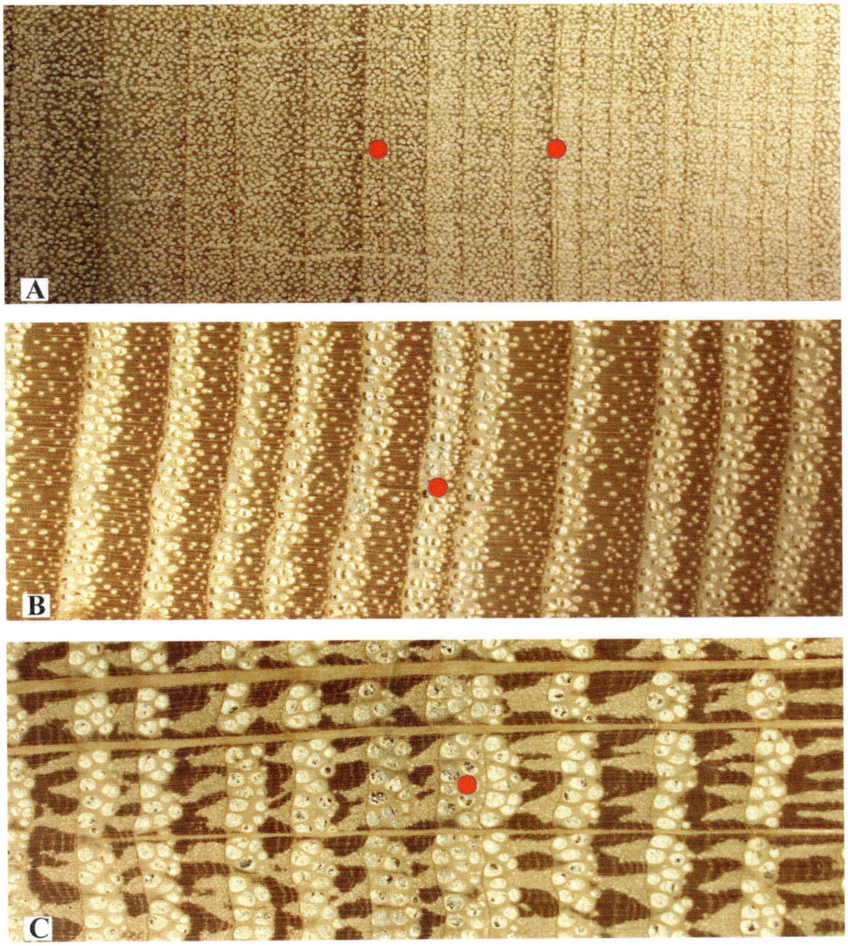

Figure 4.2
During stressful years such as times of drought, a limiting factor can cause a decrease in growth (i.e., a small tree ring). Note the one-year reduction in ring size (red dots) in these tulip poplar (A), ash (B), and white oak (C) samples. During the stressful year, the annual growth for the white oak consists of only earlywood. All images taken at the same magnification.

Source: Authors.

provide enough of all necessary resources to enable maximum growth (Barbour et al. 1999). Hence, one resource can limit the amount a plant will grow in a season. For example, if a tree is in competition for light from neighboring trees, it will produce a smaller ring even if water and nutrients are plentiful in that year. The tree may show a prolonged period of little growth (small tree rings) referred to as a suppression (Figure 4.1). A tree will, therefore record in its rings periods in which competition for some factor limited its growth rate. If the tree survives to experience less stressful growing conditions, it may exhibit a subsequent release in growth, a prolonged period marked by wide tree rings (Figure 4.1). During this release, some other factor would become a limiting factor to growth (ring width), such as water availability.

Certain environmental stresses affect many trees in a widespread geographic region simultaneously. These stresses will lead to all of the affected trees having similarities in their growth pattern. During a droughty year, for instance, trees will exhibit a sharp decline in growth for that year (Figure 4.2). Likewise, a late frost (one occurring after the growing season is well underway) can affect a large area and leave a signature in the affected trees' rings (Figure 5.20). Events such as these are very important and necessary for dating timbers from historic buildings.

These events impart a predictable signal in the growth of a tree. For example, in our study region, we can expect small rings in 1774, 1803, 1834, and 1839. Some environmental stress(es) caused a large proportion of trees to produce small (and oftentimes absent) rings during those years. We consistently observe this pattern in many different types of timber. Ultimately, these ring-width patterns make dating historical buildings and landscapes possible through the process of crossdating.

4.3 Crossdating

Now that we see why variation from year to year occurs and how growth patterns are "generated" in a series of tree rings, we turn to the principle of crossdating. Fritts (1976) states that crossdating "is the most important principle of dendrochronology"; Speer (2010) calls it "the basic tenet of dendrochronology"; and Dean (1997) says crossdating is the "one immutable principle of tree-ring science." It is through crossdating that dendrochronologists are able to confidently assign a precise calendar year to an individual tree ring.

Crossdating is performed by matching growth patterns among tree-ring samples. Without crossdating, assigning dates to tree rings found in the timbers of historical buildings would be impossible; simply put, crossdating is fundamental. When samples are taken from a building, the age of the timbers is unknown – one does not know when the tree from which the timber was made began growing or when it was felled. Dates can be determined by matching the pattern of growth in a sample of unknown age to the patterns found in accurately dated samples (Figure 4.3). Crossdating of samples can be performed using a variety of methods. We utilize skeleton plots (graphical representation of growth patterns) and statistical analysis (correlation analysis of tree-ring widths using a computer program) to crossdate the tree-ring growth pattern in timbers. An in-depth explanation of the crossdating process using skeleton plots and measured tree rings is provided in Chapter 5, and exercises and examples of crossdating are provided in Chapter 14.

Crossdating has proven to be an extremely effective method of dating wood of unknown age and has been used throughout the world to date buildings and wooden objects of unknown age (the subfield of dendrochronology known as dendroarchaeology). In order for crossdating, and, therefore, dating historic buildings and landscapes, to be possible, several conditions must be met (again, "can you date my building?").

In order for crossdating to be successful, a reference or master chronology is needed. A chronology is a set of accurately dated tree rings collected from a particular site (which can vary in size from a single forest to an entire geographic region). A chronology provides a record of the pattern of tree growth over time. Chronologies are built from multiple samples so that the "average" expected growth for a particular year can be determined. Individual trees vary genetically and will not physiologically react to environmental cues, such as stress, similarly. Additionally, though growing in the same area (even if growing in the same forested stand), individual trees experience the environment differently. Not all trees in a site receive the same amount of shade, light, moisture, nutrients, and competition in a given year. For example, trees from valleys or near streams will likely have very different levels of available moisture than those removed from bodies of water or on higher topographic positions like ridges (Anning et al. 2013). Consequently, trees, even though they are growing in close proximity, may react differently to stresses such as droughts in a given year. Through replication, collection of more and more samples, a better representation of growth can be obtained. Therefore, the most useful ref-

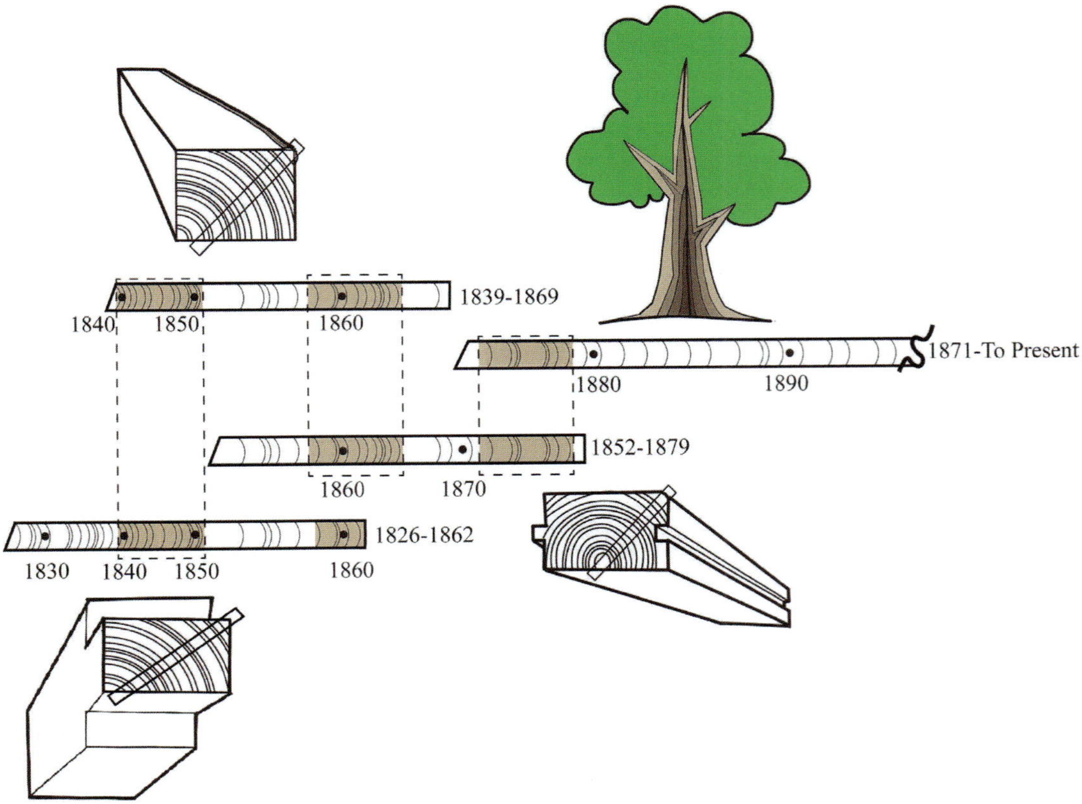

Figure 4.3
Crossdating permits dating timbers of unknown age. Crossdating is performed by matching unique growth patterns (tan) across samples. Note: the sample lengths used for this demonstration illustration are much shorter than those needed to perform an actual analysis.

Source: Authors.

erence or master chronologies should be built from numerous samples and cover a long period of time. These robust chronologies will enable reliable crossdating. See section 4.4 Creating Reference Chronologies, for an explanation of how regional chronologies can be built for reliable crossdating. For crossdating to be successful, individual timbers must have enough tree rings present to permit pattern matching. We often encounter timbers (usually sawn, dimensional timbers like floorboards, braces, or members of roof trusses) that do not have enough tree rings to allow for reliable crossdating (Figure 1.6). When samples have too few rings present, one cannot be certain that the rings in an undated sample truly match a region's expected pattern of growth. Some timbers, unfortunately, are simply not dateable.

From the preceding discussion it is evident that annual variation is key to the crossdating process. In order for crossdating and assignment of calendar dates to a timber's tree rings to be possible, trees must exhibit sensitivity or variation in growth from year to year. Compare the variability in ring widths observed in Figure 1.7. A lack of variation in growth pattern in a timber will not permit reliable crossdating – there is no pattern to match. Such a growth pattern would be considered complacent and not sensitive. Trees may exhibit complacency if their growth has not been routinely stressed by some limiting factor. Complacency may result from a tree having abundant resources (such as water and light) from year to year. Dating requires sensitive series.

Additionally, the reference chronology and the timbers being dated must overlap temporally (i.e., in time). It is not possible to crossdate a timber with tree rings ranging from 1630 to 1750 if the reference chronology spans from 1770 to 1890. Crossdating is only possible if the tree rings in a timber and the reference chronology overlap a sufficient number of years (Figure 4.3). A lack in overlap has made dating early 20th-century buildings in our study region problematic. Similarly, dating early 20th-century modifications and repairs has been hampered since our reference chronologies for certain timber types are not well-replicated in the late 19th and early 20th centuries.

4.4 Creating reference chronologies

As was stated earlier, assigning calendar dates to the tree rings in historic timbers relies on the presence of local chronologies that can be used for crossdating. Regional chronologies are initially built by taking samples from living trees. These samples are most commonly obtained by using an increment borer. An increment borer is a device that allows for the extraction of a core from a living tree without killing it (Figure 4.4; see Speer (2010) and Grissino-Mayer (2003) for a discussion of increment borers and coring techniques for living trees). Using living trees "anchors" a chronology in time since one knows when a tree-ring sample was taken. Dating living trees can be a fairly straightforward process. The tree ring adjacent to the bark is the most recently formed tree ring. If the tree is cored after the initiation of the growing season, this ring represents the current year's growth. Proceeding backwards from the bark, years can be assigned to each tree ring. By using standard dendrochronological techniques (skeleton plotting, measuring ring width, crossdating, ensuring no missing rings are present, etc.; Chapter 5), one is able to create a local chronology and to identify the growth pattern or tree-ring signal for a region.

Figure 4.4
(A) An increment borer is used to obtain cores from living trees. The boring bit (center) is inserted into the handle (blue; top). The extractor is used to remove the core from the tree by inserting it into the borer once it has been turned into the tree. The inset is the threaded tip of the increment borer that pulls the borer into the tree as the handle is turned. (B) Coring trees with an increment borer allows for the collection of tree-ring samples without the need to cut down a tree. (C) Removing a core (arrow) from a tree. (D) A mounted and dated tree core obtained from a living oak tree (1889–2014).

Source: Authors.

In addition to creating reference chronologies by sampling local, living trees, reference chronologies can be obtained from the International Tree-Ring Data Bank (ITRDB). The ITRDB is a searchable online database (www.ncdc.noaa.gov/data-access/paleoclimatology-data/datasets/tree-ring) that contains tree-ring data that have been contributed by researchers from around the world. The ITRDB is designed to help dendrochronologists share data and to archive quality tree-ring data sets (Grissino-Mayer and Fritts 1997). When we began our work in the region (mid-Ohio River Valley, United States), we were able to obtain white oak chronologies from living trees from the ITRDB. Although these chronologies were built some distance from our study area (Figure 4.5), we were able to successfully crossdate many white oak timbers with them and to begin the creation of a more regional database of measured and dated tree rings.

As we continued to sample more and more buildings and landscapes, however, we started to encounter buildings that were not fully (or predominately) built from white oak timbers. As we started to amass samples from American beech, tulip poplar, and ash timbers, we quickly realized that crossdating between different timber types was not always reliable (Chapter 3). Given the number of buildings constructed of non-oak timber we decided that the construction of non-oak chronologies was essential if we were to continue our work.

Unfortunately, in our study region (Rubino 2014), and in many regions throughout the world, finding trees of sufficient age to date historically constructed buildings is difficult if not impossible – chronologies from living trees are just not long (old) enough. Old-growth forests, forests that contain long-lived trees that

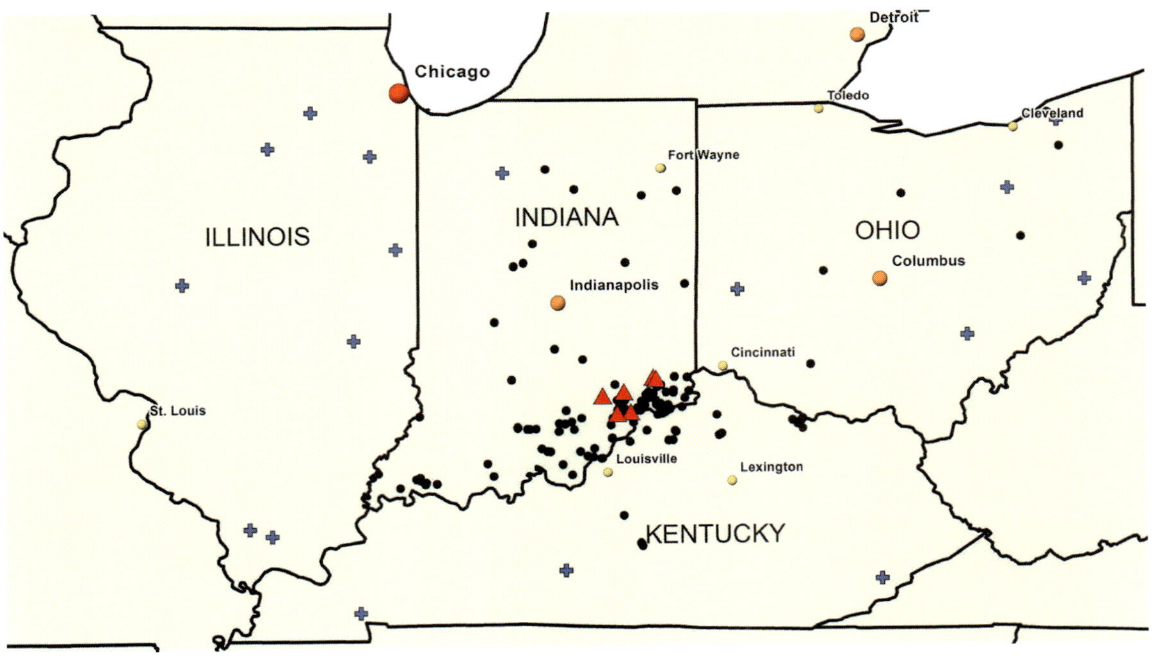

Figure 4.5
Successful crossdating necessitates reference chronologies consisting of accurately dated tree-ring series. The chronologies we use have been obtained from the International Tree-Ring Data Bank (blue crosses) and created from regional forests (red triangles) and from historically constructed buildings (black dots).

Source: Authors.

may be old enough to be used for crossdating timbers, are often off limits to tree-ring study or are found in marginal habitats where growth patterns may not be useful for crossdating since the trees are severely stressed (Stahle and Chaney 1994).

To create additional chronologies for the various timber types we encountered in historic buildings, we sampled several local forests throughout the area. However, few living trees were old enough to reach back in time far enough to be useful for crossdating the timbers from historically constructed buildings. We had identified only a handful of old oak, ash, and American beech trees – far too few for reliable crossdating since we had little replication. We turned our attention to several old-growth forests where we were given permission to sample recently fallen trees (no living stems). We created chronologies from several old-growth forests from throughout the region by chain sawing slabs from dead trees (Alexander et al. 2018). To date these slabs, we crossdated them with the living trees we had sampled (Figure 4.5). Many of the slabs dated back to the 17th and 18th centuries and were very useful in initial chronology construction.

To make these chronologies more robust, we needed to enlarge our sample size (replication) and extend the time span they covered. In order to do this, we decided to increase our sampling effort. Normally, when dating historically constructed buildings, we focus our sampling on wane- and bark-bearing timbers, since our major goal is providing a potential construction (and sometimes modification) date for a structure. We had to modify our sampling procedures. We began collecting as much wood as possible when visiting sites. We no longer simply took samples from large, *in situ* timbers. We began taking samples from floorboards, baseboards, and roof sheathing. When visiting active renovations, we gathered as many samples as possible from as many timber types as possible. Several of our chronologies were built by dumpster diving, scavenging burn piles before they were set ablaze, and sifting through piles of timbers strewn about by heavy machinery (Figure 1.8). Heavily damaged timbers (decayed and insect infested), despite their outward appearance, proved to be very fruitful in constructing regional chronologies for a number of different timber types (Figure 4.6).

Figure 4.6
Timbers with extensive decay and insect damage can still be analyzed. This white oak timber dates from 1710 to 1828.
Source: Authors.

We also collected extensively during demolitions. Structures were documented as well as possible, and we collected as many timbers as possible. We also found it very useful to build working relationships with local tradespeople that routinely or exclusively restore, renovate, and rehabilitate historically constructed buildings. We found that they have a large stock of timbers collected from previous jobs. They very generously gave us permission to take cross sections from their timber stores. For example, during the renovation of the *Musée de Venoge* (Chapter 9), workers amassed a large number of timbers from several regional buildings on the site. These timbers, whose exact provenance was unknown or uncertain, were going to be incorporated into the renovated building. We were able to collect a large number of timbers, and it was from these cross sections that we were able to construct a well-replicated ash chronology.

We also sample(d) many "mundane" sites in order to create our chronologies. Many of the structures we sampled hold seemingly little historic meaning except to that of the building's owner. We were (are) not concerned that famous politicians or civic leaders were never associated with a structure. Though many of these buildings are modest and will not qualify for landmark status by government agencies, they are important scientifically and culturally. An extant structure or landscape is important enough on its own merits. These vernacular buildings provide an opportunity to study how buildings were historically constructed – the way timbers were prepared, how joints were made, and what species of trees were used (Chapters 2 and 13). A barn is not just a structure, it is a repository of the vast forests that once covered the region, the forests European settlers encountered and harvested for their houses, barns, churches, and towns. Historic buildings are likely our only way to look into the past to study how a large number of trees grew over a several-century span.

To date, we have created well-replicated tree-ring chronologies by sampling over 200 buildings and hundreds of living trees (Figure 4.5). Over the past 15 years we have been able to produce robust ash, tulip poplar, American beech, white oak, walnut, red oak, and white elm chronologies, and we have created less well-replicated chronologies for several other timber types. We plan on expanding all of our chronologies by adding more sites and by enlarging our sampling region in the future. We now have tree rings that reach farther back in time than living trees found in extant forests. Our oldest chronology is for white elm; it extends from 1436 to 1871. By combining samples obtained from living trees, fallen trees, and historic buildings, we have been able to create continuous chronologies that span from the present to the 15th century for some timber types. For example, our tulip poplar chronology extends, uninterrupted, from 1456 to 2017. Our goals have been, and will be, to continue to crossdate more and more samples so that we can document more buildings and landscapes, better document historic tree growth, and understand how forests were utilized for timber use (Chapter 13).

4.5 Conclusion

Dendrochronology, tree-ring science, is rooted in key principles and concepts. The dating of historic structures is made possible by these scientific tenets. Accurately assigning dates to the tree rings of timbers sampled from historic buildings and landscapes requires certain conditions to be met. Reliable reference chronologies must be available, for without them there is no way to match or crossdate the growth patterns in timbers of unknown age. Additionally, samples taken from historically constructed buildings must exhibit sufficient variation (sensitivity), have enough rings to allow for pattern matching, and overlap temporally with a reference chronology. Additionally, these samples must also be of the correct timber type and, ideally, have been obtained from timbers with bark or wane so that a death date or felling date can be determined.

Let us return to our initial question proposed in the introductory chapter to this text and this chapter – "Can you date my building?" Hopefully you can better understand now why we answered "maybe." In the next chapter, we explore how tree-ring samples from timbers are obtained and elaborate on how we can accurately assign calendar dates to the individual tree rings in those timbers. The techniques used for dating historical buildings and landscapes are based on using standard dendrochronological methodology and are guided by key principles that make dendrochronology such a powerful method of historic analysis.

4.6 Literature cited

Alexander, M.R., C.R. Rollinson, D.J.P. Moore, J.H. Speer, and D.L. Rubino. 2018. Determination of death dates of coarse woody debris of multiple species in the Central Hardwood Region (Indiana, USA). *Tree-Ring Research* 74: 1–9.

Anning, A.K., D.L. Rubino, E.K. Sutherland, and B.C. McCarthy. 2013. Dendrochronological analysis of white oak growth patterns across a topographic moisture gradient in Southern Ohio. *Dendrochronologia* 31: 120–128.

Baillie, M.G.L. 1982. *Tree-Ring Dating and Archaeology*. The University of Chicago Press, Chicago, IL. 274 p.

Baillie, M.G.L. 1995. *A Slice Through Time: Dendrochronology and Precision Dating*. Routledge, London, UK. 176 p.

Bannister, B. 1963. Dendrochronology. Pp. 162–176. In: *Science in Archaeology* (D. Brothwell and E. Higgs, eds.). Basic Books, Inc., New York, NY.

Barbour, M.G., J.H. Burk, W.D. Pitts, F.S. Gilliam, and M.W. Schwartz. 1999. *Terrestrial Plant Ecology*. Benjamin, Cummings, Menlo Park, CA.

Cook, E.R. 1992. A conceptual linear aggregate model for tree rings. Pp. 98–104. In: *Methods of Dendrochronology: Applications in the Environmental Sciences* (E.R. Cook and L.A. Kairiukstis, eds.). Kluwer Academic Publishers, Dordrecht, the Netherlands.

Dean, J.S. 1997. Dendrochronology. Pp. 31–64. In: *Chronometric Dating in Archaeology*. (R.E. Taylor and M.J. Aitken, eds.). Plenum Press, New York, NY.

Fritts, H.C. 1976. *Tree Rings and Climate*. Academic Press, New York, NY. 567 p.

Grissino-Mayer, H.D. 2003. A manual and tutorial for the proper use of an increment borer. *Tree-Ring Research* 59: 63–79.

Grissino-Mayer, H.D. 2017. *The Time Is Right: Redefining the Principles in Dendrochronology*. Paper presentation, American Association of Geographers Annual Meeting, April. Boston, MA.

Grissino-Mayer, H.D., and H.C. Fritts. 1997. The International Tree-Ring Data Bank: An enhanced global database serving the global scientific community. *The Holocene* 7: 235–238.

Kaennel, M., and F.H. Schweingruber. 1995. *Multilingual Glossary of Dendrochronology: Terms and Definitions in English, German, French, Spanish, Italian, Portuguese, and Russian*. Birmensdorf, Swiss Federal Institute for Forest, Snow, and Landscape Research. Paul Haupt Publishers, Berne, Switzerland. 467 p.

Liebig, J. 1840. *Chemistry in Its Agriculture and Physiology*. Taylor and Walton, London, UK. 400 p.

Nash, S.E. 1999. *Time, Trees, and Prehistory: Tree-Ring Dating and the Development of North American Archaeology, 1914–1950*. The University of Utah Press, Salt Lake City, UT. 294 p.

Rubino, D.L. 2014. Tree-ring analysis in Indiana with special attention to dendroarchaeology. *Proceedings of the Indiana Academy of Science* 123: 7–13.

Rubino, D.L., and B.C. McCarthy. 2004. Comparative analysis of dendroecological methods used to assess disturbance events. *Dendrochronologia* 21: 97–115.

Speer, J.H. 2010. *Fundamentals of Tree-Ring Research*. The University of Arizona Press, Tucson, AZ. 333 p.

Stahle, D.W., and P.L. Chaney. 1994. A predictive model for the location of ancient forests. *Natural Areas Journal* 14: 151–158.

Trouet, V., M. Domínguez-Delmás, C. Pearson, N. Pederson, and D. Rubino. 2017. "Dendro-archeo-ecology" in North America and Europe: Re-purposing historical materials to study ancient human-environment interactions. Pp. 365–394. In: *Dendroecology: Tree-Ring Analyses Applied to Ecological Studies*. (M. Amoroso, L. Daniels, P. Baker, and J.J. Camarero, eds.). Springer, Cham, Switzerland.

Chapter 5

Tree-ring analysis methods for the field, woodshop, and lab

5.1 Introduction

Primary and secondary sources, such as diaries, personal letters, deeds, newspaper articles, legal documents, maps, tax records, wills, photographs, and oral history, provide valuable information that can be used to date the construction of historic buildings and landscapes. However, when scant, dubious, and/or contradictory historical documentation is available, tree-ring analysis offers a unique opportunity to establish and/or verify construction dates.

Providing a construction date for a building requires accurately determining the death or felling date of the trees used to fabricate a building's timbers. Death dates can be established by assigning a calendar date to the last tree ring formed by a tree – the ring that is adjacent to the bark or represents the waney surface of a timber (Figure 1.2). Wane is the uniform, rounded outer surface of a timber free of any tool marks (created by hatchet, ax, adze, chisel or saw) from which the bark has either sloughed off or was removed. If numerous timbers from throughout a structure share a similar or comparable death date, one can hypothesize a likely construction date for a structure. Similarly, when the timbers from a certain section of a building share a common felling date, one can determine when modifications or repairs were made to a building.

The goal of this chapter is to introduce you to the field and laboratory methods we use for dating historic buildings and landscapes. Just as understanding the basics of tree-ring science is necessary for truly understanding how construction dates are obtained (Chapter 4), so, too, is an understanding of how samples from timbers are collected and analyzed. Please see Chapter 6, for a thorough description of how we document (measure, sketch, etc.) historical buildings and landscapes while collecting tree-ring samples.

5.2 Field methods

Since our major goal is most often determination of a construction and/or modification date(s) of a historic building, we routinely focus sampling on wane- and bark-bearing timbers. Prior to sampling, all accessible timbers in a building are inspected to determine which have bark or a waney surface (Figure 1.2 and Figure 5.1). Once a timber has been selected for sampling, several additional characteristics are noted. First, soundness of the timber is assessed. Rotten and insect-attacked timbers (ants, termites, boring beetles) are common. Extensively damaged timbers may prove to be unusable for dating purposes and will likely be excluded from analysis because sample extraction and surface preparation require fairly sound wood. Damaged timbers can be analyzed (Figure 4.6), but they do present various sampling challenges.

Prior to sampling, we also try to determine if a sample contains enough rings to be reliably dated. When possible, we perform a very rough ring count on the end or butt of a timber. Also, when a tenon passes completely through a mortise, it is often possible to estimate the number of rings in a timber from the tenon's end.

Timbers that are selected for analysis are further assessed, and several properties are recorded. We document provenience (horizontal wall timber, beam, joist, stud, knee brace, etc.; Chapter 2), location (story, elevation), and various attributes of each sample, such as dimensions, preparation technique (hewn or sawn), and other unique timber characteristics (Figure 5.2). Multiple photographs of each sample are also taken. We use a small dry-erase board while taking photos to further document the timber's identification and location in a building (Figure 5.3).

Figure 5.1
To determine the construction date of a building, the felling date or death date of trees used to create timbers must be determined. Therefore, before sampling, we inspect the entire structure to locate wane- and bark-bearing timbers. Wane is indicated by arrows.

Source: Authors.

Figure 5.2
Field sheet used when collecting tree-ring samples.

Source: Authors.

Tree-Ring Sample Form

Sample ID: ______________________ Date: ______________

Structure: ______________________ Sample team: ________

Elevation aspect: ______________________

Floor: ______________________

Bark / wane present: ________ Collection method: ____________

Photographs:

Type of provenience:

General shape: ______________

Provenience dimensions: ________ ( ) × ________ ( ) × ________ ( )
Max Determinable

Tool markings:

Location of sample:

Provenience sketch:

Notes:

Figure 5.3
Each timber is photographed *in situ* along with a dry-erase board displaying the sample identification.

Source: Authors.

5.2.1 Obtaining tree-ring samples

Sound timbers (those showing little to no rot or insect damage) are sampled using a boring bit (Figure 5.4), since the wood in these timbers is robust enough to withstand the coring process – the core will likely not disintegrate into fragments inside the borer. Cores are obtained using a battery-powered drill (0.5 inch (1.27 cm) chuck). The borer is constructed with teeth that create a kerf (an opening made in the timber as the teeth of the spinning borer remove wood) as the borer is inserted into the timber. Holes are found along the length of the borer to minimize the amount of wood dust that accumulates inside the borer. A guide plate is screwed to the timber to keep the borer in position during the initial phases of coring (Figure 5.4). Prior to coring the timber, a permanent marker is used to color the outermost surface of the wood or bark to ensure that it remains intact during the coring process (Figure 5.4). The borer is drilled into the timber until it passes the approximate center or pith (the oldest portion of the timber) or until a void in the timber is reached. During the coring process, wood dust accumulates between the timber and borer and also around the core inside the borer. The borer must, therefore, be removed periodically and dusted (along with the timber) with compressed air; this helps to reduce friction and the possibility of breaking the core due to binding.

Cores are removed from the timber using an extractor that breaks the connection between the core and the timber (Figure 5.4). Cores are placed in labeled

Figure 5.4
(A) A borer, starting plate, and extractor (top to bottom) are used to obtain cores from timbers (inset shows the cutting teeth of the borer). (B) Once the starting plate is screwed to the timber, the outer surface of the timber is colored with a permanent marker to ensure it remains intact during the coring process. (C) Coring a timber. (D) Removing the core with an extractor. (E) Complete core removed from a timber. (F) Hole (approximately 13 mm) that remains in a timber after the core is removed.
Source: Authors.

PVC tubes capped with masking tape to protect them from mechanical damage during transport to the woodshop. A core, once properly surfaced can be thought of as a cross section (Chapter 3), since the core is taken perpendicular to the long axis of the timber (across the grain). Coring permits sample acquisition without compromising the structural integrity of a timber. Cores are taken from portions of the timber that show no branch scars (knots); wood produced in these regions will often contain irregular growth patterns that may preclude successful dating. Also, we take cores from the largest end of a beam where bark or wane is present, since this region will have more rings than others (Chapter 3).

We routinely take more than one core from a timber. A second core can help in verifying that the outer surface of a timber is, in fact, wane. Additionally, we often take a second core when working with species that are often associated with missing or locally absent rings (tulip poplar and American beech; see Chapter 3). A second sample is oftentimes helpful in locating a missing or locally absent ring. If we note growth irregularities (e.g., branch scars), rot pockets, or insect damage after removing a core we also take a second sample (Figure 7.1). More than one sample is also beneficial since stems may exhibit intra-ring variability (e.g., the 1839 ring is not 1.23 mm wide throughout the entire tree). Ideally all rings would be the same width on all radii of a stem, but this is not always the case (Figure 1.5). A second sample can assist in dating when such variation exists.

The sapwood of timbers is often riddled with boring beetle galleries (Figure 5.5). Heavily damaged areas commonly result in undateable sample material, since the wood crumbles during the coring process. In such cases, exact felling or death dates cannot be reliably attributed to the sample. Likewise, fungal degradation of sapwood is common due to the more nutrient-rich wood found in this region of the tree; heartwood is less prone to such decay due to its chemical composition (Chapter 3). To overcome these problems, we have developed a method of extracting the outer portion of damaged or degraded timbers by using an oscillating multi tool to make a series of plunge cuts in the timber (Figure 5.6). A plunge cut is made by attacking a timber with the teeth of a saw blade perpendicular to the surface to be cut. Two parallel plunge cuts are made in the timber. The cuts are made deep enough in the timber so that matching the rings between this piece and the outermost portion of a core (obtained from sound wood) may be possible. To free the wane- or bark-bearing piece, a third and fourth plunge cut is made perpendicular to the first two cuts, or a wood chisel is used when working in areas of limited clearance (Figure 5.6).

In addition to coring, samples (cross sections) are also obtained with various types of saws, including hand, reciprocating, circular, and chain saws. Sawing is most commonly performed on damaged (insect or fungal) timbers that have been or will be extracted from the structure and ultimately replaced. Sawing is beneficial since it permits rapid sampling of a large number of timbers and also allows for collection of a complete cross section of the timber. A core provides a relatively small amount of wood for tree-ring analysis, while a sawn cross section of a timber provides the largest possible surface. When timbers are greatly damaged by decay or insects, sawing is preferred over coring, since obtaining complete cores from such timbers is often not possible (Figure 5.7). To protect the cross sections and to avoid loss of fragments of deteriorated wood, we wrap them in masking tape or plastic wrap before sawing (Figure 5.8). When cutting samples, we routinely make marriage marks or carpenter's marks across breaks in the wood so that pieces can be correctly reassembled in the woodshop prior to surfacing (Figure 5.9).

Each sample (core or cross section) is assigned a unique identification containing three portions: a three-letter structure identification (FNT), a two-digit

Figure 5.5
Insect-damaged sapwood (boring beetles) in a white oak timber. Depending on the density of beetle galleries in a timber, coring may or may not be possible. Inset left: close-up view of insect-damaged sapwood. Inset right: coring timbers with heavily degraded sapwood results in an undateable sample. Note the wane portion of the timber that was colored black before coring.
Source: Authors.

provenience (individual timber) identification, and a letter indicating the individual sample obtained from a timber. For example, sample FNT03B identifies a replicate sample (B) obtained from the third provenience (03) sampled from the structure (Fant House).

As would be expected, historic buildings have often been modified over their period of service, and additions and renovations are very common. For example, barns are often expanded with additions to their gable or eave elevations. Similarly, restorations, such as replacement of rotten timbers (especially sills and floor joists) or storm-damaged roofing system are common.

In order to ensure dating of the original fabric of the building and not renovations, sampling is performed throughout an entire structure. Our goal is to obtain samples from all elevations, stories, roof, attic, and basement or crawlspace.

Determining which timbers are part of the original fabric of a building is essential, and we use several techniques to identify repairs and additions. For example, additions can often be identified by noting changes in roof pitch from the outside of a building (Figure 5.10). From the outside of a building, modifications can be hidden when buildings are re-sided. Additions to the gable end of buildings can be identified by noting exterior siding on interior walls in attics (Figure 5.11). Modifications or additions can also be identified when construction methods and

Figure 5.6
Using an oscillating multi tool to remove the outer (wane- or bark-bearing) portion of a timber damaged by decay or insect activity. (A) Parallel plunge cuts (arrows) are made into the timber. The cuts are made the width of the blade. (B) A third and fourth plunge cut is made or a wood chisel and hammer (C) are used to extract the wood section containing bark or wane (D).

Source: Authors.

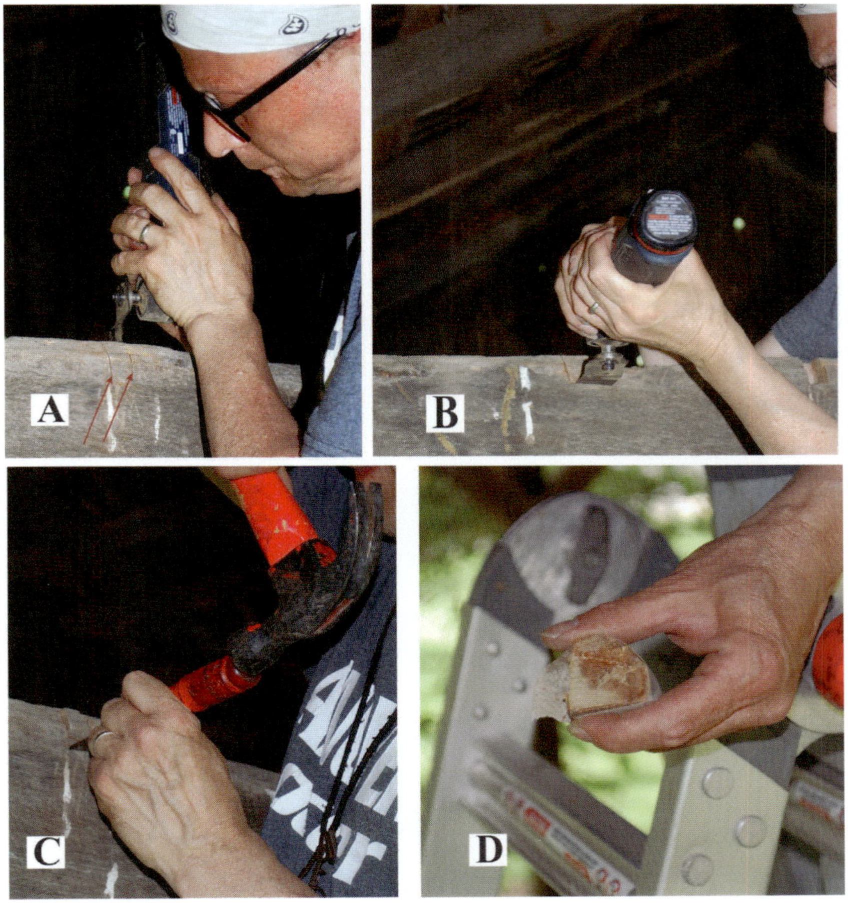

Figure 5.7
This ground-floor joist was obtained from the George Ash House (Chapter 9). The joists from this structure were severely decayed and degraded by insects. Coring of these timbers would not have provided adequate samples (too few rings for crossdating and voids in the wood). Since the joists were going to ultimately be removed during a renovation of the house (DLR actually fell through the floor while inspecting the house), the owners permitted us to saw them and remove cross sections.

Source: Authors.

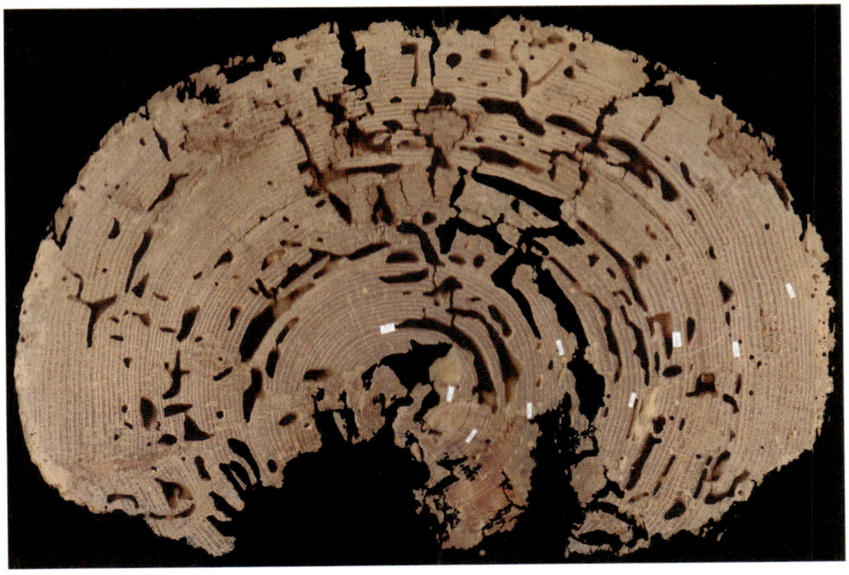

Figure 5.8
Cross sections often fragment during sawing. We use plastic wrap or masking tape to keep the pieces of wood in the proper orientation and to protect them during transport.
Source: Authors.

materials change for a particular elevation or room – timber framing with pegged mortise and tenon versus nail construction (Chapter 2).

Individual replacement timbers can often be identified by looking for several telltale signs. For example, we often suspect the presence of replacement timbers when we encounter variation in timber preparation (sawn or hewn); a replacement is likely if we encounter a sawn joist or sill when most of the timbers in the building have been hewn. The presence of a replacement timber, most likely recycled from another building, is also suggested when:

- A structural member (e.g., sill or plate) is made of several timbers that change in size and/or shape (Figure 5.12).
- Large structural timbers have mortises that are not necessary for the framing of a structure (Figure 5.13).
- Tenons do not fill the entire mortise into which they are placed (Figure 5.13).
- The patina (e.g., covering of dust/debris) of a timber does not match that of other timbers.

Figure 5.9
(A) We make marriage marks on the individual pieces of a fragmented timber so that they can be arranged in their proper configuration prior to surface preparation. (B) This cross section was glued to a base prior to sanding and subsequent analysis.

Source: Authors.

Figure 5.10
This barn has been expanded on the eave elevations as noted by the change in pitch of the roof (arrows). This is the Bakes Road Barn (Chapter 12).

Source: Authors.

Figure 5.11
From the outside of a structure, it may not be possible to identify modifications to the original footprint of a building. This structure was extended on a gable elevation, and the entire building was re-sided. We could confirm the addition by noting exterior wood siding on an interior wall while inspecting the attic. Note the location of the chimney in (A) and (C). In (B), the chimney is on the opposite side of the wall (comprised of studding and siding) and to the right of the access that was cut into the siding. This building is the Maples, a purported stop on a stage coach route that ran from New Albany to Vincennes, Indiana, in the mid-1800s.

Source: Authors.

Figure 5.12
The west sill of the *Musée de Venoge* (Chapter 9) consists of three different timbers. The central portion is a replacement timber; its felling date is 45 years after the construction of the building. The north and south parts of the sill are ash, and the central portion is white oak. A detail of the circled portion of the sill is shown in the inset.

Source: Authors.

- Modern, round wire nails are used in a building that is almost entirely fastened with historic square nails and wooden pegs.
- Timbers show striping from plaster and lath application, but the room or structure is not plastered (Figure 5.14) and would have never been (such as an outbuilding or barn).
- Presence of a type(s) of timber that is different from the remainder of the structure (e.g., a maple timber used in a building built mainly with white oak).

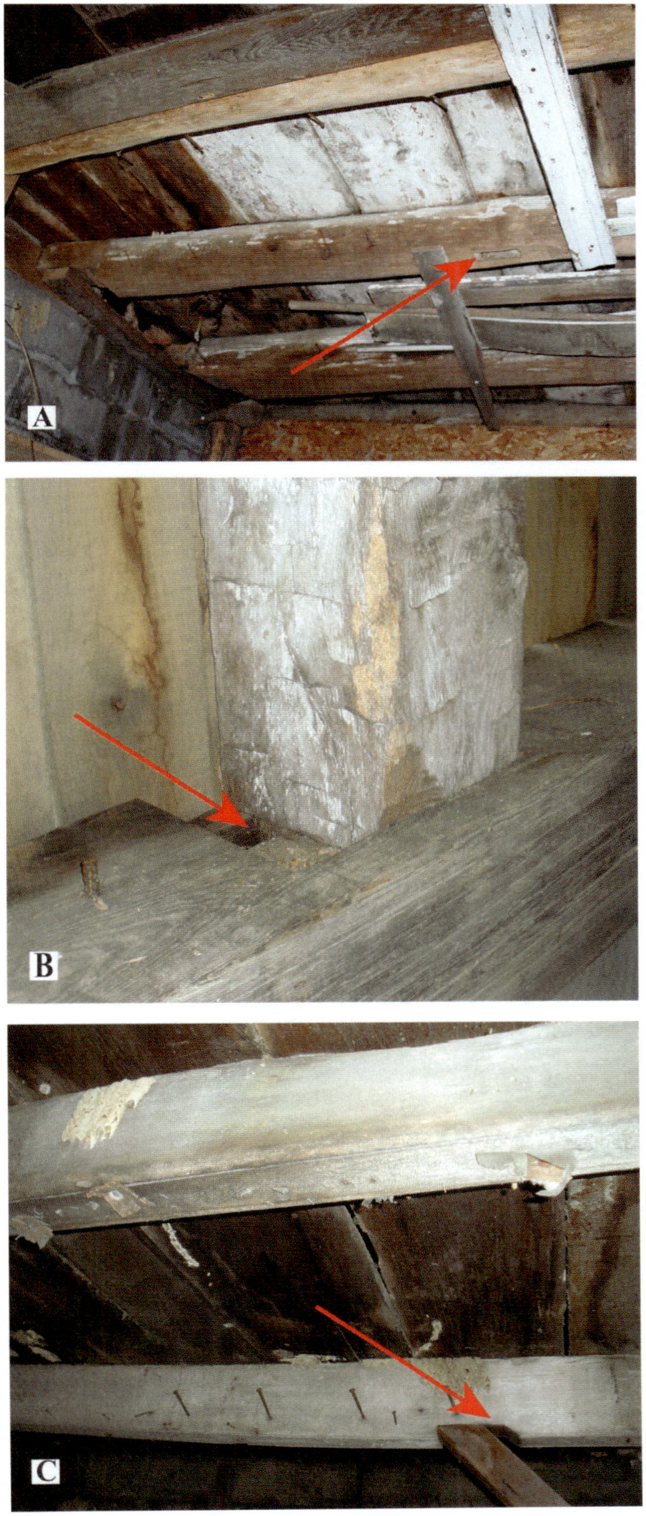

Figure 5.13
Spotting replacement timbers. (A) A mortise in this timber (arrow) is unnecessary, thus suggesting that the timber is not original to the structure and that it was likely recycled from another building.
(B) Note how the tenon of the post does not match the size of the mortise (arrow) in the sill; a block was inserted to help keep the post in place. Also note that the post is hewn whereas the sill is sawn.
(C) The diagonal brace is likely not original as it does not fill the recess cut into the floor joist. Also note that the patina of the brace is different from the two joists.

Source: Authors.

Figure 5.14
The studs show light white banding, indicating that lath and plaster were once present throughout this room. The white bands are where plaster contacted the timber, and the voids between the white bands indicate the location of the lath (thin pieces of wood used to hold plaster when finishing interior walls and ceilings). Often, we note very small nail holes or small square nails (used to affix the lath) in timbers that were once covered by lath and plaster. Intact lath with no plaster can be seen in the center of the photo. When timbers with this banding are found in structures such as barns, we conclude that the timbers are recycled.
Source: Authors.

5.2.2 Timber type determination

In addition to obtaining samples for tree-ring analysis, we also take small pieces of wood from individual timbers to determine what types (species) of timber were used to construct a building. We can often determine timber type by cutting a small nick (in essence, creating a very small cross section) and by examining the wood anatomy of the timber. We use single edge razor blades and a magnifying loupe (10× magnification) in the field. If a sample cannot be readily identified, a small sliver of wood, approximately the size of a matchstick is collected and placed in a labeled envelope. The identity of the wood type will then be determined using a microscope in the lab (see section 5.4). A list of equipment needed in the field is found in Table 5.1.

5.3 Woodshop methods

Cores are mounted into labeled, grooved mounting boards with their conductive cells (vessels and tracheids) oriented vertically (Figure 5.15). We use readily available white glue (school glue; polyvinyl acetate) for mounting the cores to the mounts and use masking tape to provide clamping pressure. Similarly, we glue fragmented sawn samples to plywood to keep them intact. We use a variety of woodworking clamps to ensure a good glue bond between the sample and plywood; a scrap piece of wood is placed between the sample and clamp to avoid damage to the sample. Once the glue has dried, the sample is ready for surface preparation and subsequent dating analysis (Figure 5.9).

Table 5.1 Field equipment typically used when sampling historic buildings and landscapes for tree-ring analysis.

Equipment	*Use/Comments*
Magnifying loupe (10×)	Wood identification
Single-edge razor blades	Wood identification
Envelopes	Collection of wood fragments for wood identification
Boring bits, starting guide, extractor	Obtaining cores from timbers
Permanent marker	Marking the outer surface of a timber; sample labeling
PVC pipe	Protection of cores during transport
Masking tape	Seal the ends of PVC tube; wrap samples
Dowel rods and rubber mallet	Remove cores from borer if the core breaks during the coring process
Rechargeable batteries and chargers	Historical buildings often lack electricity so battery-powered tools are necessary. Also working in attics and upper level of barns (often on ladders) makes the use of extension cords impractical and unsafe.
Compressed air	Boring timbers creates saw dust, which must be periodically removed from the timber and borer to avoid binding.
Camera	Photo-documentation of building, timber location, etc.
Dry-erase board, markers, and eraser	Photo-documentation of timber location in a building
Ladder	Gain access to timbers for coring. We use a multi-use/folding ladder that can be configured in different ways (extension, step, scaffold).
High-powered flashlight/ spotting light	Examination of timbers. Barns and attics are quite dark, and a bright light makes checking for wane, bark, and sound timbers easier.
Drill (battery powered)	Obtain cores using borer
Oscillating multi tool	Make plunge cuts to extract friable wane sections. We use a battery-powered tool.
Battery-powered task light	Provide light while coring and taking notes
Circular saw (battery powered)	Obtaining cross sections of *ex situ* timbers that have either fallen off a building or have been extracted during renovation
Reciprocating saw (battery powered)	Obtaining cross sections of *ex situ* timbers
Spare blades for all saws	It is very common to accidentally cut "hidden" nails in timbers. Cutting with sharp blades is safer than using dull blades.
Selection of hand tools	Saw, hammer, wood chisel, pry bars of varying size
Measuring tape and tree calipers	Measuring timber dimensions. Tree calipers are very useful to determine timber size, since *in situ* round timbers are not easily measured with a tape or ruler.
Plastic wrap	Wrap cross sections so that pieces of wood are not lost and to protect delicate wane- and bark-bearing areas of the timbers
Chain saw (fuel, bar oil, chain sawing protective equipment)	Obtaining cross sections. Used when buildings are being demolished or undergoing major renovations when major portions of the original fabric of the building are being removed
Safety equipment and first aid kit	Eye, ear, and respiratory protection equipment is essential.

Figure 5.15
A core affixed to a mounting board. The vessels and tracheids of the core need to be oriented vertically so that the woody tissue can be studied in cross section. A core can be thought of as a single radius from a cross section. Individual tree rings cannot be identified and measured if the wood is not mounted properly. In this ash sample, bark can be seen on the right end of the core. Pith is present in the upper left corner of the core.

Source: Authors.

Samples are sanded with progressively finer grits of sandpaper to remove tool marks and to produce a smooth surface for subsequent analysis. We routinely use a belt sander with ANSI 80-, 120-, 180-, and 220-grit sanding belts (Orvis and Grissino-Mayer 2002; Figure 5.16). Larger grit designation numbers are associated with finer particle size of the sandpaper; the larger the grit number, the smoother the finish left on the wood's surface. We then use a palm sander with 220-, 320-, 400-, and 600-grit sandpaper (Figure 5.16). Finally, hand sanding is performed using 30 μm and 15 μm sanding film. When samples are obtained with a chain saw or a reciprocating saw we often begin the surface preparation process by belt sanding with 50-grit belts or use an electric hand planer to remove the large gouges created by the saw's teeth. Thorough sanding of samples is essential, since dating timbers requires that individual cell and ring boundaries are clearly visible. A list of equipment needed in the woodshop is found in Table 5.2.

Figure 5.16
Surface preparation of cores and cross sections. (A) Samples are initially sanded with a belt sander (A and B) to create a level cross-sectional surface for subsequent analysis
The sander used in (A) is a 3 x 21-inch sander, and the sander pictured in (B) is a 6 x 48-inch sander. The size refers to the width and length of the sanding belt that is used. A palm sander (C) is then used to create a smooth surface that, if prepared correctly, will allow for each of the tree-ring boundaries in the sample to be identified once it is placed under a microscope.

Source: Authors.

5.4 Laboratory procedures

5.4.1 Microscopic timber type determination

The timber type of each sample is determined by using anatomical wood features and identification keys (Panshin and de Zeeuw 1980; see Hoadley 1990 for an excellent introduction to wood identification techniques). Subsamples for wood identification are obtained by removing paper-thin sections of wood with a double-edged razor blade (Figure 5.17). Subsamples are routinely taken from incomplete rings found from the innermost portion of samples or from regions of samples so that ring-width measurement is not hindered.

Table 5.2 Woodshop supplies used when dating historic buildings and landscapes.

Equipment	*Comments*
Core mounts	Grooved boards used to hold cores. We make ours using a router and ¼-inch (0.64 mm) or larger core box bit and ½-inch (1.27 cm) strips of wood.
Glue	Affix cores in the mounts and cross sections to a base.
Masking tape	Provide clamping pressure when mounting cores to mounts
Plywood or oriented strand board	Provide a base for fragmented cross sections
Wood clamps	Provide clamping pressure when gluing cross sections to a base
Belt sander	Create smooth, level cross sections and cores. We use a 6 × 48-inch (15.24 × 121.92 cm) sander and a handheld 3 × 21-inch (7.62 × 53.34 cm) sander for larger cross sections that cannot be safely placed on the larger sander.
Palm sander	Create a smooth surface suitable for microscopic analysis of cores and cross sections
Sandpaper and sanding belts	Variety of grits are used; see text for grits routinely used.
Handheld power planer	Leveling of cross sections and removing tooth marks from sawing
Clamping workbench	Hold cross sections securely while sanding and planing. See Figure 5.16.
Safety equipment	Eye, ear, and respiratory protection; anti-vibration gloves

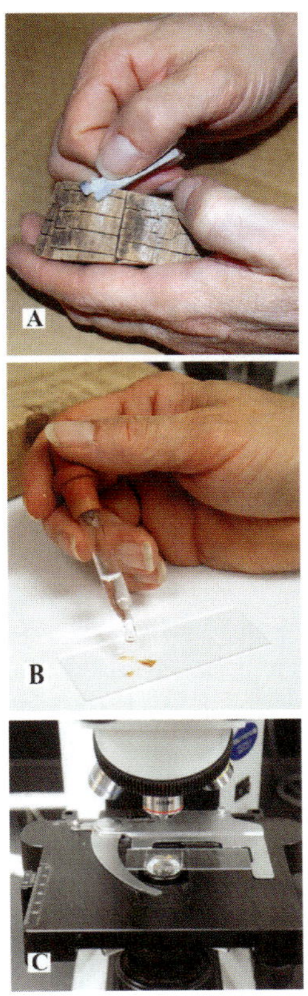

Figure 5.17
Determining timber type using microscopic wood anatomical features. (A) Thin sections of wood are obtained using double-edge razor blades. The razor blades are first broken in half. The blades can be used freehand or placed in the collet of a hobby knife. Double-edge blades are used because they are extremely sharp and relatively inexpensive. Microscopic analysis requires sections of wood thin enough to permit light to shine through them. (B) A wet mount is created by adding a drop of water to the wood on a microscope slide. (C) Once a coverslip is placed on the wood sections, it is viewed under a compound microscope at 40×, 100×, 200×, and 400× magnification.

Source: Authors.

Identifications are made by noting the types, sizes, and distribution of cells in the wood (Chapter 3 and Figure 3.7). Wood identification is performed prior to dating so that each of the different timber types is analyzed in groups; experience has shown that crossdating (growth pattern matching) proceeds best in this manor (Chapter 3). A list of equipment needed in the laboratory is found in Table 5.3.

5.4.2 Assigning years and crossdating

Once individual samples have been identified to timber type, years – not dates – are marked on each of the samples under a dissection microscope. Starting with the innermost (oldest) tree ring, years are assigned to each ring. The innermost ring is assigned year 0, the next year 1, and so on, until the outermost ring is numbered. At this stage, years, not calendar dates, are assigned to each ring since the year of a ring's formation is unknown. The last formed ring in a sample will be found on the outer edge of the sample and have the largest assigned year/ring number. To keep track of the number of rings counted, a pencil dot is marked on each decade ring (0, 10, 20, etc.). A pair of dots is marked on every 50th ring, and three dots are placed

Table 5.3 Laboratory supplies used when dating tree-ring samples from historic buildings and landscapes.

Equipment	*Comments*
Mechanical pencils	Mark rings on cores and cross sections. We recommend 0.3 and 0.5 mm pencils, since thin lead makes marking very narrow tree rings easy.
Dissecting microscope and light source	Mark years on samples and measuring individual tree rings
Double-edge razor blades	Cut extremely thin sections of wood for microscopic analysis
Microscope slides, cover slips, and dropper bottle	Prepare specimens for viewing using a compound microscope
Compound microscope	Wood identification
Graph paper	Create skeleton plots for crossdating. Paper with five lines per centimeter with every fifth line heavily marked is most commonly used. This style can be purchased or downloaded from various websites.
Transparent tape	Join skeleton plots together while crossdating
Quilter's tape (1/4-inch masking tape)	Keep track of ring count and identify radii for measuring. The tape is easy to mark with a very fine permanent marker. It must be removed after dating or it leaves adhesive residue on the wood.
Measuring device connected to a computer	Create series of measured tree-ring widths. We use a device that measures to the nearest 0.01 mm.
COFECHA computer program	Crossdating and measurement quality control. It can be downloaded for free from www.ltrr.arizona.edu/pub/dpl/COFECHA.ZIP.

Figure 5.18

Years are assigned to individual rings with the innermost (oldest) ring assigned year 0, the next year 1, etc. The outermost ring of this specimen (125) is adjacent to bark. If a calendar date can be assigned to this ring, a death or felling date for the tree from which this timber was extracted can be determined. This specimen "grew" from left (pith; not present) to right (bark).

Source: Authors.

on each 100th ring (Figure 5.18). On cross sections, we often use quilter's tape when assigning years (not dates!) to the individual rings. This thin – 1/4-inch (64-mm) wide – masking tape makes keeping track of years and identifying radii for measurement (see Section 5.4.2.2) easier (Figure 5.9).

When dating historic buildings, the date of formation of individual tree rings in a timber is unknown. These dates can be determined through a process called crossdating (e.g., Stokes and Smiley 1968; Chapter 4). Accurate crossdating is the key to assigning dates to a timber's tree rings and, ultimately, determining the date of a building's or landscape's construction and modification. Crossdating is performed by matching the pattern of tree rings in samples of unknown age with those found in samples with verified dates. Crossdating is a highly reliable and effective method for dating wood of unknown age. Dendroarchaeological techniques and crossdating have been successfully used to determine and/or verify the date of construction (and/or subsequent modification) of buildings by numerous researchers throughout the world.

5.4.2.1 CROSSDATING USING SKELETON PLOTS

At this stage of the dating process, each sample consists of a series of tree rings that is considered to be "floating" in time, since arbitrary years and not calendar dates have been assigned to the tree rings. Crossdating (Chapter 4; Figure 4.3) must be performed to assign actual calendar dates to the rings in each of the timbers. Several methods of crossdating have been developed (Speer 2010). Our experience has shown that skeleton plotting is an extremely reliable method for crossdating samples of unknown age (those with an unknown start and end date). These plots are prepared to graphically highlight the pattern of tree-ring size in a sample (Stokes and Smiley 1968). A skeleton plot is prepared for an individual sample using graph paper and assigning each growth ring in a sample a vertical line on the graph paper (Figure 5.19). This enables comparison of individual growth patterns for a large number of tree-ring samples and years. In essence, skeleton plotting allows tree-ring samples to be represented by individual strips of paper that can easily be compared and ultimately dated. Though seemingly simple to construct, a skeleton plot is a very effective tool to use in the crossdating process (Figure 5.19):

- A strip of graph paper is cut 15 squares high.
- The sample identification is recorded in the left margin of the strip.
- An arrow is placed on the skeleton plot for the first ring found in the sample (usually ring 0).
- Every tenth block (decade) is labeled (so the tree, once plotted, will "grow" from left to right; oldest ring to youngest ring). Each vertical line on the graph paper represents a year of growth.
- Lines are drawn to represent the relative amount of growth observed for individual tree rings. The length of a line assigned to an individual growth ring is inverse to the width of a ring. Larger and average-sized rings are not

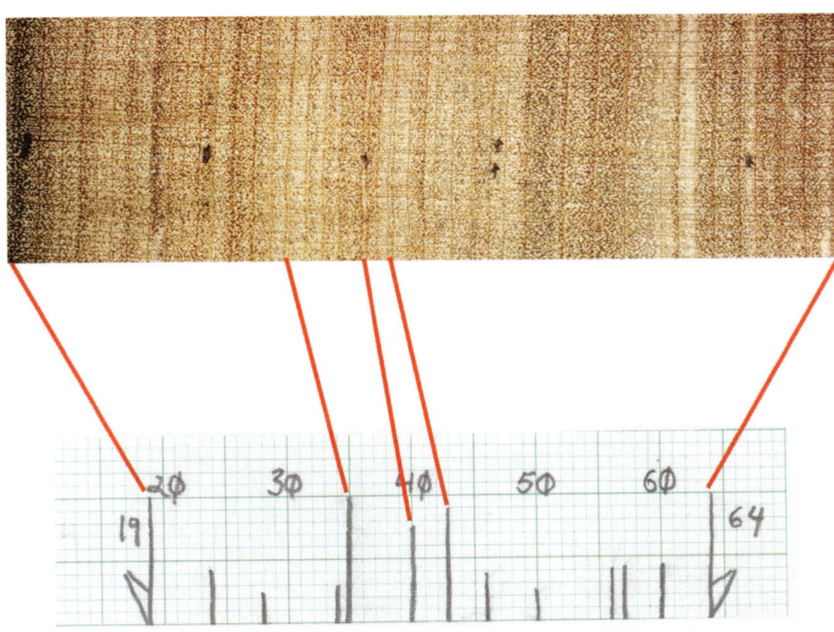

Figure 5.19
Skeleton plots are used to crossdate the tree-ring growth patterns in timbers of unknown age. See the text for a full description of how skeleton plots are constructed. In this example, we have skeleton plotted part of a series of tree rings (ring 19 to ring 64) in a tulip poplar sample. The red lines connect the rings to the individual lines drawn on the skeleton plot. The unique growth pattern in this sample will be used to crossdate it with other timbers from the building and with dated samples obtained throughout the sampling region.

Source: Authors.

assigned a line, while small rings are assigned a line. The smaller the ring, the longer the line (maximum length is ten squares on the graph paper). The length of a line is determined by the relative size of a ring when it is compared to neighboring rings (two or three rings before and after the ring of interest). By comparing to neighboring rings, and not to all of the rings in a sample, inter-annular variability in growth will be highlighted. A useful skeleton plot will not have too many or too few lines present. Small rings, as opposed to large rings, are prominently marked, since environmental stresses, such as droughts that potentially affect a large number of trees, are important in the crossdating process.

- An arrow on the line representing the last year of growth on the sample is made. If bark or wane was present on the sample, a "W" or "B" is listed by the arrow to indicate that this was the last growth ring formed by a tree and that this ring represents the death or felling date of the tree (Figure 5.19).

In addition to ring-width variation, we note unique wood anatomical characteristics for a given year while skeleton plotting. If some environmental stress affected several trees in a similar manner, a unique signature may appear in the rings of these trees and facilitate the dating process. For example, tulip poplar often exhibits variation in wood color and vessel density; we place a "P" on the line corresponding to rings that are pale in color (Figure 5.20). Similarly when working with ash (and sometimes with white oak), we routinely find anomalous earlywood formation (Figure 5.20; especially in 1774); we mark years that show an abnormal earlywood vessels with an "E." We also mark very large rings (relative to neighboring rings) by placing a plus sign on the skeleton plots. By using several lines of evidence, we are better able to accurately assign calendar dates to individual tree rings. See Speer (2010) for a more in-depth description of the skeleton plotting process. In Chapter 14, we provide exercises in which you can skeleton plot several samples in order to date timbers from a building of unknown age. You can complete the exercises on your own or as part of a class, lab, or studio.

Once skeleton plots are created, crossdating is performed by sliding plots back and forth to match the growth patterns among the timbers. While comparing skeleton plots, it is important to note both the presence/absence of a line and the length of the line for a given year. We often begin the crossdating process by looking for potential marker years (e.g., abnormally small rings represented by long lines on the plots) and align the plots to see if their pattern is consistent in each of the different samples. Note that there is not always a perfect, year-for-year signal in the plots (Figure 5.21). Again, each tree experiences a unique set of stressors even when grown in what is presumably very similar conditions (even if we assume the timbers used in construction were felled from a single forested stand).

Once patterns are satisfactorily matched, the plots are taped together into one packet (Figure 5.21). A master site chronology skeleton plot is then created by combining the signals observed in each of the trees into a single plot. This plot is constructed by "averaging" the lengths of the lines observed for each year and making a line of the appropriate length on the master site skeleton plot. The samples, and consequently, the master site skeleton plot are still floating in time. In order to assign calendar dates to the individual rings, the growth patterns found in a building's timbers must be compared to a dated master chronology. The dated master chronology may have been constructed from living or recently dead trees (of sufficient age to go back in time far enough to overlap with the tree rings

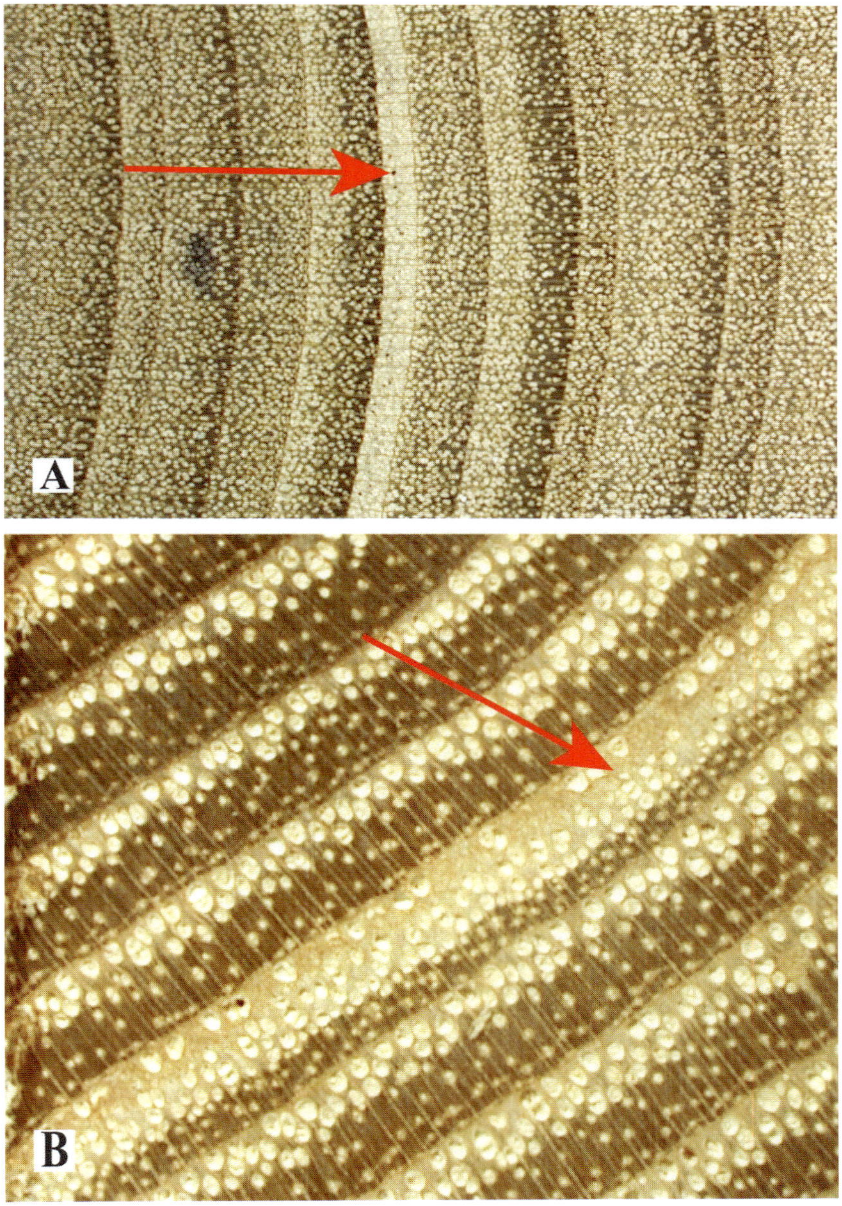

Figure 5.20
Unique wood anatomical characteristics are also recorded on skeleton plots, as these are often useful for crossdating purposes. (A) A pale ring in a tulip poplar (arrow). (B) Anomalous earlywood growth in an ash sample (arrow). Compare the earlywood of this ring to the earlywood in neighboring rings.
Source: Authors.

found in the building's timbers), from timbers dated from other historically constructed buildings, or from a combination of both (see chronology construction in Chapter 4). The crossdating process at this step is the same as above when crossdating the individual timbers from a single structure to each other, the temporal spacing of the lines and the size of the lines in the skeleton plots are matched (Figure 5.22). If the building's composite skeleton plot matches the dated regional master chronology reliably, dates can be assigned to it and, subsequently, to each of the undated samples obtained from the building.

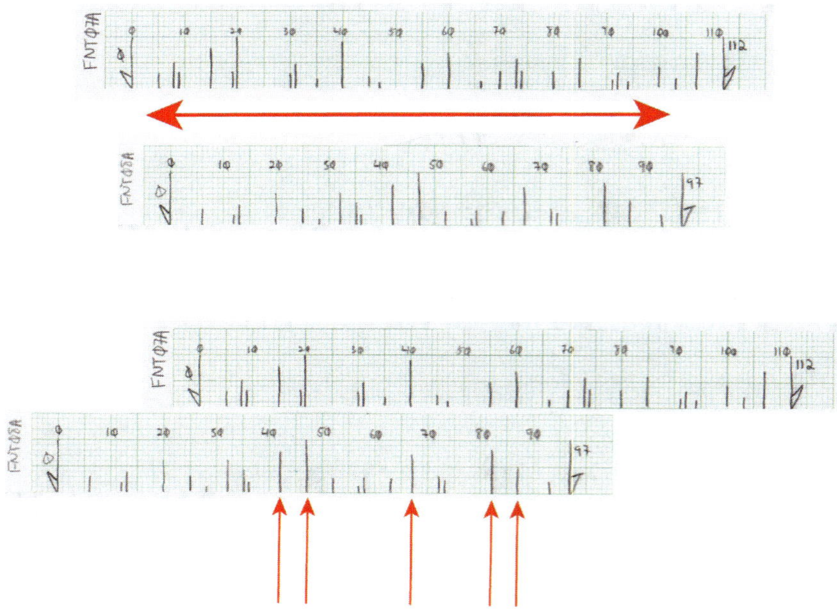

Figure 5.21
Crossdating individual samples. (Top) The skeleton plots are slid along their lengths in an attempt to match the growth patterns in each. When beginning the crossdating process we often focus on the largest lines (smallest tree rings) since these years represent years of stress (e.g., drought) that may be found in several samples. (Bottom) Skeleton plots are then taped together once their growth patterns have been crossdated. The arrows represent a common growth pattern in both of the skeleton plots.
Source: Authors.

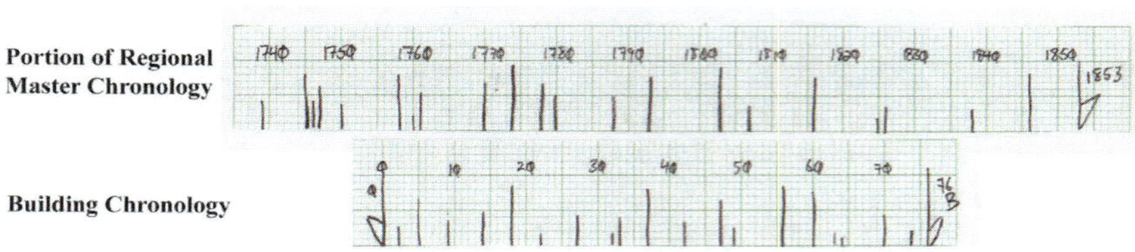

Figure 5.22
Once a site or building chronology is developed by crossdating skeleton plots from all of the timbers (Figure 5.21), a composite skeleton plot is created (lower plot). This composite skeleton plot is then crossdated against a regional master chronology skeleton plot (top plot). Based on crossdating, the hypothetical structure being crossdated in this example contains tree rings that span from 1756 to 1832. Note the similarity in line heights and spacing in the two plots.
Source: Authors.

5.4.2.2 CROSSDATING WITH MEASURED TREE RINGS

In addition to skeleton plotting, crossdating is also performed using measured tree-ring widths. We measure the widths of each ring at 45× magnification under a dissection microscope (Figure 5.23) using a measuring device with a resolution of 0.01 mm (0.00039 inches). The measuring device is connected to a computer, and at the end of measuring we have a computer file consisting of years and tree-ring measurements for each sample.

Not all of the rings in a sample are measured. The outermost ring in each sample is not measured, since it is not possible to know if the ring was fully formed (i.e., the tree could have been harvested during the growing season; Chapter 3). The innermost ring of most samples cannot be measured since sawing, hewing, cracking, or decay does not follow a ring boundary, and the ring would be incomplete. Measurement of an entire sample is not always possible, if the sample has an irregular growth pattern due to scar tissue or growth anomalies associated

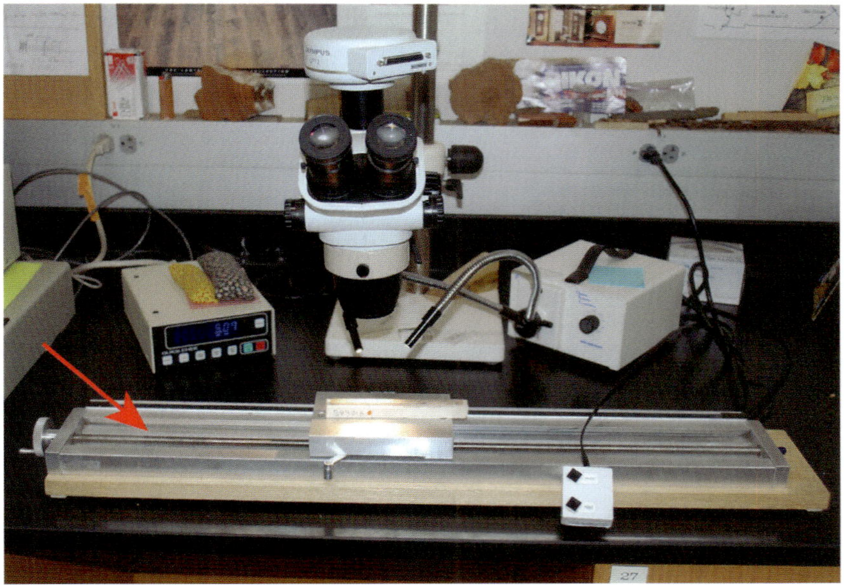

Figure 5.23
Measuring tree-ring widths. A measurement device (arrow) is used to measure each of the complete rings in a sample. As the knob on the left of the device is turned, the sample is pulled across the field of view in the microscope. Rings are measured at 45× magnification using a dissecting microscope. The measuring device is connected to a computer so that at the end of measuring a sample, a series of years and their corresponding ring widths (mm) are recorded in a computer file.

Source: Authors.

with branching (Figure 7.1). When such patterns are encountered, measuring is performed only in the region where presumably non-affected growth is observed. Inclusion of incomplete rings and abnormally formed rings in the ring-width measurements is avoided, since the "true" ring width is not determinable and subsequent inclusion of these measurements would likely hamper growth pattern analyses by including anomalous tree-ring widths.

We use the tree-ring widths and the computer program COFECHA (Holmes 1983; available at www.ltrr.arizona.edu/pub/dpl/COFECHA.ZIP) for assistance with crossdating. COFECHA is a very powerful program that aids in the dating of historic structures by detecting crossdating problems, assessing measurement accuracy, and suggesting the possible dates of undated samples (Holmes 1983; Grissino-Mayer 2001). COFECHA provides a statistical check of date assignment and also provides various data (e.g., number of rings, mean ring width) associated with each sample that is being analyzed and the master chronology (combination of all of the samples from a building).

COFECHA uses a statistical procedure known as simple linear correlation to assist in the crossdating process. In general, correlation analysis is used to explore the linear relationship between two variables. For example, one could use correlation to determine if there is a relationship between the number of leaves on a tree and a tree's height. Correlation analysis would be used to assess simultaneous changes in each variable once the heights and leaf number of several trees is measured. In a correlation analysis, a statistic, an *r*-value (Pearson product-moment correlation coefficient), is calculated. The *r*-value can range from -1 to $+1$. If an *r*-value is positive, both variables increase simultaneously; if an *r*-value is negative, one variable increases while the other decreases (Zar 2010; Figure 5.24). The magnitude of an *r*-value is a measure of the strength of the linear relationship between two variables being investigated. The closer the *r*-value is to $+1$ or -1, the stronger the relationship. An r-value of 0 indicates no relationship between two variables (Figure 5.24).

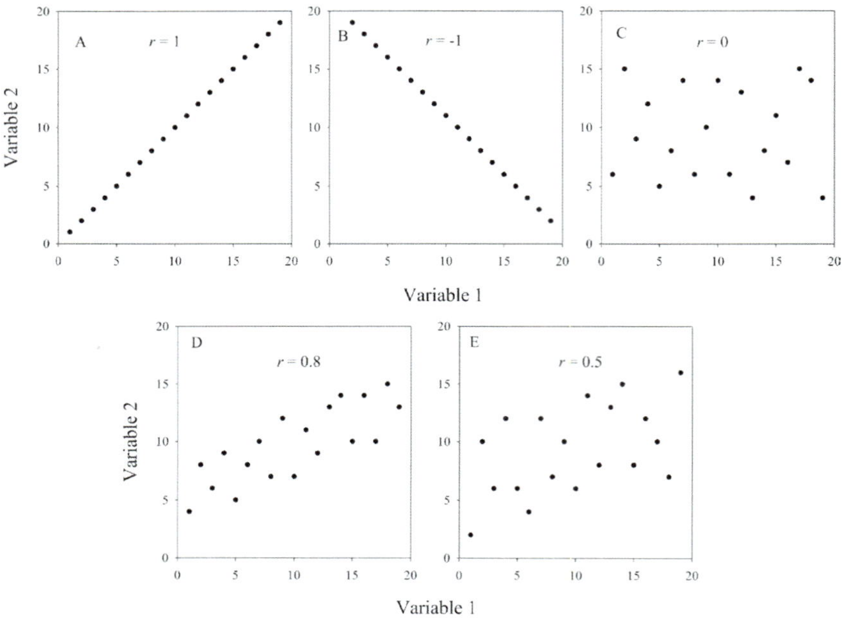

Figure 5.24
Correlation analysis allows one to explore how two variables vary simultaneously. In each of the graphs, a point represents a pair of observations made on one individual from a sample. For example, one could measure both the height (variable 1) and the number of leaves (variable 2) for an individual tree. Each tree would be a point in the plot, and the location of the point would represent an individual tree's height and number of leaves. In panel A, there is a strong (perfect) positive correlation as both variables increase simultaneously. In panel B, there is a strong (perfect) negative correlation; as one variable increases, the other decreases. In C, there is no correlation between the two variables; there is no pattern in the cloud of points, and an *r*-value of 0 is obtained. In panels D and E, the two variables are positively correlated (one variable increases as the other increases). In D, there is a stronger correlation than in E as can be seen from the more compact cloud of points and larger *r*-value. When assessing crossdating accuracy, one correlates the ring widths for each year in two samples (tree-ring widths for each year in timber A would be variable 1, and tree-ring widths for timber B would be variable 2).

Source: Authors.

COFECHA uses correlation analysis to compare sets of tree-ring measurements to assess crossdating accuracy. For example, a dendrochronologist may be interested in determining how well a set of tree-ring measurements obtained from a timber correlates or matches a verified, dated chronology. COFECHA would provide an *r*-value by running a correlation of the tree-ring widths in the two sets of data. It would do this by correlating the ring widths in the two samples for each year (1850 in the dated and timber sample, 1851 in the dated and timber sample, etc.). If the specimens are crossdated correctly, a large, positive *r*-value would be obtained; if they were crossdated inaccurately, a small or negative *r*-value would be obtained. One would ideally obtain *r*-values of +1 for all correlations, if dating was performed correctly – large ring widths for a particular year are found in both samples, and small ring widths are found in the same year for both samples. However, such a large *r*-value is not obtainable due to a variety of reasons (not all trees experience exactly the same conditions throughout their life and trees are genetically unique individuals that react differently to their environment; Chapter 4). So how large of an *r*-value is needed to conclude that crossdating is correct and that the growth patterns in two samples match each other? Statistical probability can help answer this question.

When performing a correlation analysis, the statistical probability of obtaining an *r*-value is determined. A full introduction to statistical hypothesis testing and concepts related to probability are beyond the scope of this book (we recommend Zar (2010) for an excellent explanation of hypothesis testing and probability and for further information about correlation analysis). For interpretation of the significance of the relationship between two series of tree-ring measurements, we can think about the size of the correlation coefficient obtained and the probability or likelihood of obtaining an *r*-value of this magnitude by chance. The program COFECHA uses the significance level of 0.01 (or the 1% significance level). COFECHA considers a correlation significant if the calculated *r*-value (obtained when comparing the widths of two groups of tree-ring measurements) is so large that it falls at or above the 99% confidence level. If a correlation is significant, a dendrochronologist can conclude, with a certain level of sureness, that the compared samples are very similar and that this similarity is not coincidental or caused by chance. One can, therefore, conclude that when *r* is large and has a small probability (of occurring by chance alone) associated with it that crossdating was most likely successful and accurate dates have been assigned to individual rings. COFECHA uses such a stringent significance level because crossdating needs to be precise (Grissino-Mayer 2001).

An example may make the workings of COFECHA a bit clearer. For example, after performing skeleton plotting, calendar dates have been assigned to the tree rings in a set of timbers obtained from a building. The assigned dates are based on the growth patterns observed in a regional, dated chronology. The samples would be placed in a single data file that is read by COFECHA. During a run of COFECHA, the program would create a master chronology by combining all of the individual samples. To assess how well the samples crossdated, COFECHA would correlate the ring widths in one sample with the same years in all the other samples. From each of these correlations, an *r*-value would be calculated. If dating has been performed correctly, large, positive *r*-values (that are statistically significant) would be obtained for each sample (large ring widths in one sample would be observed in the same year in the other samples, and small ring widths in one sample would be observed in the same year in the other samples). This correlation value is called the series intercorrelation. COFECHA also calculates the mean series intercorrelation. This value represents the mean correlation for each sample with the master chronology. These values are reported for each of the samples and chronologies in the case studies presented in Chapters 8 through 13.

In addition to performing correlations for entire tree-ring series (first to last measured ring), COFECHA also performs correlation of segments of individual tree-ring samples. COFECHA breaks each sample into segments of specified lengths (we normally use 50-year segments) that overlap by some user-defined number of years (we normally overlap each segment by 25 years). For example, a tree-ring series that spans from 1600 to 1850 would be broken into the segments 1600–1649, 1625–1674, 1650–1699, etc. This helps to identify portions of samples where (in time) crossdating may be incorrect by flagging segments with low correlation coefficients (*r*-values) that are not statistically significant. Again, large, positive, significant *r*-values would be found if crossdating was performed correctly. When the series intercorrelation and correlation of individual segments in each of the samples show strong correlation, we can assume that we have strong "internal" crossdating (*sensu* Henderson et al. 2009; Slayton et al. 2009). In other words, the samples obtained from a particular building crossdate against each other well and can be combined into a site or single structure composite chronology.

To further assess how well the timbers are crossdated, we perform external crossdating. In this step we take the composite chronology (composed of all of the samples of a particular timber type obtained from a building) and use COFECHA to see how well it matches other dated chronologies. Again, we assess crossdating by determining how the newly developed chronology (constructed from the tree rings of a building's timbers) correlates in its entirety and in overlapping segments with a verified chronology. Once we are satisfied that internal and external crossdating have been performed correctly, we have strong evidence that a building's timbers have been accurately dated, and calendar dates can be assigned to individual tree rings in the timbers.

During the crossdating process, COFECHA will alert a user to potentially misdated samples and measurement errors. When potential errors are reported in the COFECHA output, the skeleton plots and wood are reexamined. If errors are found (for example, a missing ring was not accounted for), adjustments are made to both the wood and skeleton plots, and the crossdating process is continued to determine if accurate calendar dates have been assigned to each ring.

Additionally, COFECHA can also be used to suggest dates for undated samples. For example, if a sample has been measured, COFECHA can be used to identify potential calendar dates for individual rings by using correlation analysis. COFECHA will break down the undated sample into segments and compare, using correlation, the ring width measurements in the sample to a dated and verified chronology (Grissino-Mayer 2001). The COFECHA output will then provide potential dates for the sample. This option of COFECHA is very useful when only a single or few timbers of a particular timber type are obtained from a building, when crossdating a sample with other timber types, and when the origin of the wood being analyzed is uncertain (Chapter 13).

Although COFECHA (and statistical analysis) is an excellent tool for assessing crossdating quality and is a very powerful program, it cannot be used alone to determine if a timber's tree rings are accurately crossdated. The dendrochronologists is guided by COFECHA, but one must still decide when a sample is or is not accurately crossdated (Holmes 1983; Grissino-Mayer 2001; Speer 2010). One must continuously revisit the wood and the skeleton plots and scrutinize date assignments. Once again, we remind the reader that crossdating is the "cornerstone" of dendrochronology (Speer 2010) and is the "most important principle and technique in dendrochronology" (Grissino-Mayer 2001).

Once crossdating (skeleton plotting and COFECHA) and quality control (COFECHA) has been completed, we consider the timbers to be dated. Skeleton plots and the wood samples are marked with actual calendar dates. The tree-ring dates are then combined with primary and secondary historical data (wills, oral history, etc.; Chapter 6) to help interpret the building or landscape. When dating historic buildings and landscapes, we do not analyze historical references until after we have dated the samples. Similarly, we ask that the owner or manager of a building not provide hypothesized construction dates. We want to perform our analysis without any potential biases.

5.5 Conclusion

To determine construction dates for historic buildings and landscapes, proper sampling, sample preparation, and sample analysis techniques must be followed. In addition to collecting wood samples in the field, we also make detailed drawings and notes of the buildings we analyze. In Chapter 6 we review our methods and goals of field documentation.

5.6 Literature cited

Grissino-Mayer, H.D. 2001. Evaluation crossdating accuracy: A manual and tutorial for the computer program COFECHA. *Tree-Ring Research* 57: 205–221.

Henderson, J.P., H.D. Grissino-Mayer, S.L. Van De Gevel, and J.L. Hart. 2009. The historical dendroarchaeology of the Hoskins House, Tannenbaum Historic Park, Greensboro, North Carolina, U.S.A. *Tree-Ring Research* 65: 37–45.

Hoadley, R.B. 1990. *Identifying Wood: Accurate Results with Simple Tools*. The Taunton Press, Newtown, CT. 223 p.

Holmes, R.L. 1983. Computer-assisted quality control in tree-ring dating and measurement. *Tree-Ring Bulletin* 43: 69–78.

Orvis, K.H., and H.D. Grissino-Mayer. 2002. Standardizing the reporting of abrasive papers used to surface tree-ring samples. *Tree-Ring Research* 58: 47–50.

Panshin, A.J., and C. de Zeeuw. 1980. *Textbook of Wood Technology*, volume 1. 3rd ed. McGraw-Hill Book Company, New York, NY. 705 p.

Slayton, J.D., M.R. Stevens, H.D. Grissino-Mayer, and C.H. Faulkner. 2009. The historical dendroarchaeology of two log structures at the marble springs historic site, Knox County, Tennessee, U.S.A. *Tree-Ring Research* 65: 23–36.

Speer, J.H. 2010. *Fundamentals of Tree-Ring Research*. The University of Arizona Press, Tucson, AZ. 333 p.

Stokes, M.A., and T.L. Smiley. 1968. *An Introduction to Tree-Ring Dating*. The University of Chicago Press, Chicago, IL. 73 p.

Zar, J.H. 2010. *Biostatistical Analysis*. 5th ed. Prentice Hall, Upper Saddle River, NJ. 944 p.

Chapter 6

Archival and scholarly sources for interpreting tree-ring analysis

6.1 Introduction

We use a variety of archival and scholarly sources, along with the results of tree-ring analysis, to interpret the sites we investigate. Chapter 6 presents the sources we have found to be the most accessible and useful. The following examples of sources are not exhaustive, but they are a great place to start a search. Each site presents a unique history and challenge requiring its own collection of supporting historic resources. Preservationists must be creative in where they seek out interpretive sources. We also encourage investigators to review National Park Service publications related to documenting historic sites, such as *A Guide to Cultural Landscape Reports: Contents, Process, and Techniques* (Page et al. 1998) and *Researching a Historic Property* (O'Donnell 1991), which are both available online.

6.2 Sanborn maps

The Sanborn Map Company published and sold bound volumes of fire insurance maps from 1867 to 1970. Sales representatives used the maps to assess the physical characteristics of a structure for the purposes of preparing a policy. The maps were routinely updated about every 10 to 15 years. However, they are limited to urban locations (so rural buildings and structures are typically not included). The maps are intricately detailed with information important to insuring buildings (Blumin 1987; Harris 1997). The scaled maps include street right-of-ways and names, property addresses, and the names of industries. The building outline, materials used for construction (e.g., brick, concrete, or wood frame), height, and the location of door openings was recorded for each structure on a property. Since the maps are snapshots of the landscape routinely published over a century, they illustrate the dramatic changes of cities and towns, and their buildings, over time as they grew and evolved (Figure 6.1).

Sanborn maps no longer in copyright are available online and in microfilm form from document archives and map libraries. Some states have developed high-resolution scans of maps that are georeferenced for use in Geographic

Figure 6.1

An 1886 Sanborn insurance map showing the Reed Case House located in Delphi, Indiana (highlighted in the red box). The map shows the house as a two-story wood frame structure with a series of smaller frame structures constructed onto the rear of the dwelling (believed to be a kitchen, carriage house, and stable). The map illustrates the house's historic context in 1886. Brick and frame residential dwellings are located to the east. A livery stable is located north of the house, and several commercial buildings are located to the northeast. Tree-ring analysis of the house is reported in Chapter 8.

Source: Indiana University, Indiana Spatial Data Portal.

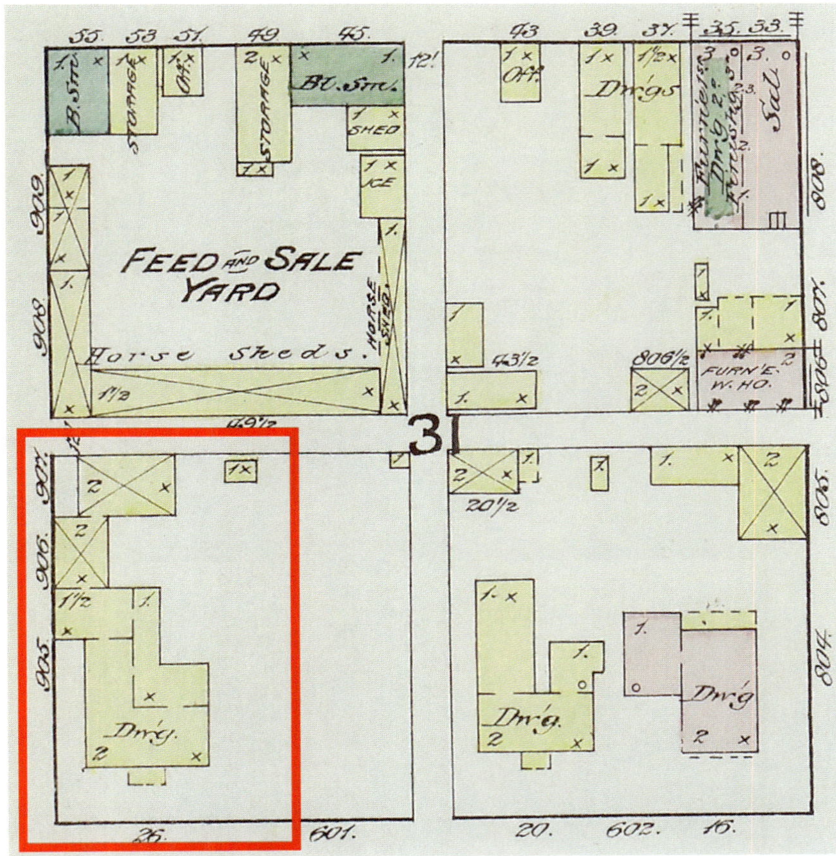

Information Systems software (Indiana Spatial Data Portal 2018). We use Sanborn insurance maps with dendrochronology to interpret multiple sites reported in this text. In Chapter 11, we demonstrate how maps illustrate alterations to Harmonist housing, and that many of the alterations had been removed prior to tree-ring analysis (see Figures 11.11, 11.14, and 11.17).

6.3 Atlases, plat maps, and landscape illustrations

An atlas is a bound collection of maps that includes interpretive information. Regional publishing companies printed atlases for most US counties in the latter half of the 19th century. They often include a county history and the biographies of notable residents. One interesting component of the county-scale atlas is how residents were able to purchase an artist's rendition of their property that was then published in the volume. While the illustrative accounts of buildings were often quite accurate, the landscape was often embellished with decorative follies such as winding drives, fountains, and gazebos (Figure 6.2; Baas 2006).

Plat maps record and illustrate the division of property, usually at the county, township, or city scale (Patton et al. 2005). They often include adver-

Figure 6.2
The top image is of Sunnyhill, the Richard M. Hazelett House located in Greencastle, Indiana. The image is from the *Illustrated Historical Atlas of the State of Indiana*, published in 1876. The image is an artist's illustration of the residence (note the family on the porch). The lower image is the house's rear façade prior to preservation. We include the image here to show the accuracy of the depiction and the artist's attention to detail. However, the house's setting of winding drives, trees rows, grand entry walk and stairs, and fountain are all contrived.

Source: Andreas 1876.

tisements for local businesses. Plat maps include a wealth of information for the study of cultural landscapes, but they are limited to being a snapshot for just the year they were published. Plat maps include the boundary lines for an individual's property. Listed with the plat is the name of the property owner and the property size (in acres; 1 acre = 0.4 hectare). Depending upon the publisher, many maps contain the location of the property's residence and other significant structures such as barns or mills. The maps identify primary topological features, such as streams, wetlands, and bluffs. Cultural features, such as towns, roads, stone quarries, treaty boundary lines, schools, and churches, are included (Figure 6.3).

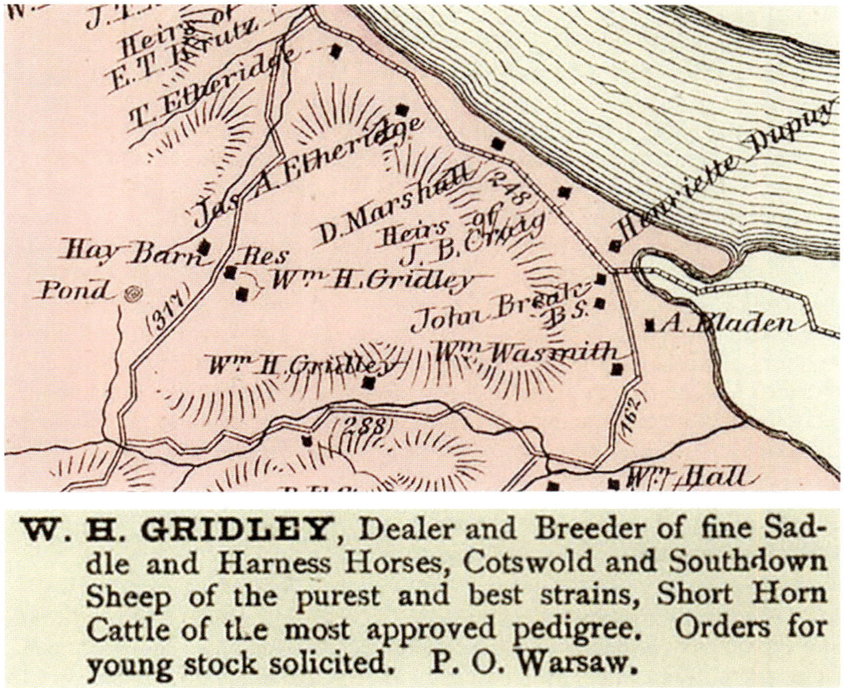

Figure 6.3
An 1883 illustration and advertisement for William H. Gridley published in *An Atlas of Carroll and Gallatin Counties, Kentucky*. The top image shows the location of Gridley's farm, residence, and a hay barn (that contains a beater hay press). The illustration also shows how the farmstead is located on the uplands adjacent to the Ohio River. The analysis of Gridley's barn is reported in Chapter 12.

Source: Lake 1883a.

When the names of property owners are cross-referenced with the population and agriculture census, broad cultural patterns can be discerned (Jordan 1966; Stanton 1985; Kamphoefner 1987; Ostergren 1988).

Plat maps were instrumental in interpreting dendrochronological results for several properties reported in this text. In our study of hay press barns in Chapter 12, a county atlas identified the James Anderson hay barn as a "Tobacco Warehouse," and it lists Anderson's position as a "Tobacco Dealer" (Lake 1883b). This information helped to explain how Anderson transitioned his agricultural production to tobacco when the hay market collapsed in the late 1870s. As a second source supporting the evolution of Anderson's occupational interests, the 1884 *Indiana Gazetteer* lists his business focus as "tobacco and grain" (Polk & Co. 1884; Baas and Rubino 2014). A plat map of the Thiebaud Farmstead demonstrates how the riverside property was ideal for participation in the regional hay economy, and identifies the expansion of the farmstead in the 1860s (Lake 1883b).

6.4 Property deeds and titles

Property deeds and titles are legal records documenting the transfer of the ownership of a property between parties (e.g., between a government and a person, between two people, the inheritance of property, etc.). These documents contain a legal description of the property, its size, names of parties involved, and the date when property changed ownership. In the United States, the records are updated and housed in a county's Recorder's and Auditor's offices. However, obsolete historic documents are often moved to a local library or historical society.

Tree-ring analysis, in conjunction with property deeds and titles, can help identify the presumed builder of a site by establishing the historic owner of a landscape or building (Grissino-Mayer et al. 2009). We use the records to help determine the owner of a property at a specific point in time, typically the construction and alteration dates established through tree-ring analysis. For instance, documents located at the Switzerland County, Indiana, Recorder's Office identify the owner of a log house in Craig Township as William Johnston (see Chapter 12). In 1820, Johnston moved to Indiana from Culpepper County, Virginia (Weakley 1885). Documents in the Recorder's office show that he paid the General Land Office in Cincinnati $197.90 for 158.4 acres. William was issued a land patent, a document showing the sale of land from the US government to an individual, and this transaction was recorded at the county office. Later, in 1865, documents in the Auditor's office show how the property was divided into two farms, and deeded to James V. Anderson and Morris McKay (Baas et al. 2017). Tree-ring analysis, along with archival documents, was able to establish Johnston as the builder of the house about 1833 and Anderson as the builder of the barn in 1870.

6.5 General Land Office survey (GLO)

In 1812, the US government created the General Land Office (GLO) to divide and sell publicly owned land (White 1991). The GLO adopted the Rectangular Land System (RLS) where the landscape was divided into square mile sections. To demarcate lines, land surveyors placed a wood post at the corner of each section. Trees near the post were blazed by chopping a flat surface in the tree's trunk with an ax to permanently mark the wood post's location (Bureau of Land Management 1980). These trees are known as "witness" trees because they officially delineated property boundaries. Government surveyors recorded the species and size of the witness trees, and described the undergrowth typically found along the section line (Figure 6.4). From GLO surveyor notes, there are first-person accounts of large sections of North America's pre-European settlement forests.

The 19th-century guidelines for GLO surveyors are available from the Bureau of Land Management's website, and copies of the field notes are available from state governmental archives or historical societies. Users of GLO data should recognize the limitations regarding a surveyor's potential bias for specific tree species (Bourdo 1956; Schulte and Mladenoff 2001; Hanberry et al. 2012).

We use GLO field notes with the results of tree-ring analysis in two distinct ways. The first is for better understanding 19th-century forest composition, a method often used by restoration ecologists to establish historic reference ecosystems (Baas 2018). The field notes supplement early 20th-century scholarly accounts of the forests surrounding our survey sites. We compare the historic forest composition with the species that comprise the buildings we have studied. As we have described in earlier chapters, historic forests are found in our historic buildings, and we attempt to connect the two data sources to describe pre-European settlement forests and species bias of builders. The second use of GLO data is demonstrating landscape preferences of Native American and European ethnic groups. Cultural geographers use GLO data to measure the cultural bias (landscape preferences) of groups for specific landscape types (for instance hardwood or softwood forests, grassland, riverine, etc.). Anderson et al. (1996) and Dilts et al. (2012) used GLO data to demonstrate anthropogenic landscape change from Native American fire regimes and agriculture. GLO data documented landscape

Figure 6.4

The 1822 General Land Office surveyor notes for the survey line between Sections 10 and 11, Township 21 North, Range 1 West (in west-central Indiana). The surveyor recorded the size and variety of several timber species: white elm, ash, beech, sugar maple, and walnut. He also noted the presence of a "handsome stream."

Source: US Bureau of Land Management.

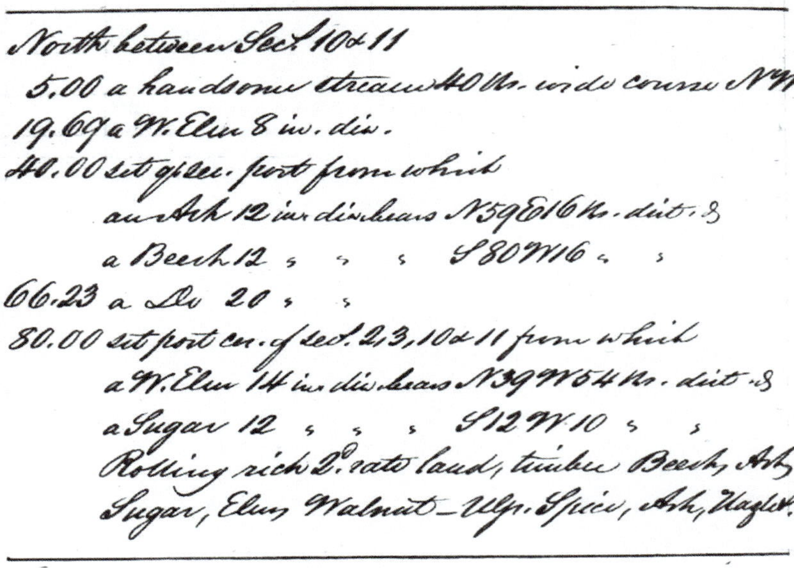
North between Sec. 10 & 11
5.00 a handsome stream 40 lks. wide course NW
19.69 a W. Elm 8 in. dia.
40.00 set qr. sec. post from which
an Ash 12 in. dia. bears N59E16 lks. dist.
a Beech 12 " " " S80W16 " "
66.23 a Do 20 " "
80.00 set post cor. of sec. 2, 3, 10 & 11 from which
a W. Elm 14 in. dia. bears N39W54 lks. dist.
a Sugar 12 " " " S12W10 " "
Rolling rich 2d rate land, timber Beech, Ash
Sugar, Elm Walnut – Undr. Spice, Ash, Hazle &c

bias of European settlement groups (Jordan 1966; Stanton 1985; Kamphoefner 1987; Ostergren 1988). In Chapter 13, we present our attempts to establish methodologies that identify or dispel bias for particular timber species in construction.

6.6 Population census

A census is a governmental counting of the population. Different countries employed different census counts that can be helpful to dating buildings and landscapes. For instance, the DECIMA project is using Geographic Information Systems to map 1420s and 1560s tax census data for Florence, Italy (Herlihy and Klapisch-Zuber 1985; Terpstra and Rose 2013, 2016). The first decennial US census took place in 1790 and is completed every ten years. The Census Bureau is required to protect the privacy of living citizens, so data for individual persons are released 72 years after they were recorded. Therefore, you can currently find the data of individual persons up to 1940 (Gauthier 2002). The US Census establishes the distribution of congressional seats, identifies when a territory's population was large enough to become a state, and reports the status of industrial and agricultural production. Census records are accessible at public libraries, universities, and historical societies through online databases (and many of these institutions still have records available on microfilm).

As archival sources used with dendrochronology, census records provide a wealth of information (National Archives 2018). The data provide the name and age of head of households and their family members, their marital status, and their relationship to the head of household (e.g., son, daughter, etc.). The census records the state or country a person was born in, as well as their parents' birthplace. This allows an individual's ethnicity to be identified. The census reports if a person is a slave or free, if they are literate, or if they have a disability (e.g., blindness).

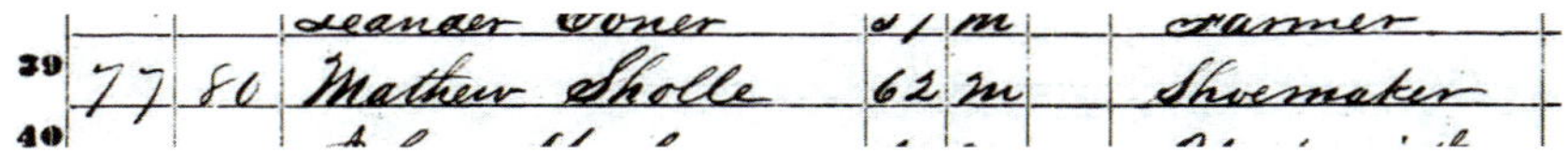

Figure 6.5
The 1850 US Population Census record for Harmonist Mattias Scholle, Economy Township, Beaver County, Pennsylvania. The record shows (Mathew) Scholle's age as 62 and identifies his occupation as a shoemaker. The census data allowed scholars to establish a 1788 birthdate for the Harmonist. The analysis of Scholle's house is reported in Chapter 11. This census entry was recorded 30 years after the Harmonists left Indiana.

Source: US Census Bureau.

It lists their occupation and the value of their property. The census paper forms completed by government enumerators has evolved over time. The earliest forms recorded the basic information of the head of household's name and the number and age of persons in a residence. Information on all residents of a household occurred in 1850 (Gauthier 2002).

We use census data to tell us about the residents of a house or landscape. The census tells us who and how many persons reside in a house, how they are related, and their occupations (e.g., laborer, carpenter, farmer, etc.; Baas 2012a). In Chapter 11, we demonstrate how the census data established the age and occupations of two Harmonists, tanner Franz Pfiel and cobbler Mattias Scholle (Figure 6.5; Strezewski 2015). In Chapter 10, we analyze Robert Glenn's farmstead landscape. The census data confirm Glenn's Scottish heritage, his occupation as a farmer, the size of his family, and that his son William Glenn inherited the farm following Robert's death.

6.7 Agricultural census

Several different types of federal and state agricultural censuses were completed throughout the 18th, 19th, and 20th centuries. For landscape research, the Census of Agriculture typically refers to the detailed enumeration completed for the years 1850, 1860, 1870, and 1880 in association with the US decennial population census. Notable cultural landscape studies have used agricultural census data to identify and compare the land preferences and agricultural production of African- and European-born residents to native-born Euro-Americans in efforts to document the cultural preferences for crops and livestock over time (Jordan 1966; Kamphoefner 1987; Ostergren 1988; Baas 2012b).

The 1850–1880 data are reported at the county scale in the compendium for each census year (National Archives 2018). However, the personal data for individual farms were distributed to individual states, and is located in state-level archives and libraries. The census enumerator collected and reported detailed information about every 19th-century farm (defined as five acres or larger). It listed the farmer's name, whether they owned or rented the farm, and the farm's value. Some of the most interesting data in the census is the recording of the number of agricultural products farmers produced and the animals they raised. The census records the amount of orchard, cereal grains, and grasses. Specialty crops, such as maple syrup, honey, hops, and grapes, were reported. The number of horses, oxen, sheep, and swine are included. The 1880 Census of Agriculture recorded the number of acres a farmer dedicated to a specific crop. For example, how many acres a farmer had in pasture or woodland, and how many acres were appropriated to a product, such as hay, rice, corn, or wheat. This data allows for the calculation of agricultural yields per acre (as in bushels of corn or pounds of cotton per acre).

The agricultural census, combined with the population census, can demonstrate New World and Old World cultural preferences. Mapping crop production demonstrates regional agricultural biases, often based on local climate, topography, and soil conditions. It can also demonstrate a cultural bias among farmers. Improved acres, the portion of a farm that had been prepared for crops or livestock, was listed along with the unimproved acreage. With this information the evolution of a farm can be tracked over time by finding a farmer's name on a plat map, locating their production in the agricultural census, and cross-referencing it with the population census to establish their (or their parents') place of birth. Understanding a farm's agricultural production can help interpret farm structures, and allow scholars to determine if they are typical to a region, outdated, or cutting edge. Did the scale of agricultural production of particular products result in distinct structures or layout of the farmstead? We used the census to show the preference of southeast Indiana and northern Kentucky farmers for producing timothy hay over tobacco (Baas and Rubino 2014). Cross-referencing the production of hay and tobacco with local market prices reported in newspapers demonstrated the financial impact of both crops for regional farmers. Census data were used to interpret a southeast Indiana hay barn that was later used as a tobacco warehouse (Baas et al. 2017). We also used the agricultural census data to interpret the structures on Robert Glenn's farmstead reported in Chapter 10, and we used it in Chapter 12 to confirm the construction and modification dates of a hay press barn on the Justi Thiebaud Farmstead.

6.8 Newspapers

Newspapers are important sources for documenting and interpreting historic landscapes and buildings. Often, from a reporter's account of the construction of newsworthy places, newspapers are the rare archival source for building dates (Baas 2012a). Additional information regarding construction dates can be garnered from various sections of the newspaper, such as the announcements for the grand opening of a new business, housing development, or public building, and in the obituaries where the biographies of the deceased are often included. We used newspapers to document the Harmon House in Chapter 14, to describe alterations to the Thrall's Opera House in Chapter 11, and to report the local hay production and the construction of press barns in Chapter 12. In Chapter 9, we are using newspaper accounts of an 1846 flood to help explain why preservationists cannot account for a 19th-century house. In Chapter 9, we use a newspaper account of a family's oral history to interpret the George Ash House.

6.9 Probate records: wills and inventories

Probate records are a listing of an individual's personal articles and their value at the time of their death. The records were assembled to help settle the deceased's outstanding debt or to divide their estate among heirs. Wills are legal documents that a person assembles to describe what happens to their property and assets when they pass. They often include descriptions for how property is to be divided and the future use of buildings. As an example, in 1872 George Tardy's will determined that his two sons and any of their descendants would equally share the use of a barn and hay press (Black 1875). Although the will does not tell us the date of the barn's construction, we do know that it was built prior to the 1872 will.

Inventories are lists of a deceased's possessions and their value. While they may not list the buildings a person owns, they may list objects specific to a building's use. For example, household items that might be typical of other geographical locations, or the tools and equipment specific to an agricultural or industrial site. Historians have also used inventories to measure the importance of material culture by identifying the types of items an ethnic group brought to North America and the items bought when they arrived (Stanton 1985).

6.10 Letters, diaries, and journals

Personnel accounts of places are often recorded in letters, diaries, and journals. These types of documents are often included in public archives, and may be transcribed and interpreted by historians. Often the documents are still in the possession of family members or property owners. They are occasionally published with interpretations. In Chapter 9, we demonstrate how the use of letters transcribed by the directors of the *Musée de Venoge*, with the results of tree-ring analysis, established that the Weaver family moved into a new house in 1828. In Chapter 7, we demonstrate how John Todd's journal account describing the construction of his new barn informed us how and when he harvested timber for the structure.

6.11 Oral histories

Oral histories collect personal memories and opinions about historical places and events (Ritchie 2005). While many oral histories are documented in text, modern histories are digital sound recordings that can be archived for use by future scholars. Like all archival sources, the histories need to make clear what events were witnessed, and what accounts and stories are passed down through a family over time. Oral histories are typically completed by genealogists or historians, but can also be reported in local histories or newspaper articles. For instance, a fair amount of information about the George Ash House analyzed in Chapter 9 comes from a newspaper article describing an interview with Ash's descendants years after his death. We also made use of an oral history that described the 1950s room layout of the William Johnston House reported in Chapter 12, and oral history is included in the teaching tree-ring analysis exercise in Chapter 14.

6.12 Directories

City and town directories are published volumes identifying an individual's address and occupation. Information regarding a person's residence or the location of their employment can be located in two ways. The first is through the alphabetical listing of names that included a person's home address, place of employment, and often their spouse's name. The second was a listing of individual residential, business, and governmental property addresses that were, again, followed by the name of the resident and employment information. Eventually, directories included a person or businesses' telephone number. Directories also include business advertisements that are often helpful in interpreting a building or landscape.

6.13 Photographic images

Photographic images, from a variety of processes, became common in the 1840s. Historic photographs document people, places, and events. In the late 19th and early 20th centuries, photographs of landscapes, monuments, landmarks, and houses were commonly used in local publications to promote cities and neighborhoods (Baas 2012a). Historic photographs are becoming more accessible through university and local libraries and historical society digital archives. The US Library of Congress houses online digital images of its collection of nationally significant sites and events.

Photographic images can be used with other archival sources, such as Sanborn maps (see section 6.2), to reconstruct a historic landscape that has been lost or altered (Seymour and Summerlin 2017). When photographic images are available for the sites we analyze, we use them to identify the original layout of a landscape or form of a building, subsequent alterations, building materials, and construction methods. We have used photographs to document 19th- and 20th-century alterations to Harmonist buildings (Chapter 11). We also used a daguerreotype, an early photographic process popular in the middle 19th century, of a timber structure related to the Fort Wayne military garrison to potentially identify the source for timbers found in a house constructed in the 1850s (Chapter 10).

6.14 Historic aerial photographs

Historic aerial photographs provide a snapshot of the landscape at specific points of time. In the United States, aerials began to be commonly used by federal and state agencies in the late 1930s to assess the state of agriculture. Aerial photographs can be used to demonstrate changes to the landscape over time by comparing historic to current photographic images (McClelland et al. 1999; United States Department of Agriculture 2016). As an example, we used historic aerial photographs of the Glenn Farmstead (see Chapter 10) to evaluate buildings that had been removed, or that were collapsed and overgrown with vegetation when they were sampled for tree-ring analysis. From the photographs we could verify fieldwork, and estimate the sizes and placement of many of the site's lost agricultural buildings (Figure 6.6).

Historic aerial photography is available in different forms from a variety of archive and online sources. More recent aerials are available from online browsers. Often, state-level Geographic Information Systems clearinghouses provide georeferenced images for use by the public. These images can be used in tandem with other map forms (e.g., atlas, plat, and Sanborn; Baas and Gibson 2011, 2015).

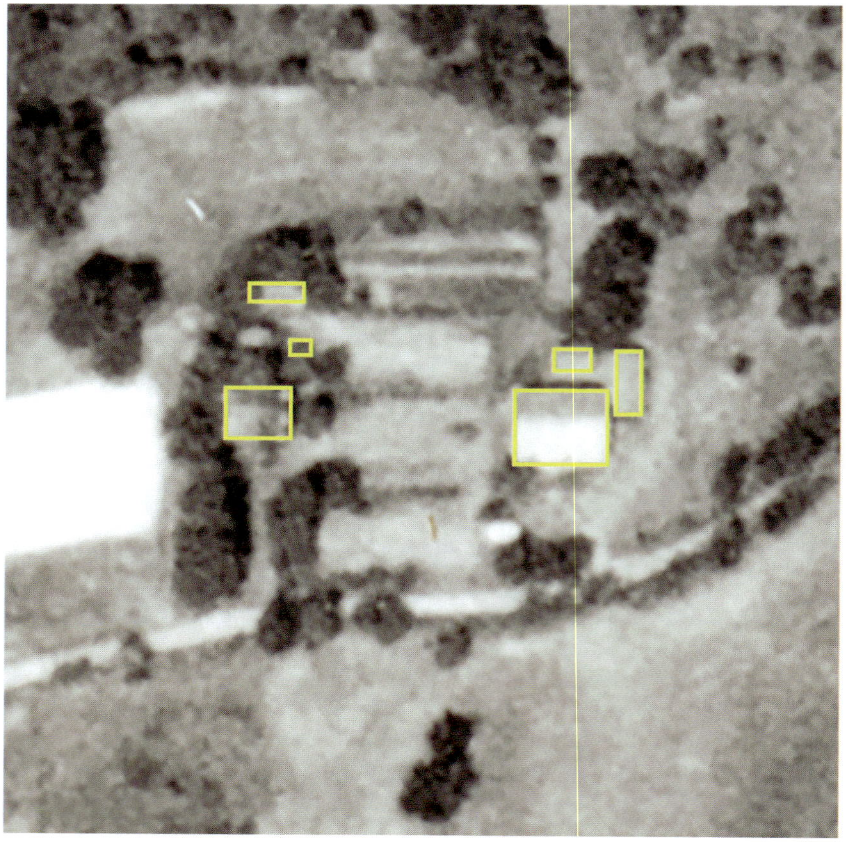

Figure 6.6
A 1955 aerial photograph of the Robert Glenn Farmstead located in Jefferson County, Indiana. The farm's houses, barn, and outbuildings are outlined in the yellow rectangles. Analysis of the farmstead is reported in Chapter 10.

Source: Indiana State Archives and Records Administration.

6.15 Secondary sources

A secondary source is scholarly information about the history of a landscape or building, reported by an author without a firsthand account of historic events. Secondary sources come in many forms, such as books, journal articles, documentaries, and websites. We use a variety of secondary sources to interpret the results of tree-ring analysis. The most common sources for the projects reported in the case studies section of this text are local histories and National Register and National Historic Landmark documentation (for instance see the New Harmony resources in Chapter 11).

We also use the information recorded by the National Park Service's Heritage Documentation Services: Historic American Building Survey (HABS), Historic American Engineering Record (HAER), and Historic American Landscape Survey (HALS). These programs use field study, archival research, photography, and measured drawings to document significant historic sites. These records start in the 1930s, are archived by the US Library of Congress, and are accessible online. Examples of how we use the records with the results of tree-ring analysis can be found in our investigations of Harmonist buildings in Chapter 11 and for William Henry Harrison's Grouseland in Chapter 8.

Archaeological investigations are often associated with significant historic sites. While the investigations often seek to locate evidence of past cultural resources (e.g., a building that no longer exists), historical archaeologists also examine existing landscapes and buildings. Investigations are typically summarized in reports and journal articles that describe and interpret artifacts. These artifacts are often construction related and may include documentation of structural foundations, nails, chinking, and window glass. Known archeological sites are recorded at a State or Tribal Historic Preservation Office (SHPO or THPO). Since archaeological sites are subject to looting, their locations are protected and scholars need to consult with the SHPO for access to a site's information.

We use the archaeological investigations of the Thiebaud Farmstead to support the existence of historic buildings on the property that we believe were recycled into the construction of a barn addition (see Chapter 12). While we were hoping archaeological investigations seeking to locate historic Fort Wayne might shed light on the timber used for its construction, the garrison's location was not identified (see Chapter 10).

6.16 Conclusion

Chapter 6 reviewed the archival and secondary sources we use to interpret the results of tree-ring analysis. Scholars need to be creative in their research of historic places since all sites are unique. In Chapter 7 we describe how we report the results of tree-ring analysis in journal articles and to property owners and managers.

6.17 Literature cited

Anderson, B., T. Crow, S.M. Lietz, and F. Stearns. 1996. Transformation of a landscape in the Upper Mid-west, USA: The history of the St. Croix River Valley, 1830–present. *Landscape and Urban Planning* 35: 247–267.

Andreas, A.T. 1876. *Illustrated Historical Atlas of the State of Indiana*. Baskin, Forster, and Company, Chicago, IL. 462 p.

Baas, C. 2006. *Richard M. Hazelett House National Register of Historic Places Registration Form*. National Park Service, Washington, DC.

Baas, C. 2012a. Concrete in the steel city: Constructing Thomas Edison's house for the working man. *Indiana Magazine of History* 108: 245–273.

Baas, C. 2012b. The effects of the civil war on land ownership and agricultural production for freedmen in St. Helena Township, Beaufort County, South Carolina: Using the Agricultural Census to evaluate and interpret material culture. *The African Diaspora Archaeology Network* 1–16.

Baas, C. 2018. Using general land office survey notes to define reference ecosystems for Ball State University's Ginn Woods, Delaware County, IN. *Proceedings of the Indiana Academy of Science* 127: 17–27.

Baas, C., and A. Gibson. 2011. 86° 10' 54" W, 39° 46' 1" N: Using geographic information systems to document historic sporting landscapes. *Journal of Sport History* 38: 475–492.

Baas, C., and A. Gibson. 2015. 86° 10' 54" O, 39° 46' 1" N: Utilizando Sistemas de Informacao Geographica Para Documentar Paisagens Esportivas Historicas. *Recorde: Revista de História do Esporte* (Brazil) 8: 1–16.

Baas, C., P. Jiang, and D.L. Rubino. 2017. Using tree-ring growth patterns to identify construction dates of two nineteenth century timber structures in Craig Township, Switzerland County, Indiana. *Indiana Archaeology* 12: 12–41.

Baas, C., and D.L. Rubino. 2014. Pressing hay in the commonwealth: Using tree-ring growth patterns to date the construction of two Kentucky beater hay press barns. *Journal of Kentucky Archaeology* 3: 2–31.

Black, J.B. 1875. *Reports of Cases Argued and Determined in the Supreme Court of Judicature of the State of Indian: Volume XLVI*. Journal Company, Indianapolis, IN.

Blumin, S.M. 1987. The Sanborn fire insurance maps, 1867–1950. *The Journal of American History* 73: 1089–1090.

Bourdo, E.A. 1956. A review of the general land office survey and of its use in quantitative studies of former forests. *Ecology* 37: 754–768.

Bureau of Land Management. 1980. *Glossary of BLM Surveying and Mapping Terms*. Denver Service Center, Denver, CO. 130 p.

Dilts, T.E., P.J. Weisberg, J. Yang, T.J. Olson, P.L. Turner, and L.A. Condon. 2012. Using historical general land office survey notes to quantify the effects of irrigated agriculture on land cover change in an arid lands watershed. *Annals of the Association of American Geographers* 102: 531–548.

D.J. Lake & Co. 1883b. *An Atlas of Switzerland and Ohio Counties*. D.J. Lake & Co., Philadelphia, PA. 46 p.

D.J. Lake & Co. 1883a. *An Atlas of Carroll and Gallatin Counties, Kentucky*. D.J. Lake & Co., Philadelphia, PA. 64 p.

Gauthier, J.C. 2002. *Measuring America: The Decennial Censuses 1790 to 2000*. US Census Bureau, Washington, DC. 148 p.

Grissino-Mayer, H.D., L.B. LaForest, and S.L. van de Gevel. 2009. Construction history of the Rocky Mount historic site (40SL386), Piney Flats, Tennessee from tree-ring and documentary evidence. *Southeastern Archaeology* 28: 64–77.

Hanberry, B.B., J. Yang, J.M. Kabrick, and H.S. He. 2012. Adjusting forest density estimates for surveyor bias in historical tree surveys. *The American Midland Naturalist* 167: 285–306.

Harris, R. 1997. Reading Sanborns for the spoor of the owner-builder, 1890s-1950s. *Perspectives in Vernacular Architecture* 7: 251–267.

Herlihy, D., and C. Klapisch-Zuber. 1985. *Tuscans and Their Families: A Study of the Florentine Catasto of 1427*. Yale University Press, New Haven, CT. 404 p.

Indiana Spatial Data Portal. 2018. *Indiana Sanborn Historic Maps 1883–1966*. http://gis.iu.edu/datasetInfo/sanborn.php. Accessed 31 August 2018.

Jordan, T.G. 1966. *German Seed in Texas Soil: Immigrant Farmers in Nineteenth-Century Texas*. University of Texas Press, Austin, TX. 237 p.

Kamphoefner, W.D. 1987. *The Westfalians: From Germany to Missouri*. Princeton University Press, Princeton, NJ. 215 p.

McClelland, L.F., J.T. Keller, G.P. Keller, and R.Z. Melnick. 1999. *National Register Bulletin: Guidelines for Evaluating and Documenting Rural Historic Landscapes*. US National Park Service, Washington, DC. 33 p.

National Archives. 2018. *Census Records: Introduction to Census Records*. www.archives.gov/research/census. Accessed 30 August 2018.

O'Donnell, E. 1991. *National Register Bulletin: Researching a Historic Property*. National Park Service, Washington, DC. 17 p.

Ostergren, R.C. 1988. *A Community Transplanted: The Trans-Atlantic Experience of a Swedish Immigrant Settlement in the Upper Middle West, 1835–1915*. University of Wisconsin Press, Madison, WI. 416 p.

Page, R.R., C.A. Gilbert, and S.A. Dolan. 1998. *A Guide to Cultural Landscape Reports: Contents, Process, and Techniques*. National Park Service, Washington, DC. 167 p.

Patton, D.K., A.K. Lobben, and B.M.C. Pape. 2005. Mapping cities and towns in the late nineteenth and early twentieth centuries: A look at plat, Sanborn, and panoramic mapping activities in Michigan. *Michigan Historical Review* 31: 93–122.

Polk & Co. 1884. *Indiana State Gazetteer and Business Directory: 1884–85, Volume III*. Indianapolis, IN. 1,264 p.

Ritchie, D.A. 2005. *Doing Oral History*. Oxford University Press, New York, NY. 320 p.

Schulte, L.A., and D.J. Mladenoff. 2001. The original US public land survey records: Their use and limitations in reconstructing presettlement vegetation. *Journal of Forestry* 99: 5–10.

Seymour, M.W., and P.R. Summerlin. 2017. Evidence and deception: A historic photo-analysis method for landscape studies. *Landscape Journal* 36: 91–109.

Stanton, G.W. 1985. *Brought, Borrowed, or Bought: Sources and Utilization Patterns of the Material Culture of German Immigrants to Southeastern Indiana, 1833–1860*. PhD Dissertation. Indiana University, Bloomington, IN.

Strezewski, M. 2015. Harmonist demography and town planning in New Harmony, Indiana. *Communal Societies* 35: 1–27.

Terpstra, N., and C. Rose. 2013. DECIMA: The digitally encoded census information and mapping archive, and the project for a geo-spatial and sensory digital map of renaissance Florence. *Journal for Early Modern Cultural Studies* 13: 156–160.

Terpstra, N., and C. Rose. 2016. *Mapping Space, Sense, and Movement in Florence: Historical GIS and the Early Modern City*. Routledge, London, UK. 220 p.

United States Department of Agriculture. 2016. *Historical Aerial Photography at the Aerial Photography Field Office (APFO): Information Sheet*. United States Department of Agriculture, Washington, DC.

Weakley. 1885. *History of Dearborn, Ohio, and Switzerland Counties, Indiana*. Weakley, Harraman & Co., Chicago, IL. 1,382 p.

White, C.A. 1991. *A History of the Rectangular Survey System*. Bureau of Land Management, Washington, DC. 774 p.

Chapter 7

Reporting the results of tree-ring analysis

7.1 Introduction

Results of tree-ring analysis are reported in a variety of ways. Chapter 7 explains how we communicate results in journal articles and to the owners and managers of historic sites. We report results with tables that organize and present the statistical analysis of our data, illustrations of landscapes and buildings that document fieldwork, and photography that records the sample locations. We include descriptions of a tree-ring analysis report, how data are reported in table form, and how to approach field documentary drawings of historic sites.

7.2 Parts of a tree-ring analysis report

Our reporting of tree-ring analysis typically contains five sections: Introduction, Methodology, Results, Conclusion, and Literature Cited. We have included a short description of each section here, followed with detailed explanations of tables and fieldwork.

Introduction. A report introduction presents a brief summary of the location, the resource, its owner, and the goals of the analysis. The goals for a project might be as simple as assigning a construction date to a landscape or building. However, they are often as complex as identifying a site's owner or builder, alterations to a site, and the temporal placing of a resource within the context of similar sites (e.g., comparing Indiana Harmonist buildings in Chapter 11 to Harmonist buildings in Pennsylvania).

Methodology. The methodology section generally explains dendrochronology theory, crossdating, fieldwork, and lab work (Chapters 4 and 5). The methodologies section will likely be used by property owners and managers with minimal technical knowledge in dendrochronology, and should be presented in layman's terms.

Results. Dendrochronology results are reported in various forms: descriptive text, data tables, graphs, illustrations, and photos. Descriptions of these forms are presented later.

Conclusion. The conclusion section, often referred to as the "discussion" section, interprets the results as they relate to the study's goals. Archival and secondary sources are essential in presenting your conclusions (see Chapter 6). In our articles and reports, we attempt to present sites within the theoretical frameworks of cultural geography and cultural landscape studies (see Chapter 2). For example, we present the Grayson Dogtrot House within discourses of Southern and Midwestern United States vernacular architecture that explain how a residence typically constructed in the southern regions of the United States is located in Indiana (Chapter 8).

Not all analysis will result in clear construction and alteration dates for a historic site, or the sequence of construction for resources in a cultural landscape. Since dendrochronological reports often influence preservation decisions for a landscape or building (i.e., the restoration or removal of all or part of a resource), the report needs to be clear when tree-ring analysis has not conclusively established construction and alterations dates.

Literature Cited. The literature cited section includes a listing of the methodologies and archival and secondary sources for the analysis.

7.3 Data tables

For the case studies reported in Chapters 8 through 13, we provide both a tree-ring sample table that provides information pertaining to each of the samples we dated or attempted to date and a summary table that provides information pertaining to each of the timber type chronologies we prepared from a building (for example, see Tables 8.1 and 8.2). The structure and content of these tables is explained here to eliminate the need to repeat them throughout the text.

7.3.1 Tree-ring sample table construction

In the tree-ring sample table, the sample identification, the first and last year present in a sample, characteristics of the outer ring, *r*-value, timber type, and provenience are provided.

- Sample identification – each timber from a building is given a unique sample identification that includes a three-letter prefix identifying the structure, a two-digit timber number referencing each timber sampled, and a letter identifying an individual sample obtained from each timber. See Chapter 5 for a more thorough description.
- First and last years present – these are the first and last years present on a sample.
- Outer ring – the characteristics of the outer ring of a sample are extremely important to the interpretation of building construction dates. We use a modified version of Nash's symbols (1999) in the tables (Table 7.1). Please see section 7.3.2 for a discussion of outer rings and their importance in determining construction and modifications dates of buildings and landscapes.
- *r*-value (series intercorrelation) – this value represents the correlation between the tree-ring widths of a timber and all other timbers of the same type (e.g., all other tulip poplars) found in a particular structure. No *r*-value is given if only one or two timbers of a particular timber type was sampled in a structure. See Chapter 5 for a discussion of *r*-values and series intercorrelation values.

- Timber type – the wood identification of a particular timber. We remind the reader that identification to species is possible for some timber types (e.g., tulip poplar and American beech), but it is not for certain groups of species (e.g., ash, red oaks, and white oaks). See Table 3.1 for a full listing of timber types encountered and the potential species that were sampled.
- Provenience – the function of a timber in a building (e.g., post, sill, or rafter; Chapter 2).

Other conventions used in the tree-ring sample tables:

- Timber type and provenience are only given once if multiple samples are obtained from a single timber.
- Cardinal directions are abbreviated by a single letter.

Table 7.1 Outer ring classification conventions. Dating the construction of a structure requires determining when the trees used to make the building's timbers were harvested. Modified from Nash (1999).

Symbol	*Interpretation*	*Harvest, cutting, or death date*
b	Bark present; outer ring represents the last year of growth for the tree. Latewood is present in the ring suggesting that the ring was fully or partially formed.	Yes
w	Wane present; outer ring represents the last year of growth for the tree. Latewood is present in the ring suggesting that the ring was fully or partially formed.	Yes
b/e	Bark present; outer ring represents the last year of growth for the tree. Only earlywood was present in the ring so the tree was harvested early in the growing season (see Chapter 3 and Figure 3.8).	Yes
w/e	Wane present; outer ring represents the last year of growth for the tree. Only earlywood was present in the ring so the tree was harvested early in the growing season (see Chapter 3 and Figure 3.8).	Yes
b#	Bark was present, but the outermost rings of the sample could not be crossdated reliably due to decay, insect damage, and/or growth suppression (Chapter 4). The number represents a best estimate of the number of rings present between the bark and the outermost ring that was reliably crossdated.	No
w#	Wane was present, but the outermost rings of the sample could not be crossdated reliably due to decay, insect damage, and/or growth suppression. The number represents a best estimate of the number of rings present between the wane and the outermost ring that was reliably crossdated.	No
s/b	Bark was present, but the outermost rings could not be crossdated reliably due to a growth suppression. The number or rings between the outermost ring that was reliably crossdated and the bark was indeterminable.	No
s/w	Wane was present, but the outermost rings could not be crossdated reliably due to a growth suppression. The number or rings between the outermost ring that was reliably crossdated and the wane was indeterminable.	No

Symbol	Interpretation	Harvest, cutting, or death date
sap	Sapwood present, but no bark or wane was present. Thus, the outermost ring was not the last ring formed by the tree prior to harvest but was near it (see Chapter 3 for an explanation of sapwood).	No
+#	No bark or wane is present. Additional rings (as noted by the "#") beyond the last dated ring were present on the sample but could not be crossdated due to decay, insect damage, and/or growth suppression.	No
blank	Non-cutting date. The last ring does not represent the last year of growth in the tree from which the timber was made. There is no way to determine when the tree was cut.	No

7.3.2 Outer rings and assigning construction dates

In the next chapters we present data from numerous case studies that we have performed throughout the mid-Ohio River Valley. For each of the structures, we report a build date and, for some of the structures, modification dates. These dates are based on assigning a calendar date to the last tree ring formed by a tree, the ring adjacent to the bark or associated with a waney outer edge of a timber (Figures 1.2 and 5.1). When multiple samples share a similar death/felling date, we interpret the date as the year of construction for the building. When timbers have comparable death dates (felling dates spread over a two- to several-year period), we interpret the construction date as the latest felling date found in any of the timbers (Nash 1999). When sampling, we concentrate our efforts on obtaining samples with bark or wane.

Commonly, the bark or wane ring has latewood present (Figure 3.8). Therefore, the tree was alive during that year, at least until the middle of the growing season (June through July). When latewood is present, it is not possible to determine when in that year it was harvested. We can only note that the tree was harvested prior to April (the beginning of the next year's growing season). Consequently, a timber with an 1856 death date could have been harvested any time from mid-growing season 1856 to April 1857. In our case studies, we would ascribe a cutting date of 1856 to the timber and an 1856 construction date to the building if other timbers from the structure exhibit a similar terminal ring. See Chapter 3 for interpretation of bark or wane rings consisting only of earlywood.

In our analysis of historic buildings, we assume that the timbers used for construction were sourced from near the construction site and that the structures were built with green or unseasoned wood, as notching and cutting to size would be much more difficult on seasoned wood (Rempel 1980; Hutslar 1986; Roberts 1996). In his construction/bookkeeping journal, John Todd (1860) describes "getting out" timber from neighboring forests for the construction of his hay press barn. Timber was cut in November of 1860, and the barn was raised about seven months later in July of 1861. Procuring timber for the barn took nine men (who spent various amounts of time on the task) 17 days to fell enough trees for the barn. Hauling timber took seven days, and framing was complete in 32.5 days. Other buildings reported in the case studies chapters were constructed over various periods of time ranging from days to years (Eleutherian College; Chapter 9 and the Huxford House; Chapter 10).

In many of the structures we have analyzed we encounter sawn timbers in addition to hewn timbers. Often the larger timbers (posts, sills, plates, floor joists) are hewn whereas smaller timbers, such as knee braces, flooring, siding, rafters, and the floor joists of the second floor and attic of houses and the second floor of barns, are sawn. The origin of sawn timbers in a building is unknown. They could have originated from trees from the construction site that were taken to a local sawmill, from trees not growing at the construction site but obtained from a sawmill, or sawn on-site with a portable sawmill (Roberts (1996) states such mills were available in southern Indiana circa 1875). The length of time between felling, sawing, hewing, and construction could potentially be quite variable. Consequently, an exact construction date is not determinable. Tree-ring analysis provides precise dates of harvest, not exact dates of construction. This is why we have come up with our convention that cutting dates equal build dates.

Not all outer or terminal rings on samples provide cutting or death dates. If bark or wane is not present on a sample, the outer ring does not represent the last year of growth for a tree. For example, when the post shown in Figure 1.3 was squared, the outer rings were removed. It is not possible to know how many rings were removed. Consequently, this timber, although dateable, will not provide a harvest date. Since 1813 was the last ring present on the timber, we can say that it was harvested sometime after 1813 at some indeterminable date. Even though this sample does not yield a harvest date, its data, its tree-ring pattern, are still very useful. It will be used in creating a more highly replicated chronology that helps in dating other timbers from the building, and it's ring-width measurements will be added to regional chronologies used to date other buildings' timbers. In the tables presented in the following chapters, we categorize the outer ring of a sample as having bark, wane, etc. (Table 7.1). When reading the tables, please realize that only those samples with a bark or wane designation provide solid evidence for harvest dates, and, consequently, potential construction dates.

7.3.3 Timbers without dates

In the tables you will notice that dates are not provided for all samples. These samples did not reliably crossdate. There are a number of reasons why a timber's tree rings could not be dated. For example, a sample could have too few years to allow for reliable pattern matching between the sample and the site (building) or regional chronologies. Also, the sample could come from a timber type that we do not have a chronology for and the sample fails to crossdate with other regional or site chronologies. For example, we encounter maple (hard or soft) so infrequently that we have not yet sampled enough historical timbers to provide a reliable chronology for crossdating; rarely are we able to successfully crossdate a maple sample. Such a chronology may be produced in the future, but it will only be made possible if a large number of samples are obtained from other buildings or we are able to date old living or recently dead trees from regional forests.

A timber may also go undated if its tree rings do not overlap temporally with existing chronologies. For example, when sawing logs into timbers, some timbers, floorboards, or studding will inevitably be sawn from the oldest heartwood present in the log. If a tree is very old, a timber made from it may contain tree rings that are too old to overlap with an existing chronology. Likewise, when renovations of buildings are made with timber harvested in the late 19th or early 20th century we are unable to crossdate them reliably with our chronologies since we lack replication or sample depth for most timber types around the turn of the 20th century.

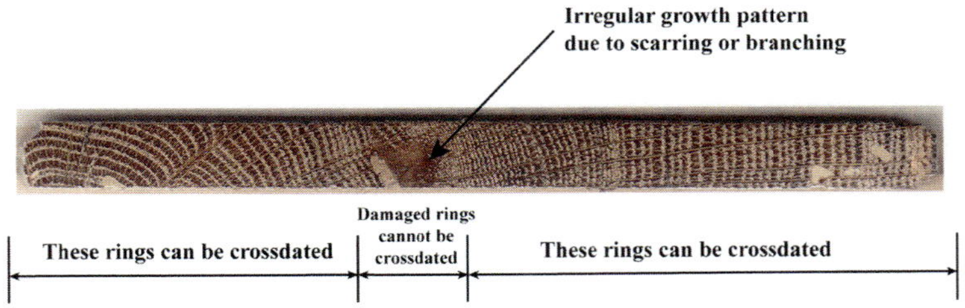

Figure 7.1
White oak core containing rings that cannot be accurately crossdated due to scarring.
Source: Authors.

Crossdating will also not be possible for a timber if a sample has an irregular growth pattern due to scarring or branching. When cores and cross sections are obtained from timbers, we select a sampling location that will hopefully produce a sample that is free of scar tissue. However, this is not always the case (Figure 7.1). In such situations, we do attempt to crossdate portions of the sample that are free of anomalous growth. Finally, if a timber was obtained from a complacent tree, dating will not be successful. A sample must exhibit variation in ring width, sensitivity, in order to be crossdated (Figure 1.7). Crossdating is only possible when rings of various widths are present – no variation, no crossdating (Chapter 4).

7.3.4 Chronology summary table construction

The chronology summary tables provide information pertaining to the chronologies (series of accurately dated and measured ring widths) created for a building. We present the chronologies for each timber type separately because we find crossdating in the region is most successful when performed with a single timber type. Each table includes:

- Samples – the number of samples successfully dated.
- Timbers – the number of timbers successfully dated. The number of samples is often greater than the number of timbers dated since multiple samples are routinely collected from an individual timber.
- Undated – the number of timbers that were not successfully dated. See section 7.3.3 for potential reasons why a timber was not reliably crossdated.
- First and last – the first and last year of the measured tree-ring chronology constructed from the building. You will note that some samples (in the sample table described earlier) have dates earlier or later than the first and last rings reported for a timber type's chronology. This discrepancy occurs because not all tree rings found in a sample are measurable (incomplete rings may be due to shaping of a timber, insect damage, etc.; Chapter 5). For some timber types there are two separate sets of first and last dates provided. These double entries are due to gaps in the final chronology for that timber type.
- Years – the number of tree rings in the final chronology.
- r – the mean series intercorrelation (Chapter 5). This value is not provided if only one or two timbers of a particular timber type were present in the building.
- Mean and SD – the mean and standard deviation (SD) ring width in millimeters.

7.4 Chronology and sample depth reporting

A graphical display showing the widths of tree rings over time and sample depth are two additional layers of data often reported with the results of tree-ring analysis. In an effort to efficiently use the space in this book for the presentation of methodologies and case studies, we have chosen not to include reporting these graphs for the case studies in the ensuing chapters. However, we do want to present examples that can be included in a tree-ring report. As discussed in Chapter 4, a chronology is a series of accurately dated tree rings. They are assembled from all of the timbers (of a particular timber type) sampled from a historic structure. Sample depth is the number of samples that had a ring present at an individual year (Figure 7.2).

7.5 Field drawings

We document our fieldwork with drawings, notes, and images. The National Park Service's Historic American Buildings Survey (HABS) and Historic American Landscapes Survey (HALS) programs have established standards for field drawings documenting historic sites (National Park Service 2011). These standards are available online and are very helpful when learning to document historic buildings and cultural landscapes. However, field documentation for dendrochronology projects rarely involves the level of detail required by these programs. For instance, HABS guidelines recommend three-person teams to record historic sites. We typically work as a pair, with one person collecting samples and the other recording building

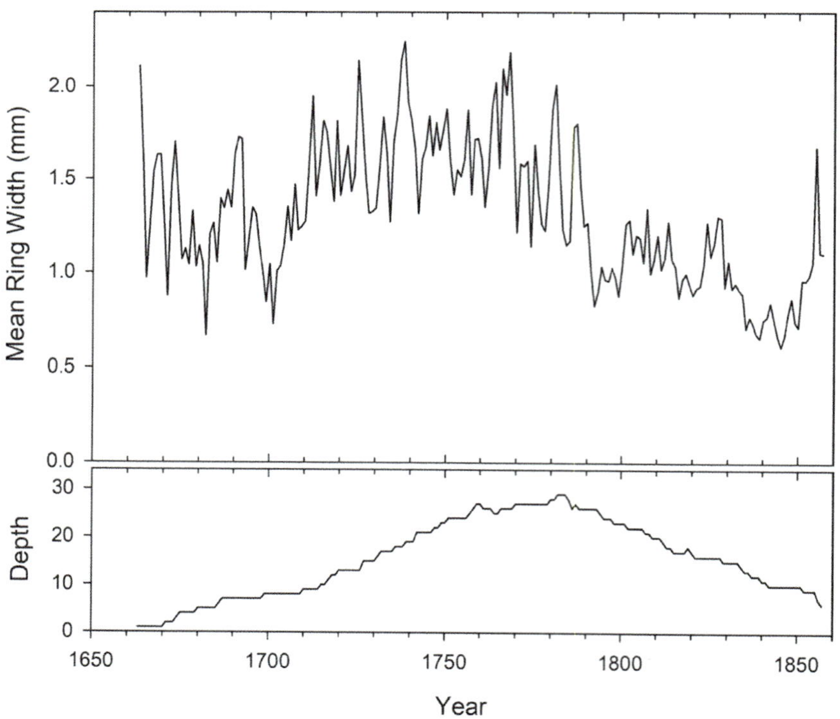

Figure 7.2
Top: Composite mean tree-ring chronology for dated ash sampled from the Frederick and Sophia Wyneken House (Chapter 8). Bottom: Sample depth for ash from the Wyneken House. Sample depth is the number of samples that had a ring present at an individual year.

Source: Authors.

attributes, measurements, and sample locations. Furthermore, HABS documentation is archived in the US Library of Congress and has specific archival grade papers and photographic requirements. Reporting tree-ring analysis does not.

There are three primary purposes for documenting dendrochronology projects. The first is to create an archival record of the sampling locations for the current analysis, and any future analysis of the site. Current and future researchers should be able to use documentary drawings and the accompanying notes to locate past sample locations (i.e., for replication of your research by others). Second, most vernacular buildings have distinct forms and patterns of interior and exterior spatial arrangements (Kniffen 1965; Glassie 1968). These forms and spatial patterns are easily recognizable in plan view when accurately represented in measured drawings (see the Grayson Dogtrot House in Chapter 8, the Bakes Road English Barn in Chapter 12, and the Harmon two-room log house in Chapter 14). More importantly, additions to a structure are more easily identified and interpreted when viewed in plan. Finally, measured drawings are used in reports and journal articles to communicate the results of tree-ring analysis.

7.5.1 Field documentation of barns

Barns are constructed from a series of bents (an assemblage of posts and beams; Figure 2.2). When we measure barns, we typically measure the bents' post locations and note the directions the bents run (i.e., bent lines run north and south, or east and west). Since the bent posts are organized in a pattern, a grid can be established to illustrate barns requiring more meticulous documentation. Once the barn's post locations have been measured and drawn, we identify exterior elevations and door and window openings. Notes should include the type and location of footings or cellar walls, and the direction of floor joists, flooring, siding, and roof. Timber frame barns are typically mortise and tenon construction. We make notes of missing or unusual mortise and tenon joints that might indicate lost timbers, alterations to a structure, or the use of recycled timbers (Figure 5.13). Interior spaces are identified (for instance, aisle, interior ramp, haymow, or cellar). Specialized barns might require identifying unique spaces (for instance a threshing floor, milking parlor, or breeding pens).

7.5.2 Field documentation of houses, churches, and commercial buildings

Measuring log and timber houses, churches, and commercial buildings is similar to measuring barns. However, when it is important to preserve historic interior finishes (e.g., plaster walls, historic wallpaper or paint, or tongue and groove flooring), sampling is limited to the cellar, crawlspace, and attic. We also note clues that suggest that interior spaces have changed over time (e.g., the addition of walls, windows, and doors). Again, the study and recognition of vernacular building types helps recognize post-construction alterations.

On occasion we will sketch a building detail if a sample is in a unique location that might make it challenging to locate in the future, or if the detail fits into a broader research interest. For instance, the bent design of a hay press barn is a key character-defining feature of the structure, and we typically show it in elevation. Also, we document the half-timber *fachwerk* construction at New Harmony, Indiana (see Chapter 11), and the *poteaux-sur-sol* (post-on-sill) construction at the *Musée de Venoge* (Chapter 9).

7.6 Equipment for field documentation

We find that time spent in the field documenting historic resources is prized, so a well-prepared surveyor or recorder is able to make efficient use of their time (Figure 7.3, Figure 7.4). We have assembled a list of documentation equipment we typically take into the field (Table 7.2).

We typically carry two types of measuring tapes. The 100-foot (or longer) reel-type tape is for measuring the long lengths of the exterior walls of buildings, and the distances between landscape features. The smaller 50-foot type is for measuring smaller lengths (such as interior rooms, or the length and width of a timber), and can fit in a pocket or hook to your belt. Often, we are in spaces where we cannot safely use a ladder, so we have created a method for approximating measurements for features such as roof peaks or the roof-level beams of barn bents. We tape a rope to a baseball, and then toss the ball over the feature we wish to measure. We then fasten the tape measure to the rope and pull it up to the height we need to measure. A laser measuring device can also be used to take various measurements throughout a building. We bring two 5-gallon buckets to the project site. The first is to carry the equipment. The second nests with the first, and is used as a seat.

7.7 Field drawings and photographs

Each field drawing should include the property name, what is being measured (e.g., a barn or house), a north arrow, drawing scale, the names of the recorders, and the date. We typically use 8.5 × 11-inch paper (ANSI A, 216 mm × 279 mm), a size we find the easiest to organize in the field, store, and reproduce (note that HABS guidelines recommends a larger 17 × 22-inch graph paper, ANSI C, 431 mm × 559 mm). We purchase large-volume packages of plastic mechanical pencils (0.7 mm or 0.9 mm) with erasers. You can never have too many pencils in the field, and unless

Table 7.2 Preservationists can optimize their time in the field by assembling a toolbox of equipment. Table 7.2 is a recommended list of equipment that we use in the field documentation of historic sites.

Hammer and nails (to anchor one end of the measuring tapes)
Graph paper
Mechanical pencils with erasers
Architectural and engineering scales
Straight edge (e.g., drafting triangles)
Clipboards
Flashlights
Headlamp
Metal measuring tapes (100-foot reel and 50-foot retractable)
Ladder
Ball and rope
Digital camera(s)
Video camera
Two 5-gallon buckets

Figure 7.3
Fieldwork equipment we use to document historic buildings and landscapes (see Table 7.2). Preparing a kit, or toolbox, of equipment optimizes field time analyzing sites.
Source: Authors.

Figure 7.4
Researchers should be prepared to collect samples in all types of conditions. This photograph is from fieldwork completed in the crawlspace of the Reed Case House (Delphi, Indiana) reported in Chapter 8. The building was moved and placed on a new concrete block foundation. The floor joists (visible in the upper left corner) are sawn white oak.
Source: Authors.

you tether one to your clipboard, they are easy to lose in the low-light spaces typical of dendrochronology sample locations. Experiment with lead sizes and hardness to see what is the most comfortable for your drafting style (Figure 7.5).

We take as many photographs as possible, including the relationship between multiple buildings within a landscape, all of a building's façades, and a resource's unique details. As described in the methodologies, photographs of a

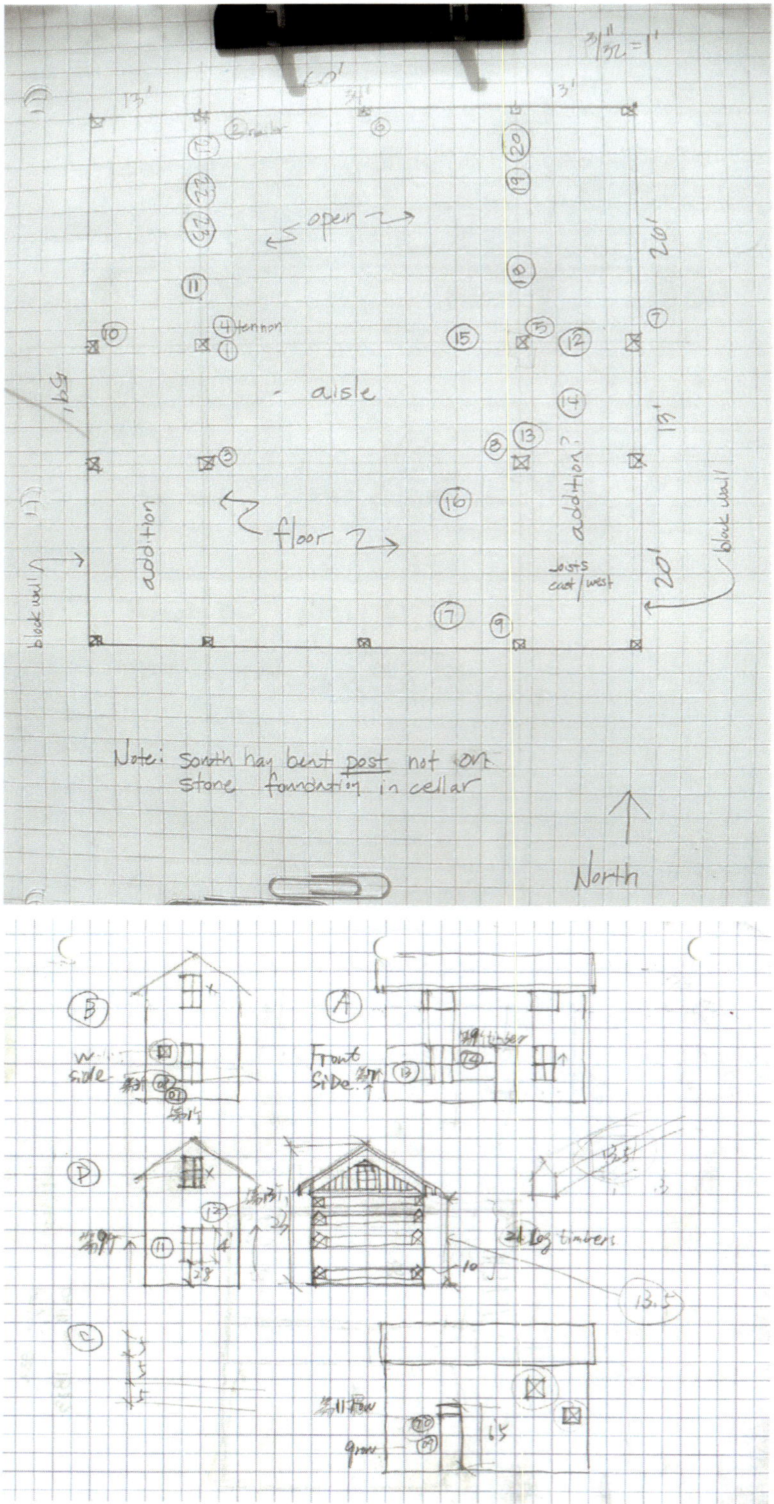

Figure 7.5
Examples of the products of the field documentation of tree-ring analysis. The top image is an example of a measured plan drawing with sample locations for the Anderson Barn reported in Chapter 12. The bottom image shows elevations and sample locations for a log house located in Markle, Indiana.

Source: Authors.

dry-erase board displaying the sample identification is vital to keeping track of your data (Chapter 5). We have recently started video recording a walk-through of the sites we analyze. This allows us to record visual and vocal observations while at the site, and allows us to document large features (such as long timbers) that are challenging to photograph with the standard lenses of digital cameras. Finally, at the day's end, we take quick digital photographs of our field drawings and notes as digital backup to the originals.

When we return from the field, we scan the field drawings as further backup. It is also wise to review and organize your photographs, for instance per building, and in subfolders representing significant spaces (e.g., cellar, first floor, etc.). If a report or journal article is to be generated for the analyzed site, the field drawings are redrawn with Computer Aided Drawing (CAD) software (Figure 7.6).

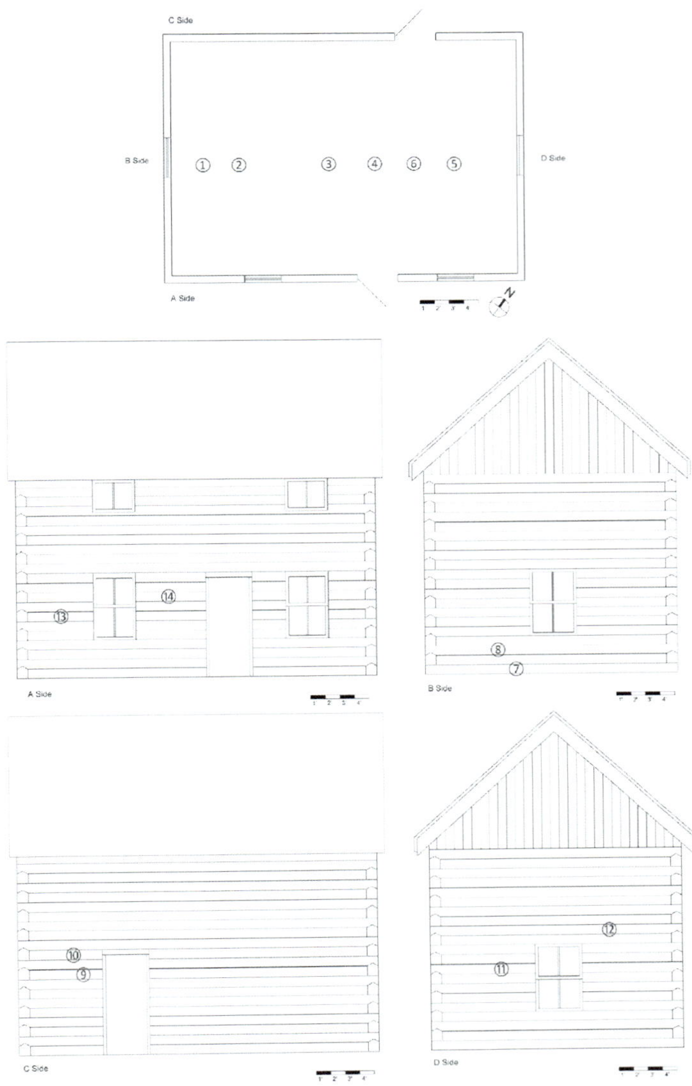

Figure 7.6
Examples of the products of the field documentation of tree-ring analysis. CAD drawing from field notes showing sample locations. This example is from a log house located in Markle, Indiana.

Source: Authors.

7.8 Conclusion

Chapter 7 explains how we communicate the results of our tree-ring analysis in journal articles and to the owners and managers of historic sites. The following chapters, Chapters 8 through Chapter 13, present a variety of case studies that report the results of tree-ring analysis. As you examine the case studies, we suggest that you continue to review Chapter 7 to optimize your comprehension of the data.

7.9 Literature cited

Glassie, H. 1968. *Pattern in the Material Folk Culture of the Eastern United States*. University of Pennsylvania Press, Philadelphia, PA. 324 p.

Hutslar, D.A. 1986. *The Architecture of Migration: Log Construction in the Ohio Country, 1750–1850*. Ohio University Press, Athens, OH. 558 p.

Kniffen, F.B. 1965. Folk housing: Key to diffusion. *Annals of the Association of American Geographers* 55: 549–577.

Nash, S.E. 1999. *Time, Trees, and Prehistory: Tree-Ring Dating and the Development of North American Archaeology, 1914–1950*. The University of Utah Press, Salt Lake City, UT. 294 p.

National Park Service. 2011. *Historic American Buildings Survey Guide to Field Documentation*. National Park Service, Washington, DC. 25 p.

Rempel, J.I. 1980. *Building with Wood and Other Aspects of Nineteenth-Century Building in Central Canada*. University of Toronto Press, Toronto, Canada. 454 p.

Roberts, W. 1996. *Log Buildings of Southern Indiana*. Trickster Press, Bloomington, IN. 228 p.

Todd, J. 1860. *Diary of Barn Building*. Private family collection.

Chapter 8

Enhancing interpretation

Case studies for open air and house museums

8.1 Introduction

Outdoor, or open air, museums are groups of collected or donated buildings moved to a museum setting. In North America, they are often arranged in simulated village landscapes that interpret colonial, frontier, or early settlement periods by displaying buildings and artifacts. Visitors often experience first- or second-person interpretation of folk life and crafts (candle making, cooking, blacksmithing, etc.; Rentzhog 2007). Proponents of outdoor museums advocate their benefits to public education and heritage tourism and explain how many significant structures would not be preserved if not for museums. Public access to "vernacular architecture perspectives in particular can help residents recognize and develop pride in neighborhoods and buildings that are not the earliest or closest to national paradigms of significance, based on high-style performance or associated with great people. Moreover, they can help the public think of buildings in terms of cultural, social, and historical issues" (Chappell 2007). However, the museums do come under criticism and "one should question if such sites have much educational value: are they teaching anything more than an imagined concept of simplicity and self-sufficiency? They often reinforce bad preservation practices, showing that significant old buildings can be bought and taken home, preferably to locations convenient to an interstate" to provide suitable access for tourists (Chappell 2007).

Our tree-ring analysis of outdoor museums seeks to establish construction dates for buildings that, due to their moving, often have minimal or lost histories. As preservationists, we struggle with the idea of moving historic structures. Historical location and context are key components to their interpretation and in defining significance. In the United States, the National Register program supports buildings remaining in their historic locations. Location, along with setting and feeling, are three of the seven aspects to determining a site's integrity (NPS 1995). However, we know of several significant structures that would not exist if they had not been moved to a museum setting (for instance, three of the hay press barns described in Chapter 12). We believe that decisions regarding the moving of buildings be made on an individual basis and as a last resort to its destruction or being left to rot.

House museums are dwellings that interpret the material culture of defined historical periods. They display artifacts, furnishings, and interior finishes representative of their owners or inhabitants. George and Martha Washington's Mount

Vernon, Thomas Jefferson's Monticello, and the Abraham and Mary Lincoln Home are examples of three well-known house museums. However, nearly every community contains a house museum celebrating local heritage. Our tree-ring analysis of house museums seeks to improve the interpretation of the resource by establishing specific dates for construction and modification. Knowing construction dates helps to identify ownership, the purpose for constructing and modifying buildings (for instance changes in family size, the economy, technology, etc.), and supports accurate and informative interpretation.

Chapter 8 presents the results for several museum settings we have analyzed (Figure 8.1). It includes a discussion about the challenges of working in significant historic settings where interpretation and education through the experience of historic spaces is the primary goal. We encourage readers that are interested in the tree-ring analysis of open air and house museums to examine the *Musée de Venoge* described in Chapter 9, the New Harmony, Indiana, houses presented in Chapter 11, and the Thiebaud House described in Chapter 12.

Figure 8.1
Open air and house museum locations analyzed in Chapter 8.
Source: Authors.

8.2 Tree-ring analysis of outdoor and house museums: opportunities and challenges

Tree-ring analysis of historic sites is becoming more common (see the recent articles by Grissino-Mayer et al. 2013; Brock et al. 2017; Druckenbrod et al. 2017; Grissino-Mayer et al. 2017; Harley et al. 2017). We have found the tree-ring analysis of historic buildings in a museum setting to be challenging. Authenticity, or the historic rigor of presenting history, varies widely among sites, based upon interpretive goals, the scale of a historical organization, marketing, and funding. The spectrum of museums ranges from highly regarded and well-funded institutions (for instance Old World Wisconsin, Conner Prairie Living History Museum, Historic Dearborn Museum) to grassroots historical societies and park departments.

Open air museum buildings are typically donated, then moved to the museum site. The buildings are often disassembled before they are moved. Frequently, we find that timber locations within a structure are improperly recorded or not recorded at all, and timbers wind up in the wrong location when reassembled. Top and bottom sill logs on timber structures are notorious for rotting because of their location near the damp ground or leaking roofs. When the building is reassembled, deteriorated logs are often replaced with timbers from other historic buildings (which is an acceptable practice as long as extant buildings in good condition are not harvested for replacement timbers). However, the replacement logs go undocumented, and over time the identity of the original timbers from the replacements is lost.

We have found that when buildings are moved, the roofs are replaced, and new doors and windows are added. Museums alter buildings based upon how they desire visitors to use a building (for instance moving visitors through a house), or to turn what were often very crude structures into more idealized forms (sometimes with no historical basis). Dendrochronology was recently performed on three 19th-century structures moved to Spring Mill State Park (Mitchell, Indiana) in the 1930s (Matheus et al. 2017). The study pointed out how logs with outlier dates and differing timber type suggest replacement at the time of the move. Also, entire buildings are often created from the timbers of multiple structures. Therefore, sampling vernacular forms of museum collections with relocated structures should rely on measured drawings and photographs of the building in its original location if they exist.

Another issue we see with log buildings in museum settings is the deterioration of timbers due to the lack of protective siding. North Americans have idealized perceptions of frontier life where images of hard-working and virtuous families reside in rustic, exposed log cabins located in the forest or on the prairie. Contrary to these ideals, historic log structures commonly had clapboard or board-and-batten siding to protect the wood and clay-based chinking (Roberts 1996). We support the accurate display of historic timber buildings, which includes the siding.

A limitation of the analysis of timber buildings in museum settings is how the invasiveness of core sampling negatively affects the historic display of the structure (see Chapter 5). Therefore, unless dendroarchaeology is part of the interpretive goals of the museum, the locations of samples need to be strategically placed and not visible to museum visitors. This may preclude accurately dating a building.

8.3 Delphi Canal Park: case study of an open air museum

The Delphi Canal Park (Delphi, Indiana) is a collection of buildings assembled to interpret the town's history. Delphi was platted in 1828 and is the Carroll County seat. The construction of the Wabash and Erie Canal began in 1832, and was completed in 1853. The corridor linked Lake Erie near Toledo, Ohio, to the Ohio River near Evansville, Indiana. Constructed from east to west, it reached Delphi in 1840. The canal was instrumental in exporting and importing agricultural and manufactured products to and from Indiana and was an economic boon for the towns along its towpath. The canal era ended in the 1870s, and was replaced by railroads, which were not as susceptible to the winter freezes, floods, and droughts of water-based transportation (Carroll County Wabash & Erie Canal, Inc. 2016; Taormina and Speer 2016). The grassroots preservation of Delphi's canal era heritage began in the 1980s. Since then the park managers have restored water to six miles (9.7 km) of the canal, constructed a museum and visitor center, and developed a museum collection of regional vernacular buildings.

8.3.1 Fouts House

The Fouts House (Figure 8.2) is an 18 × 26-foot two-story log dwelling. Family oral history says that Noah Fouts, an early settler of Cass County, Indiana, constructed it in 1839. In 1974 it was moved to a local park, and was moved again to Delphi in 2011. The house's two rooms, central fireplace, and loft suggests that it was originally "continental" in form (Glassie 1969; Jordan 1985; McAlester and McAlester 1988).

Since the house had been moved several times, we had to be diligent in collecting samples to ensure we were dating the original fabric of the building. Tree-ring analysis produced 13 dated samples from 12 timbers, resulting in 1,139 measured rings spanning from 1704 to 1868 (Tables 8.1 and 8.2). Beech, white

Figure 8.2
Image of the Fouts House located at the Delphi Canal Park (Delphi, Indiana). The house was constructed in 1869.

Source: Authors.

Table 8.1 Samples collected from the Delphi Canal Park (Delphi, Indiana). "First" and "last" refer to the first and last rings present in the sample. If no date is provided, the sample could not be reliably crossdated. "Outer ring" refers to the outermost ring of the sample: b = bark, w = wane, etc.; see Table 7.1 for a complete list of codes used for outer ring condition. "r" is the correlation coefficient obtained by correlating the ring widths of this sample with all other timbers of the same type from the building (series intercorrelation; Chapter 5). "Timber type" is the type of timber; see Table 3.1 for a list of possible tree species represented by the various timber types. "Provenience" is the timber's function in a structure; see Chapter 2 for a description of the various proveniences. Note: Timber type and provenience are only given once if more than one sample was obtained from an individual timber. An r-value is not reported if less than three samples of a particular type were obtained from a structure. A full description of the data presented in the table is provided in Chapter 7.

Sample	*First*	*Last*	*Outer ring*	r	*Timber type*	*Provenience*
				Fouts House		
DFO01A	1822	1869	w	0.70	Beech	W elevation; wall timber
DFO01B	1759	1804		0.70		
DFO02A	1759	1869	w/e	0.71	Beech	S elevation; wall timber
DFO03A					Beech	N elevation; wall timber
DFO04A	1731	1869	w	0.63	Beech	S elevation; wall timber
DFO05A	1740	1869	w	0.57	Beech	N elevation; wall timber
DFO06A	1754	1869	w/e	0.48	Beech	N elevation; wall timber
DFO07A	1732	1869	w/e	0.59	Beech	N elevation; wall timber
DFO08A	1777	1869	w	0.56	Beech	N elevation; wall timber
DFO09A	1703	1869	w	0.52	Beech	Floor joist; 1st floor
DFO10A	1812	1869	b	0.54	White oak	Rafter
DFO11A	1824	1869	b	0.63	White oak	Rafter
DFO12A	1810	1869	b	0.70	White oak	Rafter
DFO13A	1809	1866	w		Tulip	Rafter
				Loom House		
DLH01A	1739	1866	w	0.49	Beech	E elevation; wall timber
DLH02A	1769	1866	w	0.71	Beech	E elevation; wall timber
DLH02B	1778	1866	w	0.64		
DLH03A	1740	1866	w	0.57	Beech	S elevation; wall timber
DLH04A	1809	1866	w	0.59	Beech	W elevation; wall timber
DLH05A	1740	1866	w	0.57	Beech	E elevation; wall timber
DLH06A					Beech	E elevation; wall timber
				Kuns House		
DKC01A	1824	1886	w	0.67	White elm	W elevation; wall timber
DKC01B	1831	1886	b	0.59		
DKC02A	1792	1886	w	0.73	Red oak	W elevation; wall timber
DKC02B	1849	1886	w	0.77		
DKC03A	1781	1877		0.44	Red oak	E elevation; wall timber
DKC04A	1788	1886	b	0.56	Red oak	W elevation; wall timber

(Continued)

Table 8.1 Continued

Sample	*First*	*Last*	*Outer ring*	r	*Timber type*	*Provenience*
DKC05A	1786	1882		0.26	Red oak	W elevation; wall timber
DKC06A	1815	1886	b	0.68	White elm	S elevation; wall timber
DKC06B	1797	1886	b	0.64		
			Speece Warehouse			
DSW01A	1756	1855	w	0.74	Tulip	Ceiling joist; original provenience unknown
DSW01B	1727	1826		0.82		
DSW02A	1780	1855	w	0.60	Tulip	Ceiling joist; original provenience unknown
DSW02B	1651	1744		0.78		
DSW02C	1747	1815		0.81		
			Bowen House			
DBC01A					Basswood	W elevation; wall timber
DBC02A	1709	1838		0.56	Tulip	N elevation; wall timber
DBC02B	1833	1864	w/e	0.78		
DBC03A	1768	1863	w	0.62	Tulip	S elevation; wall timber
DBC04A					Basswood	W elevation; wall timber
DBC05A	1628	1864	b/e	0.71	Tulip	N elevation; wall timber
DBC05B	1782	1864	b/e	0.80		
DBC06A	1841	1864	w/e	0.73	Tulip	S elevation; wall timber
DBC06B	1756	1858		0.79		
			Reed Case House			
DRC01A	1799	1844	w	0.60	White oak	Beam; S ell
DRC02A	1794	1844	w	0.53	White oak	Beam; S ell
DRC03A					White oak	Floor joist; 1st floor; S ell
DRC04A					White oak	Floor joist; 1st floor; N ell
DRC05A					White oak	Floor joist
DRC06A	1756	1844	w	0.54	White oak	Beam; S pen
DRC07A	1647	1844		0.32	White oak	W sill
DRC07B	1773	1844		0.57		
DRC08A					White oak	Floor joist
DRC09A					White oak	Floor joist
DRC10A					White oak	Floor joist
DRC11A	1687	1811		0.26	White oak	Beam; N pen
DRC11B	1780	1844	w	0.45		
DRC11C	1811	1844	w	0.58		
DRC12A	1782	1845	w/e	0.33	White oak	E sill

Table 8.2 Summary table of the timber type chronologies prepared from Delphi Canal Park (Delphi, Indiana). "Samples" and "Timbers" are the number of successfully crossdated samples and timbers from the building. "Undated" is the number of timbers that could not be crossdated reliably. "First" and "last" refer to the first and last dates in the chronology. "Mean" and "SD" are the mean and standard deviation (mm) of the ring widths in the chronology. "r" is the mean series intercorrelation for each timber type (see Chapter 5). r *is not provided if less than three samples were obtained for a particular timber type. No total* r, *mean, or SD are provided since each chronology was prepared using only a single timber type. A full description of the data presented in the table is provided in Chapter 7.*

Structure	*Timber type*	*Samples*	*Timbers*	*Undated*	*First*	*Last*	*Years*	r	*Mean*	*SD*
Fouts House	Beech	9	8	1	1704	1868	926	0.59	1.17	0.59
Fouts House	White oak	3	3	0	1811	1868	157	0.62	1.20	0.35
Fouts House	Tulip poplar	1	1	0	1810	1865	56		0.95	0.44
Loom House	Beech	6	5	1	1740	1865	614	0.59	1.23	0.67
Kuns House	White elm	4	2	0	1798	1885	273	0.65	1.38	0.53
Kuns House	Red oak	5	4	0	1782	1885	414	0.52	1.26	0.46
Speece Warehouse	Tulip poplar	5	2	0	1729	1854	408	0.74	0.66	0.22
Bowen House	Tulip poplar	7	4	0	1629	1863	681	0.70	1.05	0.40
Bowen House	Basswood	0	0	2						
Reed Case House	White oak	9	6	6	1650	1844	694	0.42	1.31	0.33
Total		49	35	10	1629	1885	4223			

oak, and tulip poplar were sampled. Our study established a construction date of 1869, and that the roof was original (DFO10A-13A). Analysis shows that the date exceeds the family's oral history by 30 years.

8.3.2 Loom House

The Loom House (Figure 8.3) is an 18 × 24-foot log dwelling. It was originally a one and a half story structure, but the upper half-story was removed when it was moved from a farm north of Burrows, Indiana, and reconstructed at the park. The rear (north) wall of the house was rebuilt by volunteers with recycled logs form another structure, and was not sampled. Tree-ring analysis yielded six samples from five beech timbers, resulting in 614 measured rings ranging from 1740 to 1865 (Tables 8.1 and 8.2). The house was constructed in 1866.

8.3.3 Kuns House

The Kuns House (Figure 8.4) is a 15 × 24-foot one and a half story timber dwelling. It was moved to the museum from a farm northeast of Flora, Indiana. Historic photographs show that its unique gable end entry is original. At one time, as many as 13 family members lived in the house (Carroll County Wabash and Erie Canal, Inc. 2016).

Figure 8.3
Image of the Loom House located at the Delphi Canal Park (Delphi, Indiana). The house was constructed in 1866.

Source: Authors.

Figure 8.4
Image of the Kuns House located at the Delphi Canal Park (Delphi, Indiana). The house was constructed in 1886.

Source: Authors.

Tree-ring analysis produced nine samples from six timbers, resulting in 687 measured rings ranging from 1782 to 1885 (Tables 8.1 and 8.2). White elm and red oak timbers were sampled. A construction date of 1886 was established for the house. Analysis supported the oral history that the house was about a century old when it was moved to Delphi in 1981.

8.3.4 Speece Shelter

The Speece Shelter contains sawn lumber from the Speece Warehouse, a business that was located on the Wabash & Erie Canal's towpath in Carrollton, Indiana. The structure was believed to have been constructed about 1850, but it is unclear from what part of the warehouse the lumber originates and if it was part of an original structure or a later addition. Tree-ring analysis allowed for crossdating five samples from two timbers, resulting in 408 measured rings ranging from 1729 to 1854. A death date of 1855 was established for both timbers, but we are hesitant to present it as a conclusive construction date of the original warehouse without knowing the timbers' provenience and with a limited number of sampled timbers.

8.3.5 Bowen House

The Bowen House (Figure 8.5) is a 19 × 28-foot, two-story log house with an 11 × 28-foot porch that was enclosed to add interior museum space. It was used as tenant housing on the Bowen Farm in central Carroll County. The house was rotated when it was moved in 1990 to bring visitors into the rear of the structure rather than through the porch.

Sampling was limited because overzealous chinking from the restoration limited access to wane and bark over most of the building. Other potential sampling locations were avoided because they were visible to visitors. Tree-ring

Figure 8.5
Image of the Bowen House located at the Delphi Canal Park (Delphi, Indiana). The house was constructed in 1864.

Source: Authors.

analysis yielded samples from four timbers, resulting in 681 measured rings ranging from 1629 to 1863 (Tables 8.1 and 8.2). Tulip poplar and basswood were represented. A construction date of 1864 was established for the house.

8.3.6 Reed and Mariah Case House

The Case House (Figure 8.6) is a two-story, Federal Style, I-house built by the contractor that constructed the Delphi section of the canal. It is believed to have been built in 1844 and then enlarged about 1853 for use as an inn (Carroll County Wabash & Erie Canal, Inc. 2016). It was moved to the park from downtown Delphi in 1986, and is the crown jewel of the park's collection.

The term I-house was established by geographer Fred Kniffen (1986) to describe a dwelling type mostly found in rural Indiana, Illinois, and Iowa. It is a folk architectural form with British roots, and it was transplanted to the Midwestern United States along migration routes. While it had slight material and layout variations over geographic regions, it is consistently at least two rooms in length, one-room deep, and two full stories in height. The I-house's persistence as a housing type ranged from the late 18th into the early 20th centuries. Its plain exterior allowed builders to add decorative details from various national architectural styles. The Case House was constructed in the Federal Style, but it also displays Greek Revival details in the gable returns of the ell, suggesting a modification to the original structure (Peat 1962; Lewis 1975; McAlester and McAlester 1988).

Samples were only collected in the house's cellar because the interior walls of the house are plaster and the attic was not accessible for sampling. Tree-ring analysis produced nine dated samples from six white oak timbers, resulting in 694 rings ranging from 1650 to 1844 (Tables 8.1 and 8.2). Logs for the house were cut in both 1844 and 1845 (DRC12A), so analysis shows a construction date of 1845 (a year later than the 1844 estimated by historians). It was thought that the house's ells (the wings off the rear of the dwelling) were both constructed in the 1850s. Analysis shows that the south ell was constructed in 1845 as part of the original

Figure 8.6
Image of the Case House located at the Delphi Canal Park (Delphi, Indiana). The house was constructed in 1845.

Source: Authors.

house. Dates could not be established for the north ell, and no samples dating to the 1850s were collected, so at this time dendrochronology cannot verify expansion of the house in the 1850s. However, the Greek Revival style gable returns suggest a later addition to the Federal Style house, and further analysis of attic timbers at a future time might provide answers.

8.3.7 Summary of the tree-ring analysis of Delphi Canal Park

The six Delphi Canal Park structures we sampled produced 49 samples from 35 timbers. Chronologies spanned from 1629 to 1885, and 4,223 rings were measured. The results of our analysis are the reverse of what we expected. In the US Midwest, heritage museums often portray the simple and rustic qualities of frontier life, where log houses are associated with the first wave of settlers in the opening decades of the 19th century. In the case of the Delphi building collection, the opposite is true. The oldest building we sampled is the high style Case House built in 1845. Ironically, the collection's log houses date to the latter half of the 19th century, decades after first-wave frontier settlement. Three of the log houses were constructed in the 1860s, and the Kuns House shows that log dwellings were still being built in the region as late as 1886.

8.4 John Hay Center

The John Hay Center is an open air museum located in Salem, Indiana, that interprets and celebrates the history of Washington County. The complex has a simulated village containing the John Hay House (in its original location) and log buildings moved to the site. Dendrochronology was performed on the National Register-listed Hay House, and the one-room, log Haganman House.

8.4.1 John Hay House addition

John Hay was born in Salem in 1838, and moved to Illinois when he was three years old. He is renowned as President Abraham Lincoln's secretary and military aide and Secretary of State under presidents William McKinley and Theodore Roosevelt (Boomhower 1995). The John Hay House is a one-story, brick, Federal Style house with a rear addition. The house was originally constructed in 1824 as the Salem Grammar School. The property was purchased by Hay's father in 1833 (Abstract n.d.), and they lived there until 1842 (Figure 8.7). It was restored and opened as a museum in 1971 (Hay Center n.d.; Burns et al. 1970; Salem Leader 1971).

Dendrochronology was completed on the house's addition to determine if it existed when Hay resided there as a child. Historians concluded that it was constructed after the Hay's family move to Illinois in 1843 as a breezeway, summer kitchen, and smokehouse and that it was later enclosed for interior rooms (Hay Center n.d.). Fifteen samples from nine white oak timbers resulted in the measurement of 934 rings spanning from 1692 to 1833 ($r = 0.45$; mean ring width ± SD was 0.95 ± 0.37 mm). A construction date of 1834 was determined for the addition, suggesting it was built by the Hay family, and affirming that it was present when John Hay resided at the property as a young boy.

Figure 8.7
Images of the John Hay House located at the John Hay Center open air museum (Salem, Indiana). The top image is the front (west) and south façades. The bottom image shows the north façade of the addition. The house was constructed in 1824. Tree-ring analysis established a constructed date of 1834 for the addition, confirming that it was present when Hay resided in Salem.

Source: Authors.

Table 8.3 Samples (all white oak) collected from the Hay House addition (Salem, Indiana). See Table 8.1 for a description of the table contents; a full description of the data presented is provided in Chapter 7.

Sample	*First*	*Last*	*Outer ring*	r	*Provenience*
HAY01A	1759	1834	w	0.50	Ground floor floor joist
HAY01B	1769	1834	w	0.48	
HAY01C	1735	1789		0.41	
HAY02A	1737	1834	w	0.59	Ground floor floor joist
HAY03A	1767	1821		0.42	Ground floor floor joist
HAY03B	1758	1807		0.34	

Sample	First	Last	Outer ring	r	Provenience
HAY04A	1756	1833	w	0.42	Ground floor floor joist
HAY05A	1725	1780		0.71	Ground floor floor joist
HAY05B	1727	1797	w28	0.48	
HAY06A	1748	1796		0.30	Ground floor floor joist
HAY06B	1747	1781		0.26	
HAY06C	1737	1802		0.40	
HAY07A	1691	1763		0.35	S sill
HAY08A	1707	1784		0.40	E sill
HAY09A	1713	1783		0.54	W sill

8.4.2 Haganman House

The Haganman House (Figure 8.8) is a 20 × 24-foot, one and a half story log dwelling built with V-notch corner notching. The house is reported to have been built in 1830 by John and Anna Overman who were local farmers and craftspeople. It was purchased by the Haganman family in 1915 and donated to the museum in 1979 (Hay Center n.d.).

Analysis of the Haganman House produced a tulip poplar chronology comprised of 18 samples from 12 timbers and 1,437 measured rings (r = 0.70; 1.57 ± 0.84 mm). We also dated one red oak timber; one chestnut timber did not date (Table 8.4). Analysis established an 1822 construction date, roughly a decade earlier than estimated by historians. The house was restored with logs harvested in the 1840s and 1850s. Log "patches" filled the fireplace and door openings. These alterations reinforced the need to thoroughly scrutinize logs while sampling, and the fact that good field notes are essential.

Figure 8.8
Image of the Haganman House located at the John Hay Center open air museum (Salem, Indiana). The house was constructed in 1822.

Source: Authors.

Table 8.4 Samples collected from the Haganman House (Salem, Indiana). See Table 8.1 for a description of the table contents; a full description of the data presented is provided in Chapter 7.

Sample	*First*	*Last*	*Outer ring*	r	*Timber type*	*Provenience*
HAG01A	1750	1850	w	0.75	Tulip	1st floor; E elevation; wall timber; patch
HAG01B	1747	1850	w	0.79	Tulip	
HAG02A	1728	1822	w	0.82	Tulip	1st floor; E elevation; wall timber
HAG02B	1737	1822	w	0.74	Tulip	
HAG03A	1763	1843	w		Red oak	1st floor; E elevation; wall timber; patch
HAG04A	1736	1794		0.57	Tulip	1st floor; E elevation; wall timber; patch
HAG04B	1736	1849	sap	0.65	Tulip	
HAG05A	1730	1841	sap	0.43	Tulip	1st floor; E elevation; wall timber; patch
HAG06A	1716	1818	s/w	0.77	Tulip	2nd floor; S elevation; top wall timber
HAG06B	1720	1821	s/w	0.70	Tulip	
HAG07A	1799	1822	w	0.65	Tulip	2nd floor; W elevation; top wall timber
HAG07B	1791	1822	w	0.52	Tulip	
HAG08A	1731	1793		0.76	Tulip	2nd floor; W elevation; wall timber
HAG08B	1801	1820	w	0.73	Tulip	
HAG09A	1713	1818	w	0.76	Tulip	2nd floor; W elevation; wall timber
HAG10A	1721	1792		0.75	Tulip	2nd floor; S elevation; wall timber
HAG11A	1743	1850	w	0.64	Tulip	1st floor; S elevation; wall timber
HAG12A					Chestnut	1st floor; S elevation; wall timber; patch
HAG13A	1703	1821	s/w	0.82	Tulip	1st floor; S elevation; wall timber
HAG14A	1701	1822	w	0.65	Tulip	1st floor; S elevation; wall timber

8.5 Jeremiah Sullivan House

The Jeremiah Sullivan House (Figure 8.9) is located in the Ohio River town of Madison, Indiana. It was documented by the Historic American Buildings Survey (HABS) in 1934 (HABS 1934a), and it is included as a contributing resource to the Madison Historic District National Historic Landmark (Fife et al. 2007). Sullivan (1794–1870), a lawyer, moved to Madison from Virginia in 1816 to start a business in the young town. He served in the Indiana State Legislature, was an Indiana Supreme Court judge, and was one of the organizers of nearby Hanover College (Indiana's oldest private college, founded in 1827; Schauinger 1941). According to the National Park Service, Sullivan suggested “Indianapolis” as the name for the new state capital (NPS 2018).

Sullivan’s house was constructed in the Federal Style, the dominant design and vernacular architectural style in the United States between 1780 and 1820 when the country’s population was expanding and spreading into the Midwest (McAlester and McAlester 1988). The house is unique for its gable end double chimney, which is typically found in 18th-century Georgian architecture (Figure 8.9; Peat 1962; Fife 2007). National Historic Landmark documentation reports that the house was constructed prior to 1820, and the HABS documentation describes the house being built in 1818.

Tree-ring analysis of the residence came about following a misfortunate event. In 2008, a street tree fell on the house damaging the roof and east wall. While the event provided abundant timber for analysis, determining provenience for several pieces was difficult as the wood had already been removed from the building. The roof sheathing was readily determined by dimensions (wide, thin timbers) and the large number of small square-headed nails used to secure shingles. The timbers came from the ceiling (third floor), east wall (third floor), or roof framing.

Tree-ring analysis allowed for accurate crossdating of 33 tulip poplar samples from 21 timbers, resulting in a chronology consisting of 2,774 measured rings ranging from 1648 to 1821 (r = 0.67; 0.88 ± 0.39 mm). Samples SUL19A

Figure 8.9
Image of the front (south) and east façades of the Judge Jeremiah Sullivan House (Madison, Indiana). This photo was taken following the house’s repair of damage to the east façade and roof caused by a falling street tree. The house was constructed between 1820 and 1822.

Source: Authors.

and SUL39A do not overlap with the chronology but do crossdate with other regional tulip poplar chronologies (Table 8.5). Analysis established that construction of the house likely started about 1820 (SUL78 and SUL80) and was completed in 1822, just a few years later than estimated by historians. In addition to the hardwood (tulip poplar and ash) timbers found in the building, we collected many white pine timbers from the building. Interestingly, white pine is not native to the region; see Chapter 13 for an interpretation and dating of these softwood timbers.

Table 8.5 Samples collected from the Jeremiah Sullivan House (Madison, Indiana). See Table 8.1 for a description of the table contents; a full description of the data presented is provided in Chapter 7.

Sample	*First*	*Last*	*Outer ring*	r	*Timber type*	*Provenience*
SUL17A	1658	1729		0.64	Tulip	Wall or ceiling framing timber
SUL17B	1657	1725		0.71	Tulip	
SUL18A	1714	1767		0.50	Tulip	Framing timber
SUL19A	1502	1618			Tulip	Framing timber
SUL20A					Tulip	Roof truss
SUL21A	1725	1822	b	0.81	Tulip	Rafter?
SUL22A	1707	1822	b	0.81	Tulip	Roof truss
SUL23A	1708	1765		0.47	Tulip	Framing timber
SUL24A	1678	1811	b10	0.83	Tulip	Roof truss
SUL24B	1680	1822	b	0.61	Tulip	
SUL26A	1716	1822	b	0.87	Tulip	Framing timber
SUL26B	1713	1822	w	0.86	Tulip	
SUL27A	1699	1822	b	0.86	Tulip	Framing timber
SUL27B	1703	1822	b	0.85	Tulip	
SUL28A	1682	1781		0.73	Tulip	Framing timber
SUL29B	1650	1726		0.71	Tulip	Framing timber
SUL30A	1671	1738		0.77	Tulip	Framing timber
SUL32A	1724	1822	b	0.79	Tulip	Roof truss
SUL33A					Tulip	Roof truss
SUL34A	1715	1822	b	0.82	Tulip	Framing timber
SUL35A	1684	1773		0.77	Tulip	Framing timber
SUL37A					Tulip	Framing timber
SUL38A					Tulip	Framing timber
SUL39A	1515	1628			Tulip	Framing timber
SUL40A					Tulip	Framing timber
SUL78A	1713	1820	b	0.68	Tulip	Floor joist; 1st floor
SUL78B	1739	1820	w	0.61	Tulip	
SUL78C	1721	1802	+18	0.70	Tulip	
SUL79A	1749	1791		0.64	Tulip	Beam; 1st floor
SUL79B	1749	1804		0.66	Tulip	
SUL80A	1713	1820	w	0.57	Tulip	Floor joist; 1st floor
SUL81A	1759	1820		0.64	Tulip	Beam; 1st floor
SUL81B	1761	1822	w	0.65	Tulip	
SUL81C	1757	1816	+6	0.68	Tulip	

Sample	*First*	*Last*	*Outer ring*	r	*Timber type*	*Provenience*
SUL82A	1761	1822	w	0.60	Tulip	Floor joist; 1st floor
SUL82B	1653	1711		0.70	Tulip	
SUL82C	1647	1711		0.68	Tulip	
SUL82D	1761	1822	w	0.53	Tulip	
SUL91A					Ash	Ceiling joist; 3rd floor
SUL92A					Ash	Ceiling joist; 3rd floor

8.6 Grayson House

The Grayson House (Figure 8.10) is a dogtrot house located in the historic town of New Harmony, Indiana. The village was founded on the Wabash River by German Harmonists in 1814 and later became a short-lived utopian experiment when industrialist Robert Owen purchased the town in 1825. The house is part of a collection of log structures that were moved to a museum site within the town. Unfortunately, very little is known about the house's historical location (Indiana State Historic Sites 2013). The interpretive signage for the dwelling identifies 1775 as the construction date. Our knowledge of the region's settlement patterns caused us to question the date as much too early, thereby making the house a worthy candidate for dendroarchaeological study (Rubino and Baas 2014).

The house is a 20 × 42-foot, single-story structure composed of an open, central hallway flanked by two rectangular rooms. Chimneys are located at each end, and a roof covers the entire structure (Figure 8.10). The corner notching is half-dovetail. The hallway runs the depth of the house and is referred to as a dogtrot (Montell and Morse 1995). A family would have slept in one room and cooked in the other (Glassie 1968; Jordan 1985; Kniffen 1986). The trot provided a covered outdoor living space for warm southern climates.

The dogtrot form has European roots (Glassie 1968; Jordan 1985; McAlester and McAlester 1988; Roberts 1996) and was brought to North America by colonists. It was distributed into the American interior over generations via the

Figure 8.10
Image of the Grayson House (New Harmony, Indiana). The house is a one-story dogtrot constructed in 1852.

Source: Authors.

18th- and 19th-century Middle Atlantic migration stream that ran from Pennsylvania through eastern Tennessee, then across northern Mississippi and Alabama. Settlers then brought it north into western Kentucky and the southern regions of Indiana, Illinois, and Ohio (Kniffen 1986; Roberts 1996). The open trot allowed breezes to cool the structure in southern climates and is the reason the highly functional house form was used over a wide geographical area and for an extended length of time. While the house form is common in southern states, it is rare in Indiana. Tree-ring analysis was completed on a two-story dogtrot house in Spring Mill State Park, and an 1820s construction date was determined (Matheus et al. 2017).

Analysis produced 27 dated tulip poplar samples from 16 timbers, resulting in a chronology consisting of 1,774 measured rings ranging from 1682 to 1851 (Tables 8.6 and 8.7). Timber types sampled were white elm, white oak, hickory, and tulip poplar. Analysis identified an 1852 build date, 77 years later than thought. The analysis also shows the presence of the dogtrot house form in Indiana into the middle decades of the 19th century.

Table 8.6 Samples collected from the Grayson House (New Harmony, Indiana). See Table 8.1 for a description of the table contents; a full description of the data presented is provided in Chapter 7.

Sample	*First*	*Last*	*Outer ring*	r	*Timber type*	*Provenience*
MAC01A	1803	1851	w	0.66	Tulip	Wall timber; W pen; E wall
MAC01B	1774	1851	b	0.69		
MAC01C	1738	1792		0.51		
MAC01D	1713	1738		0.82		
MAC01E	1738	1757		0.43		
MAC02A	1761	1851	w	0.64	Tulip	Wall timber; W pen; E wall
MAC02B	1681	1761		0.65		
MAC03A	1772	1834		0.68	Tulip	Wall timber; W pen; E wall
MAC03B	1703	1851	w	0.66		
MAC03C	1709	1772		0.55		
MAC03D	1834	1851	w	0.21		
MAC04A	1780	1852	w	0.36	Tulip	Wall timber; W pen; E wall
MAC05A	1730	1852	w	0.62	Tulip	Wall timber; W pen; E wall
MAC06A	1760	1851	w	0.59	Tulip	Wall timber; E pen; W wall
MAC07A					Hickory	Ceiling beam; W pen
MAC08A					White oak	Ceiling beam; W pen
MAC09A					White elm	Ceiling beam; breezeway
MAC10A					White elm	Ceiling beam; breezeway
MAC11A	1797	1827		0.49	Tulip	Ceiling beam; E pen
MAC12A	1719	1812		0.54	Tulip	Ceiling beam; E pen
MAC12B	1799	1852	w	0.49		
MAC13A					White oak	Ceiling beam; breezeway
MAC14A	1770	1852	b	0.66	Tulip	Wall timber; E pen; W wall
MAC14B	1808	1852	b	0.74		
MAC15A	1767	1808		0.40	Tulip	Wall timber; E pen; N wall
MAC16A	1758	1852	b	0.61	Tulip	Wall timber; E pen; E wall
MAC17A	1759	1852	b	0.55	Tulip	Wall timber; E pen; S wall
MAC18A	1745	1824		0.72	Tulip	Wall timber; E pen; E wall

Sample	First	Last	Outer ring	r	Timber type	Provenience
MAC18B	1827	1852	b	0.81		
MAC19A	1783	1852	w	0.37	Tulip	Wall timber; E pen; E wall
MAC20A	1802	1852	w	0.60	Tulip	Wall timber; W pen; W wall
MAC21A	1708	1806		0.74	Tulip	Wall timber; E pen; N wall

Table 8.7 Summary table of the timber type chronologies prepared from the Grayson House (New Harmony, Indiana). See Table 8.2 for a description of table contents; a full description of the data presented is provided in Chapter 7.

Timber type	Samples	Timbers	Undated	First	Last	Years	r	Mean	SD
Hickory	0	0	1						
Tulip poplar	27	16	0	1682	1851	1774	0.60	1.06	0.63
White elm	0	0	2						
White oak	0	0	2						
Total	27	16	5	1682	1851	1774			

8.7 Frederick and Sophia Wyneken House

The Frederick and Sophia Wyneken House is located in rural Adams County, Indiana. Frederick Conrad Dietrich Wyneken's (1810–1876) biography as a 19th-century Lutheran pastor and frontier circuit rider is consistently told in a variety of sources (Hageman 1926; Webber 1957; Wolf 1997; Smith 1999). Born in Verden, Germany, he came from a family of local religious leaders. Finding his calling in missionary work, he migrated to Baltimore, Maryland, in 1838 and was sent to northeast Indiana to preach and organize German Protestant settlers. As a survey missionary he assessed the needs of the church in post-frontier northwestern Ohio, southeastern Michigan, and areas of Indiana extending as far west as Lafayette. Following this work, he became the pastor for parishes in the growing city of Fort Wayne and the neighboring rural town of Decatur.

Wyneken's significance as a religious leader comes from his efforts to find Lutheran pastors to organize and preach to German Protestant immigrants. Germans of all faiths came to Indiana in great numbers in the early and middle decades of the 19th century. Settlement of the young state was promoted by Indiana's governmental leaders, and the draw of affordable farmland, commercial opportunities in growing Hoosier cities, and chain migration resulted in the establishment of German urban communities and rural settlements. German families scattered throughout developing rural counties had limited access to local German faith churches, priests, or pastors. Throughout the range of his travels, Wyneken found isolated Germans yearning for organized congregations and leadership. His fear was that other Protestant denominations were converting underserved Lutherans (Wolf 1997).

In 1841, Wyneken returned to Germany to organize missionary societies and to recruit pastors for service in America. About this time, the pastor authored *The Distress of the German Lutherans in North America*, a pamphlet that had "far-reaching effects in attracting German pastors to America and establishing a strong Lutheran presence" (Taylor and McBirney 1996). Wyneken left Indiana to pastor a congregation in Baltimore, Maryland, in 1845, then moved to St. Louis, Missouri, in 1850 where he was elected president of the Missouri Synod, a position he held for 14 years. Through these efforts he contributed to keeping the Lutheran faith relevant in North America.

In the years leading up to the Civil War the synod headquarters and seminary were moved out of the border state of Missouri to Fort Wayne, Indiana. Preservationists believed that the house was newly constructed for the Wyneken family, and that they resided in it until about 1864 when Frederick retired as synod president and accepted the job of pastor of the Trinity Lutheran Church in Cleveland, Ohio.

The Friends of Wyneken, a not-for-profit organization preserving the house and Adams County's German heritage, purchased and moved the dwelling in 2005. It is being restored as a museum and interpretive center. We were invited to sample and analyze the house to identify its construction date and to confirm if it was associated to Wyneken's initial time in Indiana in the 1830s, or if it was connected to his return to Indiana in the late 1850s.

The house's vernacular form is commonly known as a "side hall." The main, or front, façade of a side-hall house is the gable end (Figure 8.11). This gives the residence a temple-like appearance, which fit in with the Greek Revival style of the mid-19th century (Peat 1962; McAlester and McAlester 1988). While a more upscale house might display classically inspired decorative features, the Wyneken house is quite plain, with the only ornamentation being the diamond shaped attic louver in the gable's peak. In plan, the house is two rooms deep with a side hall that runs its length. A one-story wing, or "ell," is located on the rear of the house and served as the kitchen (Figure 8.12).

Tree-ring analysis resulted in 59 crossdated samples from 53 timbers, yielding chronologies consisting of 4,599 measured rings ranging from 1663 to 1857 (Tables 8.8 and 8.9). Seven timber types were sampled: ash, basswood, red

Figure 8.11
Image of the front (east) and north façades of the Reverend Frederick and Sophia Wyneken House (Adams County, Indiana). The house was moved to this location in 2005, and the siding was replaced as part of the restoration. Samples were collected in the second floor and attic spaces. The house was constructed in 1858.

Source: Authors.

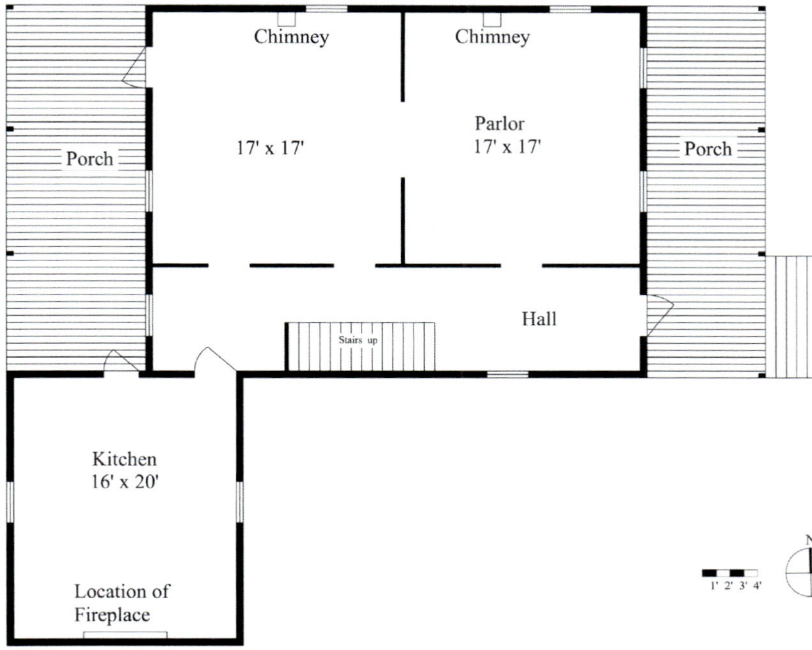

Figure 8.12
First-floor plan of the Reverend Frederick and Sophia Wyneken House (Adams County, Indiana). The house is a gable-front, side-hall form with Greek Revival style details.

Source: Authors.

oak, sycamore, tulip poplar, white elm, and white oak. The analysis established an 1858 construction date for the house, indicating the residence was built new for the Wynekin family following their St. Louis residency. The samples include a large number of ash tongue and groove floorboards. A portion of the second floor was stolen while the house was being moved. The floor was cut out of the house, leaving the small ends of the planks (WYN60-WYN87) nailed in place. One plank contained bark (WYN82) allowing us to confirm the floor was original to the house. The first floor joists were also stolen and are missing from the data.

Table 8.8 Samples collected from the Wyneken House (Decatur County, Indiana). See Table 8.1 for a description of the table contents; a full description of the data presented is provided in Chapter 7.

Sample	*First*	*Last*	*Outer ring*	r	*Timber type*	*Provenience*
WYN01A	1828	1858	w	0.60	White oak	Floor joist; attic
WYN02A	1765	1858	w	0.41	Ash	Floor joist; attic
WYN03A	1752	1858	b	0.61	Tulip	Floor joist; attic
WYN03B	1744	1858	w	0.64	Tulip	
WYN04A	1758	1858		0.33	Ash	Floor joist; attic
WYN05A					Red oak	Stud; 2nd floor; S elevation
WYN06A	1740	1847		0.49	White oak	Floor joist; attic
WYN06B	1810	1858	w	0.41	White oak	
WYN06C	1796	1858	w	0.35	White oak	
WYN07A	1746	1858	w	0.49	Ash	Floor joist; attic

(Continued)

Table 8.8 Continued

Sample	*First*	*Last*	*Outer ring*	r	*Timber type*	*Provenience*
WYN08A	1733	1858	w		White elm	Beam supporting rafter supports; attic
WYN09A					Basswood	Loose board; attic
WYN10A					Basswood	Loose board; attic
WYN11A	1777	1858	b	0.48	Tulip	Floor joist; attic
WYN12A	1779	1858	b	0.46	Ash	Floor joist; attic
WYN13A	1729	1810		0.35	White oak	Floor joist; attic
WYN14A	1727	1845		0.42	White oak	Floor joist; attic
WYN14B	1825	1858	b	0.74	White oak	
WYN15A					Tulip	Loose board; attic
WYN16A	1716	1781		0.44	Ash	tongue and groove flooring; loose in attic
WYN17A					Basswood	Loose board; attic
WYN18A					Sycamore	Loose board; attic
WYN19A	1750	1829		0.65	Ash	tongue and groove flooring; loose in attic
WYN30A	1728	1858	w/y		White elm	Beam supporting rafter supports; attic
WYN31A	1735	1825		0.54	White oak	Plate; W elevation
WYN31B	1827	1858	w	0.53	White oak	
WYN32A	1818	1858	w	0.49	Ash	Post/rafter support W elevation
WYN33A					Sycamore	Loose board; attic
WYN34A	1732	1850		0.55	White oak	Floor joist; attic
WYN34B	1834	1858	w	0.78	White oak	
WYN35A	1757	1858	w	0.38	Ash	Floor joist; attic
WYN36A	1738	1858	w	0.34	Ash	Plate; E elevation
WYN37A	1808	1858	b	0.89	Red oak	Rafter support
WYN38A	1807	1858	b	0.84	Red oak	Rafter support
WYN39A	1835	1858	w	0.64	Red oak	Rafter
WYN40A	1808	1854		0.92	Red oak	Rafter support
WYN41A	1786	1858	w	0.31	Tulip	Stud; 2nd floor; S elevation
WYN42A	1808	1848		0.76	Red oak	Plate; S elevation
WYN43A	1773	1858	b	0.40	White oak	Post; ground floor
WYN60A	1719	1786		0.69	Ash	tongue and groove flooring; 2nd floor; NW room
WYN61A	1760	1842		0.68	Ash	tongue and groove flooring; 2nd floor; NW room
WYN62A	1726	1786		0.58	Ash	tongue and groove flooring; 2nd floor; NW room
WYN63A	1673	1798		0.77	Ash	tongue and groove flooring; 2nd floor; NW room
WYN64A	1685	1807		0.75	Ash	tongue and groove flooring; 2nd floor; NW room

Sample	*First*	*Last*	*Outer ring*	r	*Timber type*	*Provenience*
WYN65A	1662	1766		0.56	Ash	tongue and groove flooring; 2nd floor; NW room
WYN66A	1748	1837		0.40	Ash	tongue and groove flooring; 2nd floor; NW room
WYN67A	1686	1764		0.70	Ash	tongue and groove flooring; 2nd floor; NW room
WYN68A	1735	1821		0.46	Ash	tongue and groove flooring; 2nd floor; NW room
WYN69A	1698	1761		0.52	Ash	tongue and groove flooring; 2nd floor; NW room
WYN70A	1769	1853		0.35	Ash	tongue and groove flooring; 2nd floor; NW room
WYN71A	1781	1857		0.73	Ash	tongue and groove flooring; 2nd floor; NW room
WYN72A	1714	1796		0.73	Ash	tongue and groove flooring; 2nd floor; NW room
WYN73A	1717	1813		0.42	Ash	tongue and groove flooring; 2nd floor; NW room
WYN74A	1670	1794		0.79	Ash	tongue and groove flooring; 2nd floor; NW room
WYN75A	1741	1835		0.58	Ash	tongue and groove flooring; 2nd floor; NW room
WYN76A	1679	1762		0.74	Ash	tongue and groove flooring; 2nd floor; NW room
WYN77A	1729	1787		0.57	Ash	tongue and groove flooring; 2nd floor; NW room
WYN78A	1674	1788		0.74	Ash	tongue and groove flooring; 2nd floor; NW room
WYN79A	1780	1841		0.63	Ash	tongue and groove flooring; 2nd floor; NW room
WYN80A					Ash	tongue and groove flooring; 2nd floor; NW room
WYN81A	1731	1820		0.53	Ash	tongue and groove flooring; 2nd floor; NW room
WYN82A	1786	1857	b	0.63	Ash	tongue and groove flooring; 2nd floor; NW room
WYN83A	1756	1816		0.74	Ash	tongue and groove flooring; 2nd floor; NW room
WYN84A	1741	1812		0.74	Ash	tongue and groove flooring; 2nd floor; NW room
WYN85A	1726	1809		0.60	Ash	tongue and groove flooring; 2nd floor; NW room
WYN86A	1765	1834		0.69	Ash	tongue and groove flooring; 2nd floor; NW room
WYN87A	1709	1802		0.63	Ash	tongue and groove flooring; 2nd floor; NW room

Table 8.9 Summary table of the timber type chronologies prepared from the Wyneken House (Decatur County, Indiana). See Table 8.2 for a description of table contents; a full description of the data presented is provided in Chapter 7.

Timber type	*Samples*	*Timbers*	*Undated*	*First*	*Last*	*Years*	r	*Mean*	*SD*
Ash	36	36	1	1663	1857	3026	0.58	1.36	0.47
Basswood	0	0	3						
Red oak	5	5	1	1808	1857	202	0.83	1.62	0.40
Sycamore	0	0	2						
Tulip poplar	4	3	1	1745	1857	328	0.54	0.94	0.69
White elm	2	2	0	1729	1856	246		1.39	0.55
White oak	12	7	0	1728	1857	797	0.48	1.01	0.50
Total	59	53	8	1663	1857	4599			

8.8 Grouseland

Grouseland was the home of William Henry Harrison, the ninth president of the United States, while he was governor of the Indiana Territory. Harrison constructed the residence at Vincennes, Indiana, on a 300-acre (121-hectare) plot along the Wabash River (Figure 8.13). The 26-room house, described as the White House of the West, was the location of many of Harrison's treaty negotiations with Native Americans (HABS 1934b). The Federal Style, brick, two-story house measures 60 × 75 feet. Its construction is impressive because the craftsmanship required to build the high-style mansion would have been rare on the western frontier. The site was documented by HABS in 1934 and was designated a National Historic Landmark in 1983. A construction date for the house can be estimated based on Harrison's appointment in 1800 and departure from Vincennes in 1814.

Figure 8.13
Image of Grouseland, William Henry Harrison's home while he was governor of the Indiana Territory (Vincennes, Indiana).

The photo was taken after 1933 and shows the front (south) and west façades. Limited tree-ring analysis was performed in the cellar, and two joists had 1802 death dates. Historians report that construction of the house was completed in 1804. Source: Library of Congress Digital Archives, Historic American Building Survey Documentation: William Henry Harrison House, Park & Scott Streets, Vincennes, Knox County, IN; HABS IND, 42-VINC.

Documentation of the house records it as being built between 1801 and 1804 (HABS 1934a; Mendinghall 1976).

Since Grouseland is a National Historic Landmark and is operated as a house museum, tree-ring analysis was limited to two white oak floor joists in the cellar. Two samples were collected from two timbers, and 177 rings were measured spanning from 1712 to 1801. A death date of 1802 for the cellar joists falls within the range of the 1801 start date and the 1804 completion of the structure and supports the dates established by historians. Grouseland is likely the oldest brick house in Indiana (see also the George Ash House in Chapter 9).

8.9 Conclusion

Chapter 8 presented our analysis of open air and house museums, and reported the challenges of sampling and interpreting buildings that have been moved and modified and currently used for interpretation. In Chapter 9 we report our analysis of diverse cultural landscapes that also include several significant houses.

8.10 Literature cited

Abstract of Title. n.d. *Abstract of Title William Rodman to Charles Hay*. Hay Center collections, Salem, IN.

Boomhower, R. 1995. The John Hay home. *Traces* 7: 36–37.

Brock, D.W.H., B.J. Heath, and H.D. Grissino-Mayer. 2017. Dendroarchaeology at the Tipton-Haynes state historic site: Contextualizing the landscape of an upland South farmstead in East Tennessee, USA. *Dendrochronologia* 43: 41–49.

Burns, C.M., L. Davis, and E.S. Dean. 1970. *John Hay Birthplace National Register of Historic Places Inventory-Nomination Form*. National Park Service, Washington, DC.

Carroll County Wabash & Erie Canal, Inc. 2016. *Volunteer Notebook*. Carroll County Wabash & Erie Canal, Inc, Carroll County, IN. 68 p.

Chappell, E.A. 2007. Viewpoint: Vernacular architecture and public history. *Buildings & Landscapes: Journal of the Vernacular Architecture Forum* 14: 1 – 12.

Druckenbrod, D.L., L.A. Stachowiak, E.A. Schneider, D.M. Graves, and H.D. Grissino-Mayer. 2017. Dendrochronological dating of the Graves Mill Gristmill, Madison County, Virginia, USA. *Dendrochronologia* 43: 27–32.

Fife, C.B., M. Buchanan, and E. Auxier. 2007. *Madison Historic District National Historic Landmark Application Form*. US National Park Service, Washington, DC.

Glassie, H. 1968. *Pattern in the Material Folk Culture of the Eastern United States*. University of Pennsylvania Press, Philadelphia, PA. 344 p.

Glassie, H. 1969. A central chimney continental log house. *Pennsylvania Folklife* 18: 32–39.

Grissino-Mayer, H.D., J.T. Maxwell, G.L. Harley, N.A. Garland, D.H. Holt, C. Absher, B.J. Beale, M.S. Boehm, K.A. de Graauw, A. Rautio, and A.W. Dye. 2013. Dendrochronology reveals the construction history of an early 19th century farm settlement, Southwestern Virginia, USA. *Journal of Archaeological Science* 40: 481–489.

Grissino-Mayer, H.D., E.A. Schneider, M.L. Rochner, L.A. Stachowiak, and M.E. Dennison. 2017. Tree-ring dating of timbers from Sabine Hill, home of General Nathaniel Taylor, Elizabethton, Tennessee, USA. *Dendrochronologia* 43: 33–40.

HABS. 1934b. *Grouseland*. Library of Congress (HABS IND, 42-VINC), Washington, DC.

HABS. 1934a. *Judge Jeremiah Sullivan House*. Library of Congress, Washington, DC (HABS IND, 39-MAD).

Hageman, G.E. 1926. *Frederick Konrad Dietrich Wyneken: Pioneer Lutheran Missionary of the Nineteenth Century. St. Louis, Missouri*. Concordia Publishing House, St. Louis, Missouri. 55 p.

Harley, G.L., J.T. Maxwell, D. Holt, and C.B. Speagle. 2017. Construction history of the Deason House, Jones County, Mississippi. *Dendrochronologia* 43: 50–58.
Hay Center. n.d. *The Pioneer Village of the Stevens Memorial Museum*. Tour brochure. 16 p.
Indiana State Historic Sites. 2013. *Interpretive Notes and Newspaper Clippings from the New Harmony Office*.
Jordan, T.G. 1985. *American Log Buildings: An Old World Heritage*. University of North Carolina Press, Chapel Hill, NC. 193 p.
Kniffen, F.B. 1986. Folk housing: Key to diffusion. Pp. 3–26. In: *Common Places: Readings in American Vernacular Architecture*. (D. Upton and J.M. Vlach, eds.). University of Georgia Press, Athens, GA. 560 p.
Lewis, P.F. 1975. Common houses, cultural spoor. *Landscape* 19: 1–28.
Matheus, T.J., J.T. Maxwell, J. Oliver, M. Thornton, M. Hess, and G.L. Harley. 2017. A dendrochronological evaluation of three historic pioneer cabins at Spring Mill Village, Indiana. *Dendrochronologia* 43: 12–19.
McAlester, V., and L. McAlester. 1988. *A Field Guide to American Houses*. Alfred A. Knopf, New York, NY. 525 p.
Mendinghall, J.S. 1976. *William Henry Harrison Home. National Register of Historic Places Inventory and Nomination Form*. National Park Service, Washington, DC.
Montell, L.W., and M.L. Morse. 1995. *Kentucky Folk Architecture*. University Press of Kentucky, Lexington, KY. 120 p.
NPS. 1995. *National Register Bulletin: How to Apply the National Register Criteria for Evaluation*. National Park Service, Washington, DC.
NPS. 2018. *Judge Jeremiah Sullivan House*. www.nps.gov/nr/travel/madison/judge_jeremiah_sullivan_house.html. Accessed 13 August 2018.
Peat, Wilbur D. 1962. *Indiana Houses of the Nineteenth Century*. Indiana Historical Society, Indianapolis, IN. 195 p.
Rentzhog, S. 2007. *Open Air Museums: The History and Future of a Visionary Idea*. Ostersund, Jamtli, Sweden. 530 p.
Roberts, W.E. 1996. *Log Buildings of Southern Indiana*. Trickster Press, Bloomington, IN. 228 p.
Rubino, D., and C. Baas. 2014. Using Tree-ring growth patterns to date the construction of a nineteenth century dogtrot house in Posey County, Indiana. *Proceedings of the Indiana Academy of Science* 123: 14–22.
Salem Leader. 1971. *Historically Speaking*. July 28. Newspaper article.
Schauinger, J.C. 1941. Jeremiah C. Sullivan, Hoosier Jurist. *Indiana Magazine of History* 37: 217–236.
Smith, Robert E. 1999. Wyneken as missionary: Mission in the life & ministry of Friedrich Conrad Dietrich Wyneken. Pp. 321–340. In: *Let Christ Be Christ: Theology, Ethics & World Religions in the Two Kingdoms*. Tentatio Press, Huntington Beach, CA. 378 p.
Taormina, R., and J.H. Speer. 2016. Forgotten waterways: Analyzing beams from the Wabash and Erie Canal. *Tree-Ring Research* 72: 78–90.
Taylor, Robert M., and Connie A. McBirney. 1996. *Peopling Indiana: The Ethnic Experience*. Indiana Historical Society, Indianapolis, IN. 703 p.
Webber, F.R. 1957. *A History of Preaching in Britain and America: Part Three*. Northwestern Publishing House, Milwaukee, WI. 682 p.
Wolf, C. Umhau 1997. *Nineteenth Century Lutheranism in Northwestern, Ohio and Southeastern, Michigan: Part Two 1851–1866*. Manuscript. 144 p.

Chapter 9

Case studies

Dating and interpreting diverse cultural landscapes

9.1 Introduction

Cultural groups "socially construct landscapes as reflections of themselves" where the "social, cultural, and natural environments are meshed and become part of the shared symbols and beliefs of members of the groups" (Evans et al. 2001). Therefore, culturally influenced changes to the natural environment (e.g., patterns of settlement, landownership, building arrangements, foodways, etc.) have unique meanings specific to a cultural group. Archaeologists', historians', and preservationists' ability to recognize these cultural traits is challenging and requires an attentive approach to examining historic sites (Hardesty 2000; Egan 2003). In this chapter, we present five case studies of eight structures that demonstrate how dendrochronology allows diverse cultural landscapes to be situated in time, a level of information typically unavailable to preservationists. (Figure 9.1). The short histories presented with each study examine archival resources associated with a variety of distinct cultural groups. Many of these groups initially struggled to garner much attention by historians, but are now included as central themes in the interpretation of historic sites.

9.1.1 Dendrochronology and interpreting historic sites

A goal of this book is to promote use of dendrochronology, archival resources, and field analysis to inform preservation decisions. Scholarly documentation of the use of dendrochronology in the preservation of North American historic sites is not common. An internet and literature search did reveal notable examples. Dendrochronology influenced the restoration of a Newtown History Center's dwelling in Stephens City, Virginia, by confirming that the log addition to a 1760s stone house was added in 1786, and that the height of the addition's exterior walls was raised in 1804. The museum leaders struggled with how to approach preserving the property. The house's patron desired that it be restored to its late-Colonial roots, however, that would require removing the timber addition. The museum leaders chose to restore and interpret the house to an 1830s time

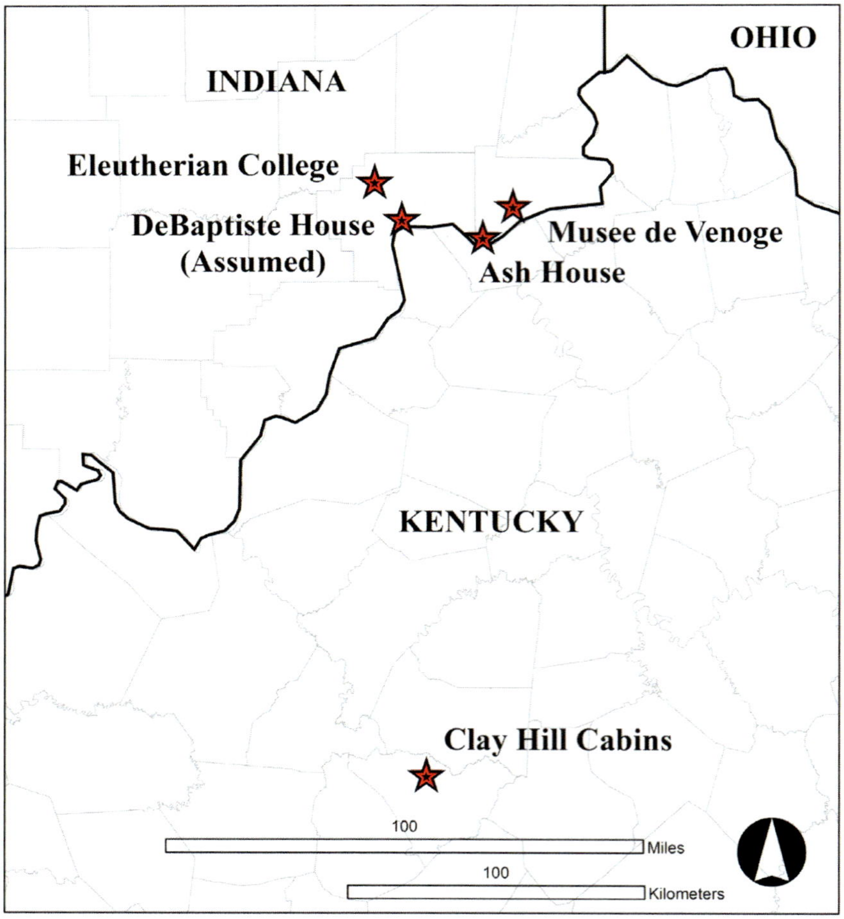

Figure 9.1
The Indiana and Kentucky locations of the cultural landscapes reported in Chapter 9.
Source: Authors.

period, instead of the Colonial era, rather than remove the additions that were key parts to the history of the building (Newtown History Center 2018). Tree-ring analysis informed the restoration of the 1828 Rice-Upshaw House, a log dwelling restored by the Black River Technical College in Randolph County, Arkansas. Historically, its builder Reuben Rice used the house as a trading post. The house was restored to the 1840s era (when it was expanded to serve as a home), requiring the removal of later (post-1840s) additions to the structure (Project Reach 2018). Dendrochronology determined that the historic office and store for the Belle Grove Plantation in Middletown, Virginia, was constructed in 1788 and is the property's oldest structure. The building was rehabilitated to interpret its historic uses as a plantation office and store. The site managers also chose to introduce the contemporary use of housing for visiting scholars. This required the modernization of a restroom and the installation of a kitchenette (Belle Grove Plantation 2018).

Most of our tree-ring analysis has been used to interpret historic sites. Sometimes analysis results in no date. Often, it does not corroborate written histories or the understanding, or beliefs, of a site's owners or managers. Three sites reported in this chapter turned out not to be what preservationists thought they were. These types of results should cause owners and managers to rethink the significance of a historic site as significant examples of a period landscape or building type rather than time-based connections to important people or events.

9.2 Using dendrochronology to identify historic slave dwellings

There is increased interest and activity in the documentation of slave landscapes (Dumas et al. 2017). This interest is motivated by the inclusion of slave life in the interpretation of high profile historic sites such as Mount Vernon, Monticello, and Colonial Williamsburg (Mount Vernon 2018; Monticello 2018; Williamsburg 2018). In describing slave-related dwellings in Alabama, Dumas et al. explains how "slave houses are artifacts. Their attributes can be recorded, analyzed, and broad patterns about their historical context, their builders, and inhabitants revealed. The methods of construction, for instance, often indicate climate, economy, and available technology. Ethnographic and archaeological studies demonstrate that houses, yards, and landscapes reflect cultural values and social relationships" (Dumas et al. 2017).

Dendroarchaeology of slave-related buildings is rare. It has been used to determine if a dwelling was constructed for enslaved or free African-Americans (before the 1863 Emancipation Proclamation, or the 1865 end of the US Civil War), or if it is slave-related at all. A construction date of 1805 was established for a Tennessee slave cabin at the Belle Meade Plantation (Stachowiak 2016). The study also found that elm (*Ulmus* spp.) and ash (*Fraxinus* spp.) were the primary species used in its construction, and that this choice of timber for constructing dwellings was not typical in the southeastern United States. Dendroarchaeology was used to examine the cabin of Alfred Jackson, a slave born at The Hermitage, Andrew Jackson's plantation near Nashville, Tennessee (Lewis et al. 2009). The study sought to discover if the cabin was constructed while he was a slave or after he was freed. The results established a pre-Civil War 1840s construction date confirming it was built as a slave cabin.

9.2.1 The Sanders House and the Clay Hill Cabin

The Sanders House and Clay Hill Cabin are timber houses located in Taylor County, Kentucky, and are purported to be slave dwellings. The buildings are associated with the historic 4,000-acre (1,619-hectare) Clay Hill Farm first settled by Thomas James Sanders in 1795. The farm was one of the largest in Kentucky and was famous as a retreat for John Hunt Morgan, a cavalry general during the US Civil War (1861–1865) who made raids into northern states.

In 1860, approximately 35 slaves resided on the farm, and oral history describes how they were housed in as many as 19 cabins (Clay Hill Forest 2018). The cabin locations are now part of a 305-acre (123-hectare) woodland that was historically part of the farm. The site is now managed by Campbellsville University (located in Campbellsville, Kentucky). The University sought the use of dendrochronology, along with the input of historians and archaeologists, to determine the significance of the cabins within the context of Clay Hill's plantation history.

The Sanders House was originally a one and a half story single-pen log house measuring 20 × 25 feet. The house has V-notch corner notching and sections have board-and-batten siding. It was later modified with the addition of an interior wall, stairs, and a loft. A frame addition was added to its south wall, and interior flues and stoves were installed. Through field documentation, historians concluded that the cabin was constructed about 1850, and the interior modifications were completed in the 1870s. The frame addition, believed to be late 19th or early 20th century, had been removed prior to tree-ring analysis (Kentucky Heritage Council 1996). The historian's Antebellum date of 1850 suggested that it was a slave cabin, although they pointed out that the claim was only supported by oral history and not field or archival evidence (Belson 1999). As a potentially significant resource, the cabin was encased in a metal barn to protect it until it could be properly restored (Figure 9.2).

Tree-ring analysis resulted in the successful crossdating of 26 series from 19 timbers, with a total of 1,571 measured rings (1706–1882). The samples were obtained from four timber types: tulip poplar, American chestnut, red oak, and white oak. Dendrochronology established a build date of 1883 for the dwelling (Tables 9.1 and 9.2).

The Clay Hill Cabin (Figure 9.3) is a one and a half story single-pen log house measuring 18 × 20 feet. The house demonstrates frugal and often crude construction. Features such as the loft's floor joists and the roof sheathing and purlins are hand riven (i.e., split) rather than sawn. The building is constructed with square corner notching, a method that is common, but not the predominant notching type south of the Ohio River (Figure 9.4; Kniffen and Glassie 1966; Montell and Morse 1976; Roberts 1996). While the simple notch requires minimal skills

Figure 9.2
Sanders House (Clay Hill Forest, Taylor County, Kentucky). The dwelling was enclosed in a small barn after historians identified it as a potential slave cabin. It turned out the cabin was constructed in 1883 and was never a slave dwelling.

Source: Gordon Weddle.

Table 9.1 Samples collected from the Sanders House (Taylor County, Kentucky). "First" and "last" refer to the first and last rings present in the sample. If no date is provided, the sample could not be reliably crossdated. "Outer ring" refers to the outermost ring of the sample: b = bark, w = wane, etc.; see Table 7.1 for a complete list of codes used for outer ring condition. "r" is the correlation coefficient obtained by correlating the ring widths of this sample with all other timbers of the same type from the building (series intercorrelation; Chapter 5). "Timber type" is the type of timber; see Table 3.1 for a list of possible tree species represented by the various timber types. "Provenience" is the timber's function in a structure; see Chapter 2 for a description of the various proveniences.

Sample	*First*	*Last*	*Outer ring*	r	*Timber type*	*Provenience*
CHF01A	1854	1883	w	0.84	White oak	Floor joist; ground floor; N pen
CHF01B	1709	1878		0.50		
CHF02A	1810	1883	b	0.47	White oak	Floor joist; ground floor; N pen
CHF03A	1705	1849	w	0.48	White oak	Wall timber; N elevation
CHF04A	1841	1881	w	0.63	Chestnut	Wall timber; N elevation
CHF05A	1804	1830	w	0.58	Tulip	Wall timber; N elevation
CHF05B	1736	1829	sap	0.47		
CHF06A					Tulip	Wall timber; N elevation
CHF07A	1872	1883	w	0.74	Tulip	Wall timber; N elevation
CHF07B	1852	1883	w	0.53		
CHF08A	1857	1883	w	0.55	Tulip	Wall timber; E elevation
CHF08B	1858	1883	w	0.82		
CHF09A	1830	1879	sap	0.58	Tulip	Wall timber; E elevation
CHF10A					Red oak	Chinking; E elevation
CHF11A	1813	1871	sap	0.61	Tulip	Wall timber; E elevation
CHF11B	1821	1883	w	0.78		
CHF12A	1800	1883	w	0.61	Tulip	Wall timber; E elevation
CHF12B	1855	1883	w	0.71		
CHF13A					Chestnut	Chinking; E elevation
CHF14A	1766	1846		0.38	Red oak	Chinking; E elevation
CHF15A	1741	1829		0.45	White oak	Chinking; E elevation
CHF16A	1836	1883	b	0.56	Chestnut	Chinking; E elevation
CHF17A	1840	1883	b	0.61	Tulip	Wall timber; E elevation
CHF18A	1803	1883	w	0.61	Tulip	Wall timber; W elevation
CHF18B	1850	1883	w	0.77		
CHF19A	1855	1883	w	0.79	Chestnut	Wall timber; N elevation
CHF21A	1733	1883		0.44	Red oak	Loose
CHF22A	1761	1882		0.60	Red oak	Shingle
CHF23A	1771	1876		0.69	Red oak	Shingle

Note: Timber type and provenience are only given once if more than one sample was obtained from an individual timber. An *r*-value is not reported if less than three timbers of a particular type were obtained from a structure or if the sample did not overlap temporally with the other samples obtained for a particular timber type. A full description of the data presented in the table is provided in Chapter 7.

Table 9.2 Summary table of the timber type chronologies prepared from Sanders House (Taylor County, Kentucky). "Samples" and "Timbers" are the number of successfully crossdated samples and timbers from the building. "Undated" is the number of timbers that could not be crossdated reliably. "First" and "last" refer to the first and last dates in the chronology. "Mean" and "SD" are the Mean and standard deviation (mm) of the ring widths in the chronology. "r" is the mean series intercorrelation for each timber type (see Chapter 5). r is not provided if less than three timbers were obtained for a particular timber type. No total r, mean, or SD are provided since each chronology was prepared using only a single timber type. A full description of the data presented in the table is provided in Chapter 7.

Timber type	*Samples*	*Timbers*	*Undated*	*First*	*Last*	*Years*	r	*Mean*	*SD*
Chestnut	3	3	1	1839	1882	110	0.64	2.85	0.91
Red oak	4	4	1	1735	1882	450	0.54	1.39	0.39
Tulip poplar	14	8	1	1738	1882	621	0.64	1.55	0.62
White oak	5	4	0	1706	1882	390	0.50	1.16	0.43
Total	26	19	3	1706	1882	1571			

to create, it fails to lock the timbers together as effectively as other types of corner notching (e.g., dovetail and half-dovetail), and this is the only house we have seen with this type of notching. The site's archaeology report estimates that the house's construction was in the first half of the 19th century, and based on certain features (exposed and beaded floor joists, square corner notching, and nail type) likely dated between 1820 and 1830 (Schlarb 2014; Kentucky Heritage Council 2015).

Tree-ring analysis produced 88 series from 82 timbers, totaling 8,384 measured rings spanning from 1546 to 1877. The samples were obtained from five timber types: tulip poplar, American beech, walnut, red oak, and white oak.

Figure 9.3
Clay Hill Cabin (Clay Hill Forest, Taylor County, Kentucky). The dwelling was constructed in 1843 and was likely a slave dwelling for the Clay Hill Farm.

Source: Authors.

Figure 9.4
Clay Hill Cabin (Clay Hill Forest, Taylor County, Kentucky). The building is constructed with square corner notching. While the simple notch requires minimal skills to create, it fails to lock the timbers together as effectively as other types of corner notching.

Source: Authors.

Dendrochronology established a build date of 1843 for the Clay Hill Cabin, placing it within the era of slavery at the plantation (Tables 9.3 and 9.4).

Architectural historians and archaeologists documented each of the Clay Hill buildings, but their estimated dates for construction were only partially accurate. Dendroarchaeology was able to establish that the Sanders House was constructed 18 years after the US Civil War, and was not a slave dwelling. The Clay Hill Cabin was constructed 22 years before the war's end, and is likely a slave cabin associated with the Clay Hill Plantation. Our analysis of the Clay Hill Cabin also suggests that it may have originally had a dirt floor. An examination of the floor construction, and the dating of its joists, indicates that a timber floor was added in the 1850s (FTR27A, FTR30A, and FTR31A and B). Unfortunately, the Clay Hill Cabin was beyond repair and was removed.

Table 9.3 Samples collected from the Clay Hill Cabin (Taylor County, Kentucky). See Table 9.1 for a description of the table contents; a full description of the data presented is provided in Chapter 7.

Sample	*First*	*Last*	*Outer ring*	r	*Timber type*	*Provenience*
FTR01A	1746	1843	w	0.53	Tulip	Wall timber; S elevation
FTR02A	1733	1843	b	0.78	Tulip	Wall timber; S elevation
FTR03A	1681	1843	b	0.72	Tulip	Wall timber; S elevation
FTR04A	1728	1843	b	0.65	Tulip	Wall timber; S elevation
FTR05A	1726	1843	b	0.64	Tulip	Wall timber; W elevation; gable
FTR06A	1677	1843	w	0.67	Tulip	Wall timber; N elevation
FTR07A	1708	1843	b	0.68	Tulip	Wall timber; N elevation
FTR08A	1725	1843	w	0.66	Tulip	Wall timber; E elevation
FTR08B	1725	1843	w	0.53		
FTR09A	1698	1812		0.60	Tulip	Wall timber; E elevation
FTR10A	1793	1843	b	0.52	Tulip	Wall timber; E elevation; gable
FTR11A	1743	1843	b	0.78	Tulip	Wall timber; E elevation; gable
FTR12A	1704	1843	w	0.68	Tulip	Wall timber; S elevation
FTR13A	1726	1843	b	0.59	Tulip	Wall timber unknown elevation
FTR14A	1619	1751		0.56	Tulip	Floor joist; loft
FTR15A	1674	1763		0.72	Tulip	Floor joist; loft
FTR16A	1614	1730		0.59	Tulip	Staircase
FTR17A	1621	1726		0.63	Tulip	Floor joist; loft
FTR18A	1647	1749		0.67	Tulip	Floor joist; loft
FTR19A	1657	1772		0.63	Tulip	Floor joist; loft
FTR20A	1548	1682		0.83	Tulip	Staircase
FTR21A					Tulip	Floor joist; loft
FTR22A	1685	1852	b	0.63	Tulip	Stair tread
FTR23A	1670	1761		0.70	Tulip	Floor joist; loft
FTR24A					Tulip	Staircase
FTR25A	1740	1813		0.68	Tulip	Staircase
FTR26A	1676	1792		0.77	Tulip	Staircase
FTR27A	1718	1853	b	0.66	White oak	Floor joist; ground floor
FTR28A	1770	1844	w	0.39	White oak	Floor joist; ground floor
FTR29A	1545	1654		0.77	Tulip	Loose
FTR30A	1728	1853	b	0.58	White oak	Floor joist; ground floor
FTR31A	1769	1852	w	0.64	White oak	Floor joist; ground floor
FTR31B	1695	1852	b	0.57		
FTR32A	1702	1843	b	0.71	Tulip	Wall timber; W elevation
FTR33A	1600	1669		0.67	Tulip	Subfloor; ground floor
FTR34A					Tulip	Subfloor; ground floor
FTR35A					Tulip	Subfloor; ground floor
FTR36A	1626	1733		0.75	Tulip	Subfloor; ground floor
FTR37A	1738	1828		0.78	Tulip	Subfloor; ground floor

Sample	First	Last	Outer ring	r	Timber type	Provenience
FTR38A	1730	1788		0.85	Tulip	Subfloor; ground floor
FTR39A	1595	1717		0.63	Tulip	Subfloor; ground floor
FTR40A					Tulip	Subfloor; ground floor
FTR41A	1764	1821		0.79	Tulip	Subfloor; ground floor
FTR42A	1642	1741		0.69	Tulip	Subfloor; ground floor
FTR43A	1740	1807		0.70	Tulip	Subfloor; ground floor
FTR44A	1709	1764		0.74	Tulip	Subfloor; ground floor
FTR45A	1601	1727		0.63	Tulip	Subfloor; ground floor
FTR46A	1598	1714		0.77	Tulip	Subfloor; ground floor
FTR47A	1742	1833		0.72	Tulip	Subfloor; ground floor
FTR48A	1660	1769		0.63	Tulip	Subfloor; ground floor
FTR49A	1637	1754		0.78	Tulip	Subfloor; ground floor
FTR50A	1674	1779		0.74	Tulip	Tongue and groove flooring; loft
FTR51A	1692	1830		0.68	Tulip	Winder; staircase
FTR51B	1687	1755		0.58		
FTR52A					Beech	Purlin; replacement
FTR53A					Red oak	Purlin; replacement
FTR54A					Tulip	Staircase
FTR55A	1680	1837		0.62	Tulip	Staircase
FTR56A	1732	1765		0.67	Tulip	Staircase
FTR57A	1689	1775		0.75	Tulip	Stringer; staircase
FTR58A	1815	1844	w	0.63	Tulip	Rafter
FTR59A					Red oak	Purlin
FTR60A	1791	1843	w		Red oak	Purlin
WHI12A	1614	1726		0.78	Tulip	Tongue and groove flooring; loft
WHI13A	1660	1758		0.74	Tulip	Floorboard; loft
WHI14A	1652	1756		0.79	Tulip	Tongue and groove flooring; loft
WHI15A	1678	1751		0.67	Tulip	Floorboard; loft
WHI16A	1684	1780		0.74	Tulip	Floorboard; loft
WHI17A	1740	1809		0.69	Tulip	Tongue and groove flooring; loft
WHI18A	1713	1788		0.71	Tulip	Floorboard; loft
WHI19A	1669	1715		0.72	Tulip	Tongue and groove flooring; loft
WHI20A	1547	1676		0.88	Tulip	Tongue and groove flooring; loft
WHI21A	1655	1736		0.66	Tulip	Floorboard; loft
WHI22A	1617	1748		0.76	Tulip	Tongue and groove flooring; loft
WHI23A	1638	1799		0.73	Tulip	Floorboard; loft
WHI24A	1712	1782		0.65	Tulip	Tongue and groove flooring; loft
WHI25A	1620	1751		0.76	Tulip	Floorboard; loft

(Continued)

Table 9.3 Continued

Sample	*First*	*Last*	*Outer ring*	r	*Timber type*	*Provenience*
WHI26A	1686	1733		0.47	Tulip	Tongue and groove flooring; loft
WHI27A	1658	1765		0.65	Tulip	Tongue and groove flooring; loft
WHI28A	1579	1685		0.73	Tulip	Subfloor; ground floor
WHI29A					Tulip	Subfloor; ground floor
WHI30A	1640	1722		0.85	Tulip	Subfloor; ground floor
WHI31A	1747	1821		0.75	Tulip	Subfloor; ground floor
WHI32A	1707	1748		0.62	Tulip	Subfloor; ground floor
WHI33A	1739	1818		0.81	Tulip	Subfloor; ground floor
WHI34A	1703	1762		0.68	Tulip	Subfloor; ground floor
WHI35A	1638	1785		0.69	Tulip	Subfloor; ground floor
WHI36A	1801	1848		0.65	White oak	Subfloor; ground floor
WHI37A	1787	1857		0.51	White oak	Subfloor; ground floor
WHI38A	1805	1879		0.60	White oak	Subfloor; ground floor
WHI39A					Beech	Subfloor; ground floor
WHI40A					Beech	Subfloor; ground floor
WHI41A					Beech	Subfloor; ground floor
WTE03A	1722	1843	w	0.67	Tulip	Wall timber; S elevation
WTE04A	1819	1843	b	0.56	Tulip	Wall timber; S elevation
WTE04B	1724	1843	b	0.68		
WTE05A	1792	1844	w	0.68	White oak	Floor joist; ground floor
WTE05B	1790	1844	w	0.70		
WTE06A					White oak	Floor joist; ground floor
WTE08A					Walnut	Filler in mortise for loft joist
WTE09A	1724	1782		0.83	Tulip	Filler in mortise for loft joist
WTE10A					Walnut	Filler in mortise for loft joist
WTE11A	1778	1831	b	0.61	Tulip	Wall timber; E elevation
WTE11B	1736	1831	w	0.71		

Table 9.4 Summary table of the timber type chronologies prepared from the Clay Hill Cabin (Taylor County, Kentucky). See Table 9.2 for a description of table contents; a full description of the data presented is provided in Chapter 7.

Timber type	*Samples*	*Timbers*	*Undated*	*First*	*Last*	*Years*	r	*Mean*	*SD*
Beech	0	0	4						
Red oak	1	1	2	1792	1842	51		2.42	0.42
Tulip poplar	77	73	7	1546	1851	7508	0.70	1.38	0.71
Walnut	0	0	2						
White oak	10	8	1	1696	1877	825	0.59	0.97	0.38
Total	88	82	16	1546	1877	8384			

9.3 The George DeBaptiste House

George DeBaptiste (1814–1875) was an African-American abolitionist that resided in Madison, Indiana, from 1837 until 1846. DeBaptiste was born a free black in Fredericksburg, Virginia. He was General William Henry Harrison's (1773–1841) personal valet, and when Harrison was elected president DeBaptiste was named a White House steward.

DeBaptiste's Madison barbershop was a center for Underground Railroad activity that helped move escaping African-American slaves from Southern states to freedom in the North. He is credited with assisting the escape of 180 slaves (Madison Courier 2011; Furnish 2014, 2016). When his life was threatened for his anti-slavery actions, he moved to Detroit, Michigan, in 1846. He continued his work helping fellow African-Americans until his death in 1875 (Detroit Free Press 1875; Detroit Historical Society 2018).

What was thought to be the DeBaptiste House for the time he lived in Madison is located in the city's Georgetown District, a neighborhood housing free blacks in the years leading up to the US Civil War (Indiana Department of Natural Resources 2018; National Park Service 2018). We were invited to analyze the house in an effort to establish a construction date concurring with DeBaptiste's time in Madison. The Federal Style brick dwelling was in disrepair, and grassroots efforts to save it confirmed that it was located on property previously owned by DeBaptiste (Figure 9.5; Madison Courier 2011).

Figure 9.5
House believed to be owned by George DeBaptiste (Madison, Indiana). This house was constructed in 1846.

Source: Authors.

Analysis produced 18 tulip poplar samples from 16 timbers, resulting in 3,350 measured rings spanning from 1546 to 1845 ($r = 0.77$, 0.89 ± 0.52 mm; mean ± SD; Table 9.5). Since the house was in poor shape, provenience is not absolute. Most of the pieces were identified as joists based on their dimensions. An 1846 build date was established, the year DeBaptiste had moved to Detroit.

We believe DeBaptiste's house was likely destroyed in the September 1846 Crooked Creek flash flood that drowned 11 people and swept dwellings off their foundations (Madison Republic 1846; Woollen 1883). The creek originates in the bluffs behind the city, and the effects of the flood compounded when a railroad culvert became plugged with houses and debris creating a large pool. When the culvert broke, a torrent of water destroyed buildings and bridges along the stream bank. The assumed location of DeBaptiste's house lies within meters of the creek and would have been affected by the water. A new brick house, perhaps one more resistant to future floods, was likely constructed the following year. We point out here that a timber with an 1846 death date (GDB11A and GDB99A) could have been harvested any time from mid-growing season 1846 to April 1847. In our case studies, we would ascribe a death date of 1846 to the timber and, consequently, an 1846 construction date to the building if other timbers from the structure exhibit a similar terminal ring. However, a new owner could have constructed the building in 1847. Tree-ring analysis was able to assign a construction date for the house, but the results did not establish a connection with DeBaptiste. As a consequence, interest in preserving the structure has waned.

Table 9.5 Samples collected from the George DeBaptiste House (Madison, Indiana). See Table 9.1 for a description of the table contents; a full description of the data presented is provided in Chapter 7.

Sample	*First*	*Last*	*Outer ring*	r	*Timber type*	*Provenience*
GDB01A	1553	1753		0.87	Tulip	Floor joist
GDB02A	1661	1778		0.71	Tulip	Sill
GDB03A	1659	1803		0.63	Tulip	Brace
GDB04A	1551	1770		0.86	Tulip	Floor joist
GBD05A					Hickory	Brace
GDB06A	1590	1763		0.83	Tulip	Floor joist
GDB07A	1549	1662		0.87	Tulip	Floor joist
GDB07B	1549	1786		0.86	Tulip	
GDB07C	1662	1776		0.85	Tulip	
GDB08A	1558	1783		0.89	Tulip	Floor joist
GDB09A	1545	1793		0.77	Tulip	Floor joist
GDB10A	1551	1805		0.74	Tulip	Floor joist
GDB11A	1640	1846	w	0.69	Tulip	Floor joist
GDB12A	1610	1790		0.63	Tulip	Door jamb
GDB13A	1602	1789		0.68	Tulip	Door jamb
GDB96A	1604	1828	sap	0.68	Tulip	Floor joist
GDB97A	1644	1795		0.71	Tulip	Floor or wall panel
GDB98A	1635	1833	sap	0.77	Tulip	Floor joist
GDB99A	1643	1846	b	0.78	Tulip	Floor joist

9.4 Eleutherian College

Eleutherian College was a mid-19th-century institution known for promoting gender and racial equality (the Greek word *eleutheros* means freedom). The roots of the college are based in the region's anti-slavery ideals and the activities of the abolitionists centered in the rural village of Lancaster, Indiana. In 1846, the local Neil's Creek Abolitionist Baptist Church convinced Thomas Craven, a minister from Oxford, Ohio, to establish his educational institute near their town. Church members, who became early leaders of the college, were active in the Underground Railroad that moved escaping African-American slaves from Southern states to freedom in the North. The Ohio River, located just nine miles (14.5 kilometers) south of Lancaster, was the boundary between free and slave cultures (Bennett 1996).

The school opened with 15 students in 1848, with classes taught in a local community hall (Bennett 1996). In 1849, Craven purchased 80 acres (32 hectares) of land for the college. Later, in 1855, he and his wife Rebecca conveyed seven and a half acres (3 hectares) of the property to the school (Indiana Historical Bureau 2018). An 1853 promotional brochure described the campus landscape as an "elevated and healthful site of six and a half acres [2.6 hectares] of land, on which was a comfortable log house, well, spring, and orchard" (Thompson 1923). The elevated parcel, chosen for its symbolic ascension to knowledge, resulted in the nickname College Hill (Department of Public Instruction 1917). The college's most successful years were from 1855 through 1861 when 150 students were enrolled. Following enrollment declines during and after the US Civil War, the building was sold in 1888 to Lancaster Township for use as a public school. It closed in 1938 (Department of Public Instruction 1917; Bennett 1996).

The Eleutherian campus landscape was a collection of structures that included the classroom and chapel building, a dormitory, a log house, and a caretaker's house (now the visitor center). There were likely additional support buildings, such as stables and privies. A promotional brochure reported that a two-story stone boarding house containing 15 rooms had been constructed by 1853. It also describes that a log house existed in 1853. The massive limestone classroom and chapel building is reported to have been started in 1854 and finished in 1856 (Thompson 1923; Bennett 1996). Beyond the campus landscape were the farms and houses of the college's supporters, including the Lyman and Asenath Hoyt House (Thompson 1923).

9.4.1 The classroom and chapel building

The classroom and chapel building is a rectangular, three-story, gable-front, Greek Revival limestone structure measuring 42 × 65 feet in size (Figure 9.6). Its gable front is topped by a bell tower. The primary interior space is a two-story chapel. Classrooms are located on the third floor (Bennett 1996).

Tree-ring analysis produced 69 series from 57 timbers, resulting in 5,180 measured rings ranging from 1530 to 1854. Five timber types are represented: tulip poplar, sweet gum, American beech, black gum, and white oak. Dendrochronology confirmed the 1856 construction date estimated by archival sources (Tables 9.6 and 9.7). The results suggest that the timber construction of the building may have started as early as 1852, two years earlier than estimated.

Figure 9.6
Classroom and chapel building at Eleutherian College (Lancaster, Indiana). The building was constructed in 1856.
Source: Authors.

9.4.2 Caretaker's House

What is now the visitor center was originally a caretaker's house (Figure 9.7). Multiple additions to the dwelling have obscured its historic form. We crossdated four samples from three timbers, resulting in 265 measured rings ranging from 1762 to 1854. Two timber types are represented: tulip poplar and sweet gum (Tables 9.6 and 9.7). An estimated construction date for the visitor center appears to be in the mid-1850s. We cannot be any more precise since only two timbers yielding a harvest date were available to sample. In the multitude of modifications to the timber building, dendrochronology was able to discover datable timbers that provide essential information to the interpretation of the college. However, the limitations of dendrochronology did not allow a definitive construction date to be identified at this time.

Figure 9.7
Caretaker's house at Eleutherian College (Lancaster, Indiana). The house has been modified many times, but portions of the original were located and analyzed. The building was constructed about 1852.
Source: Authors.

9.4.3 The Lyman and Asenath Hoyt House

The Lyman (1804–1857) and Asenath (1810–1897) Hoyt House is a two-story Greek Revival stone residence believed to have been constructed about 1850 (Figure 9.8). Its limestone construction, rare for a rural residence, and its similarities in style and artisanship to the classroom and chapel building suggest they were built by the same mason. Although it is on a different property, we consider the Hoyt House part of the Eleutherian College landscape because of the Hoyt's and the college's shared abolitionists connections. The Hoyts were members of the Neil's Creek Abolitionist Baptist Church and active in the Underground Railroad (Narayanan 2003).

Analysis of the Hoyt House resulted in 18 crossdated series obtained from 15 timbers; the various chronologies contained 1,617 measured rings spanning from 1625 to 1852. Seven species were represented: tulip poplar, sweet gum, American beech, black gum, white oak, red oak, and black locust (Tables 9.6 and 9.7). Dendrochronology established an 1848 construction date for the house based on the date of wall and roof timbers (HOY09A, HOY11A, HOY17A, and HOY18A), and it appears the timber portions of the house were started as early as 1846. The house predates the classroom and chapel building by eight years. Many samples obtained from the Hoyt House were not datable (Tables 9.6 and 9.7). Many of these samples had few tree rings that were very large. We hypothesize that many of these timbers were replacements harvested from a second-growth forest since they did not exhibit the small, ring-dense growth patterns we usually find in the timbers of historic houses. Given the disrepair and renovation (likely multiple) of the building, understanding the evolution of the building's form and absolute age is difficult.

9.4.4 Tree-ring analysis of Eleutherian College

The preservation of the Eleutherian College landscape is in its early stages, and the results of dendrochronology are informing the interpretation of the historic campus. Tree-ring analysis of the site was very productive. Analysis resulted in 91 series from 75 timbers, and 7,062 measured rings ranging from 1530 to 1854.

Figure 9.8
Lyman and Asenath Hoyt House (Lancaster, Indiana). The Hoyt's were part of the Lancaster Underground Railroad, and supporters of Eleutherian College. The building was constructed in 1848.

Source: Authors.

Table 9.6 Samples collected from the Eleutherian College campus (Lancaster, Indiana). See Table 9.1 for a description of the table contents; a full description of the data presented is provided in Chapter 7.

					Eleutherian College	
Sample	*First*	*Last*	*Outer ring*	r	*Timber type*	*Provenience*
ELT01A	1681	1838	s/b	0.58	Sweet gum	Stud; 3rd floor; E room; N wall
ELT02A	1692	1839	s/b	0.55	Sweet gum	Stud; 3rd floor; E room; N wall
ELT03A	1722	1853	b	0.66	Sweet gum	Stud; 3rd floor; E room; N wall
ELT04A	1676	1852	w	0.54	Sweet gum	Roof truss
ELT05A	1817	1853	b	0.53	Tulip	Diagonal roof brace
ELT06A	1753	1853	w	0.68	Tulip	Horizontal beam between attic and bell tower
ELT07A	1655	1772		0.62	Tulip	Stud; 3rd floor; W room; N wall
ELT08A	1671	1754		0.54	Tulip	Stud; 3rd floor; W room; N wall
ELT09A	1562	1637		0.54	Tulip	Stud; 3rd floor; W room; N wall
ELT10A	1694	1792		0.55	Sweet gum	Stud; 3rd floor; W room; N wall
ELT11A					Sweet gum	Stud; 3rd floor; W room; N wall
ELT12A					Beech	Stud; 3rd floor; W room; N wall
ELT13A	1748	1852		0.70	Sweet gum	Stud; 3rd floor; W room; N wall
ELT14A	1703	1838	s/b	0.60	Sweet gum	Stud; 3rd floor; W room; N wall
ELT15A	1758	1852	w	0.61	Beech	Diagonal roof brace
ELT16A	1738	1850	w	0.66	Sweet gum	Diagonal roof brace
ELT17A					Beech	N-S roof support
ELT18A	1769	1820	s/b	0.71	Sweet gum	Stud; 3rd floor; N room; S wall
ELT19A	1807	1854	b		White oak	Stud; 3rd floor; N room; S wall
ELT20A	1782	1842		0.40	Sweet gum	Stud; 3rd floor; N room; S wall
ELT21A	1650	1790		0.51	Sweet gum	Stud; 3rd floor; N room; S wall
ELT21B	1796	1838	s/b	0.70	Sweet gum	
ELT22A	1775	1855	w	0.68	Beech	Stud; 3rd floor; W room; E wall
ELT23A	1826	1855	w	0.63	Beech	Stud; 3rd floor; W room; E wall
ELT23B	1786	1817		0.77	Beech	
ELT24A	1810	1852	b	0.83	Sweet gum	Stud; 3rd floor; N wall; E office
ELT24B	1813	1852	b	0.81	Sweet gum	
ELT24C	1732	1776		0.71	Sweet gum	
ELT24D	1724	1800		0.74	Sweet gum	
ELT24E	1777	1809		0.63	Sweet gum	
ELT25A	1746	1852	b	0.42	Sweet gum	Stud; 3rd floor; E office; N wall
ELT26A	1704	1849	b	0.55	Sweet gum	Stud; 2nd floor; N room; W wall
ELT27A	1721	1852	w		Black gum	Floor joist; 3rd floor
ELT27B	1836	1852	w		Black gum	
ELT27C	1643	1784			Black gum	
ELT28A	1719	1807	s/b	0.68	Sweet gum	Scab; 2nd floor; N room; S wall closet
ELT29A	1762	1815		0.50	Sweet gum	Stud; 2nd floor; N room; S wall
ELT29B	1793	1849	w	0.34	Sweet gum	

Eleutherian College						
Sample	*First*	*Last*	*Outer ring*	r	*Timber type*	*Provenience*
ELT30A	1794	1853	w	0.66	Tulip	Stud; 2nd floor; N room; W wall
ELT31A	1831	1855	w	0.83	Beech	S stretcher; 2nd–3rd floor stairs
ELT31B	1806	1855	w	0.61	Beech	
ELT32A	1827	1855	w	0.55	Beech	N stretcher; 2nd–3rd floor stairs
ELT33A	1797	1848		0.68	Tulip	W stretcher; E stairwell; 1st–2nd floor stairs
ELT33B	1806	1853	w	0.60	Tulip	
ELT34A	1824	1853	w	0.62	Tulip	E stretcher; E stairwell; 1st–2nd floor stairs
ELT34B	1819	1851		0.57	Tulip	
ELT35A	1810	1853	w	0.60	Tulip	Kick plate; E stairwell; 1st–2nd floor stairs; 7th step from bottom
ELT36A	1760	1832		0.70	Tulip	Tongue and groove flooring
ELT37A	1529	1606		0.44	Tulip	Tongue and groove flooring
ELT38A	1744	1814		0.78	Tulip	Tongue and groove flooring
ELT39A	1743	1810		0.68	Tulip	Loose timber
ELT40A	1716	1790		0.66	Tulip	Loose timber
ELT41A					Sweet gum	Loose timber
ELT42A					Sweet gum	Loose timber
ELT43A					Sweet gum	Loose timber
ELT44A	1701	1799		0.60	Tulip	Tongue and groove flooring
ELT45A	1725	1795		0.76	Tulip	Tongue and groove flooring
ELT46A	1685	1733		0.73	Tulip	Tongue and groove flooring
ELT47A	1714	1789		0.66	Tulip	Tongue and groove flooring
ELT48A	1553	1648		0.64	Tulip	Tongue and groove flooring
ELT49A	1731	1804		0.74	Tulip	Tongue and groove flooring
ELT56A	1681	1747		0.40	Tulip	Tongue and groove flooring
ELT57A	1590	1730		0.48	Tulip	Tongue and groove flooring
ELT58A	1679	1756		0.66	Tulip	Tongue and groove flooring
ELT59A	1659	1740		0.70	Tulip	Tongue and groove flooring
ELT60A	1656	1704		0.66	Tulip	Tongue and groove flooring
ELT61A					Tulip	Tongue and groove flooring
ELT62A	1812	1853	w	0.59	Tulip	Tongue and groove flooring
ELT63A	1672	1741		0.75	Tulip	Tongue and groove flooring
ELT64A	1668	1807		0.75	Tulip	Tongue and groove flooring
ELT65A	1670	1740		0.80	Tulip	Tongue and groove flooring
ELT66A	1694	1789		0.64	Tulip	Tongue and groove flooring
ELT67A	1568	1627		0.48	Tulip	Tongue and groove flooring
ELT68A	1741	1802		0.62	Tulip	Tongue and groove flooring
ELT69A	1606	1658		0.48	Tulip	Tongue and groove flooring
ELT70A	1662	1802		0.68	Tulip	Tongue and groove flooring

(Continued)

Table 9.6 Continued

Hoyt House						
Sample	*First*	*Last*	*Outer ring*	r	*Timber type*	*Provenience*
HOY01A					Black locust	Sleeper
HOY02A					Beech	Floor joist; ground floor
HOY03A					Black locust	Sleeper
HOY04A					White oak	Floor joist; ground floor
HOY05A					Red oak	Floor joist; ground floor
HOY06A	1822	1847	b	0.45	Tulip	Floor joist; 2nd floor
HOY06B	1809	1847	b	0.47	Tulip	
HOY07A					Tulip	Floor joist; 2nd floor
HOY08A	1765	1841	s/w		Sweet gum	Beam; 2nd floor support
HOY08B	1759	1841	s/w		Sweet gum	
HOY09A	1721	1848	w	0.65	Beech	Stud; ground floor
HOY10A					Tulip	Scab; ground floor; likely for lath attachment
HOY11A	1741	1848	w	0.69	Beech	Stud; ground floor
HOY12A	1744	1847	w	0.66	Tulip	Diagonal roof support
HOY13A	1786	1847	w	0.74	Tulip	Diagonal roof support
HOY13B	1787	1847	w	0.75	Tulip	
HOY14A	1765	1847	b		Black gum	Loose
HOY15A	1782	1841	s/w		Sweet gum	Floor joist; attic
HOY16A	1780	1846	w	0.49	Tulip	Floor joist; 2nd floor
HOY17A	1723	1848	w	0.74	Beech	Stud; ground floor
HOY18A	1746	1848	w	0.71	Beech	Stud; ground floor
HOY19A					Red oak	Floor joist; ground floor
HOY20A					Beech	Floor joist; ground floor
HOY21A	1661	1846	w		White oak	Loose
HOY22A	1660	1846	w		White oak	Loose
HOY23A					Black gum	Loose
HOY24A	1624	1687			Tulip	Loose
HOY25A	1761	1853	w	0.54	Tulip	Loose
HOY26A					Tulip	Loose
HOY27A					Black gum	Loose
HOY28A					Red oak	Loose
HOY29A					Tulip	Loose

Caretaker's House (Visitor Center)						
Sample	*First*	*Last*	*Outer ring*	r	*Timber type*	*Provenience*
VEL01A	1780	1822			Tulip	Floor joist; ground floor
VEL01B	1783	1852	w		Tulip	
VEL02A	1783	1852	w		Sweet gum	Floor joist; ground floor
VEL03A	1760	1855	w		Sweet gum	Floor joist; ground floor

Table 9.7 Summary table of the timber type chronologies prepared from the Eleutherian College campus (Lancaster, Indiana). See Table 9.2 for a description of table contents; a full description of the data presented is provided in Chapter 7.

Eleutherian College									
Timber type	*Samples*	*Timbers*	*Undated*	*First*	*Last*	*Rings*	r	*Mean*	*SD*
Beech	7	5	2	1759	1854	319	0.66	1.28	0.49
Black gum	3	1	0	1646	1851	281		0.66	0.21
Sweet gum	22	16	4	1658	1852	1970	0.59	0.76	0.38
Tulip poplar	36	34	1	1530	1852	2565	0.64	1.33	0.55
White oak	1	1	0	1809	1853	45		1.75	0.53
Total	69	57	7	1530	1854	5180			
Hoyt House									
Timber type	*Samples*	*Timbers*	*Undated*	*First*	*Last*	*Rings*	r	*Mean*	*SD*
Beech	4	4	2	1722	1847	455	0.70	1.56	0.68
Black gum	1	1	2	1766	1846	81		0.72	0.31
Black locust	0	0	2						
Red oak	0	0	3						
Sweet gum	3	2	0	1760	1840	214		1.17	0.71
Tulip poplar	7	5	4	1754	1852	428	0.61	1.00	0.49
	1	1	0	1625	1686	72		1.71	0.63
White oak	2	2	0	1662	1845	367		1.39	0.48
Total	18	15	13	1625	1852	1617			
Caretaker's House (Visitor Center)									
Timber type	*Samples*	*Timbers*	*Undated*	*First*	*Last*	*Rings*	r	*Mean*	*SD*
Sweet gum	2	2	0	1762	1854	161		0.59	0.30
Tulip poplar	2	1	0	1785	1851	104		0.88	0.56
Total	4	3	0	1762	1854	265			
Totals	91	75	20	1530	1854	7062			

9.5 *Musée de Venoge*

The *Musée de Venoge* is a house museum located on Indian Creek in Switzerland County, Indiana. The museum interprets and celebrates the region's French-speaking Swiss immigrants. Swiss began settling the Indiana Territory region about 1800 to join John James Dufour's efforts to establish a commercial winery in America (Dufour 1826; Schelbert 1996).

The house was the residence of Jacob Weaver and Charlotte Golay. The couple joined her father, David Golay, in New Switzerland in 1813 and pursued a career in farming. *Venoge* is an example where interpretation of the house was assisted by personal written documents. A collection of 16 letters from Jacob to his father describes the couple's travel to New Switzerland, his attempts at a career in farming, and life in the house (Weaver 2011; Indiana Historical Society Library 2018). An 1814 letter explains how the couple "remain living at my father-in-laws as yet, but have rented a piece of ground of my brother-in-law, Mr. Jay (Gex), [w]here I intend to move in a fortnight or three weeks" (Weaver 2011). The couple, however, did not make the move, perhaps because David Golay died, and the couple remained at his residence to help operate the family farm. In a letter dated 1828, 14 years following Weaver's initial attempt to move to Gex's property, he writes, "I will also inform you that I am preparing to leave the place where I now live and go on a small place I had wrote you before I had bought of Mr. Jex (Gex) Oboussier about one and a half miles [2.4 kilometers] from where I now live containing forty one acres [17 hectares] of land" (Weaver 2011).

The couple moved to a small, 18 × 38-foot, one and a half story building with a covered front porch (Figure 9.9). The lower story contains a single room with a fireplace. The second story, accessed from the porch, contains a storage space and a bedroom. The house is *poteaux-sur-sol* (post-on-sill) construction, a vernacular building method typical of French colonial architecture in North America (Figure 5.12; Figure 2.10; Kniffen and Glassie 1966). *Poteaux-sur-sol* buildings were constructed with vertical posts inserted and pegged into horizontal timber sills. The spaces between the posts are infilled with brick. Historians assumed the house was constructed sometime after 1805, the year the property was purchased (Weaver 1996), and was existing when Weaver first mentions moving out of his father-in-law's residence in 1814.

Tree-ring analysis allowed for the dating of 32 series from 24 timbers with 2,591 measured rings (1584–1882). The samples were obtained from seven timber types: ash, tulip poplar, American beech, walnut, hard maple, sweet gum, and white oak (Tables 9.8 and 9.9). Thirteen wane-bearing timbers identify 1828 as

Figure 9.9
Musée de Venoge (Craig Township, Switzerland County, Indiana). The house was constructed in 1828 and displays *poteaux-sur-sol* construction typical of French colonial architecture. Personal letters were used with tree-ring analysis to interpret the dwelling.
Source: Authors.

the year of construction. The remaining wane-bearing timbers indicate modifications to the building later in the 19th century. The timbers with 1828 harvest/death dates are mostly American beech (9 of 13), and the remaining are ash. As we discuss in Chapter 13, the Weaver's choice of American beech to construct their residence, while other settlers of the region tended to construct timber buildings of oak and tulip, suggests a cultural bias for the species. Our analysis shows that the house was newly constructed when the Weavers moved to the property in 1828 as described in the letter to Jacob's father. Interestingly, the 1828 date also documents French colonial construction influences well into the 19th century when the region's initial Swiss settlement had been overwhelmed with settlers from southern states who brought their own cultural bias for building forms and timber species.

Table 9.8 Samples collected from the Musée de Venoge *(Craig Township, Switzerland County, Indiana). See Table 9.1 for a description of the table contents; a full description of the data presented is provided in Chapter 7.*

Sample	*First*	*Last*	*Outer*	r	*Timber type*	*Provenience*
VEN01A	1714	1841	w	0.33	Ash	Beam; W addition
VEN01B	1735	1841	w	0.35	Ash	
VEN02A					Walnut	Beam; W addition
VEN03A	1740	1835	w		White oak	Beam; W addition
VEN04A					Ash	Loose timber; W addition
VEN05A	1708	1828	w	0.55	Ash	W sill; north member
VEN05B	1812	1828	w	0.70	Ash	
VEN06A	1734	1873	w		White oak	W sill; central member; appears recycled
VEN07A	1755	1793		0.79	Ash	W sill; south member
VEN07B	1714	1780		0.63	Ash	
VEN07C	1714	1780		0.62	Ash	
VEN08A	1802	1828	w	0.53	Beech	Stud; ground floor; W elevation
VEN09A	1763	1811		0.63	Beech	Floor joist; ground floor
VEN10A					White oak	Loose beam or joist
VEN11A	1734	1828	w	0.66	Beech	Stud; W elevation
VEN12A					Sweet gum	Scabbed to joist under window; E elevation
VEN13A	1801	1874	w		Tulip	Scabbed to 2nd floor joist
VEN14A					Beech	Stud E elevation; south of door
VEN15A	1730	1802		0.68	Ash	E sill
VEN17A	1647	1720		0.52	Tulip	Stud; 2nd floor
VEN18A	1705	1828	w	0.50	Beech	Floor joist; 2nd floor
VEN19A	1779	1883	w	0.50	Ash	Likely joist or beam; loose
VEN19B	1772	1883	w	0.56	Ash	
VEN33A	1699	1828	w	0.54	Beech	Floor joist; ground floor
VEN34A	1765	1828	w	0.81	Beech	Stud W elevation
VEN35A	1684	1828	w	0.72	Beech	Stud N elevation
VEN36A	1758	1828	w	0.67	Beech	Stud S elevation

(Continued)

Table 9.8 Continued

Sample	*First*	*Last*	*Outer*	r	*Timber type*	*Provenience*
VEN37A					Hard maple	Beam; 2nd floor; E elevation
VEN38A	1703	1828	w	0.56	Beech	Floor joist; 2nd floor
VEN39A					Hard maple	Floor joist; 2nd floor
VEN40A	1719	1768		0.61	Ash	S sill
VEN40B	1777	1828	w	0.42	Ash	
VEN40C	1728	1775		0.70	Ash	
VEN77A	1764	1818		0.45	Beech	Loose
VEN90A	1728	1828	w	0.59	Ash	Likely floorboard; loose under house
VEN94A					Walnut	Porch post
VEN95A	1608	1714		0.73	Tulip	Likely part of door jamb; 2nd floor
VEN96A	1582	1693		0.79	Tulip	Likely part of door jamb; 2nd floor
VEN97A	1744	1822		0.75	Beech	Stud; S elevation
VEN97B	1743	1828	w	0.83	Beech	

Table 9.9 Summary table of the timber type chronologies prepared from the Musée de Venoge *(Craig Township, Switzerland County, Indiana). See Table 9.2 for a description of table contents; a full description of the data presented is provided in Chapter 7.*

Timber type	*Samples*	*Timbers*	*Undated*	*First*	*Last*	*Rings*	r	*Mean*	*SD*
Ash	14	7	1	1709	1882	1038	0.53	1.19	0.45
Beech	12	11	1	1701	1827	977	0.63	1.08	0.50
Black walnut	0	0	2						
Hard maple	0	0	2						
Sweet gum	0	0	1						
Tulip poplar	3	3	0	1584	1719	286	0.70	1.05	0.47
	1	1	0	1803	1872	70			
White oak	2	2	1	1741	1872	220		1.01	0.29
Total	32	24	8	1584	1882	2591			

9.6 Tree-ring analysis of the George Ash House

The science of dating buildings using tree rings was largely established through the study of Native American pueblo settlements in the American southwest (Nash 1999 and included references). However, the dendrochronology of Native American timber structures in the Midwest and eastern regions of the United States is uncommon because surviving resources are rare. Tree-ring analysis of the (Cherokee) Chief John Ross House in Rossville, Georgia, established an 1816 build date, 19 years later than the 1797 date estimated by historians (DeWeese et al. 2012). A 43-year-long hickory and a 35-year-long maple chronology was developed from charcoal (burnt wood) excavated from an archaeological site near Fort Ouiatenon, an 18th-century French and Native American trading post in Tippecanoe County, Indiana (Strezewski and Rubino 2017).

The George Ash House is a two-story, brick, Federal Style I-House (Figure 9.10). Archival documents led historians to determine that it was constructed about 1798, making it the oldest house in Indiana (Historic Landmarks 2006). When Ash (c.1770–1850) was ten years of age, Shawnee Indians kidnapped him from his family farm near Bardstown, Kentucky. His story is accounted in several regional and national histories (Turner 1836; Frost 1854; Weakley 1885). Ash remained with the Shawnee tribe for 17 years, and he fought with Native Americans in battles against the United States. He left the tribe about 1797 and worked as a guide and interpreter. Delaware and Shawnee chiefs gifted him four square miles (1,036 hectares) of land in Indian territory for his services. Ash's gift was located immediately west of the Greenville Treaty Line that separated the territory of the United States to the east from Native American lands to the west. The treaty line was created by a 1795 agreement, the Treaty of Greenville, between the United States and native tribes, and ran from Greenville, Ohio, southwest to the Ohio River where the village of Lamb, Indiana, is now located.

Archival resources supporting construction dates for the house are unique. Published histories place Ash in Lamb sometime between 1797 and 1810, and the house's construction as early as 1798 (Turner 1836; Weakley, Harraman & Co. 1885; Historic Landmarks 2006). The most interesting archival source in documenting the construction of the house is Ash's application to the US Congress for ownership of the property gifted by the chiefs. In February of 1802 Ash asked the "United States to permit him to receive as a present" land owned

Figure 9.10
George Ash House (Lamb, Indiana). The house was constructed in 1808 on land gifted to Ash by Native American chiefs. Congressional records were used with tree-ring analysis to interpret the dwelling.
Source: Authors.

by Native Americans "across from the mouth of the Kentucky River" (American State Papers 1832). The congressional committee reviewing the application identified that the Treaty of Greenville only permitted Native Americans to sell or gift land to the US government, and that awarding land to Ash would set a precedent for similar claims by dishonest individuals (American State Papers 1832). Nevertheless, the committee proposed awarding the sale of one square mile (one township section, or 640 acres (259 hectares)) to Ash beginning at the treaty line and extending west along the Ohio River into Indian Territory. The phrase in the document "subsequent to the Treaty of Greenville" in the 1807 response to the application indicates Ash settled the property after the treaty's signing in 1795 (American State Papers 1832). "Including his improvement" in the 1807 response to the application, suggests the property might have contained the house, but improvements might also have meant Ash's river ferry operations or agricultural fields. Congress approved the sale in March of 1807 (American State Papers 1832).

Another interesting clue about the construction of the house was a 1922 newspaper article that interviewed Oliver and Nicholas Ash, George's descendants. The article reports an 1809 construction date, and that it is "probably the oldest brick house in Indiana" (Indianapolis News 1922). The article, again, sets a post-treaty construction date.

We were invited to analyze the house to see if dendrochronology could determine if the Ash House was the oldest brick building in the state. Tree-ring analysis produced eight series from five timbers totaling 785 measured rings ranging from 1613 to 1807 (Tables 9.10 and 9.11). The samples were obtained from three timber types: ash, tulip poplar, and butternut. Sampling focused on the floor joists and the roof (that has pegged rafters and joists). Our analysis established an 1808 build date indicating construction started after Congress' 1807 approval of the land sale to Ash, and not the 1790s gift by Native American chiefs. The 1922 newspaper reporting the Ash family's oral history differed from the results of analysis by only one year. The Ash House is an excellent example of how tree rings date the cutting of timber, not the construction of a building. If trees were cut in 1808 but before April 1809 the descendants are correct. While the Ash House is not the oldest brick residence in Indiana, it is close. It appears Grouseland, Governor Henry Harrison's home in Vincennes, Indiana, predates Ash's house by three years (see Chapter 8).

Table 9.10 Samples collected from the George Ash House (Lamb, Indiana). See Table 9.1 for a description of the table contents; a full description of the data presented is provided in Chapter 7.

Sample	*First*	*Last*	*Outer ring*	r	*Timber type*	*Provenience*
HUF01A	1618	1808	b	0.35	Ash	Floor joist; ground floor
HUF01B	1654	1808	b	0.38		
HUF02A	1623	1693			Tulip	Floor joist; ground floor
HUF02B	1612	1711				
HUF03A	1713	1808	w	0.41	Ash	Floor joist; ground floor
HUF03B	1731	1808	w	0.51		
HUF05A	1727	1805			Tulip	Floor joist; ground floor
HUF97A	1765	1800	w8		Butternut	Rafter

Table 9.11 Summary table of the timber type chronologies prepared from the George Ash House (Lamb, Indiana). See Table 9.2 for a description of table contents; a full description of the data presented is provided in Chapter 7.

Timber type	*Samples*	*Timbers*	*Undated*	*First*	*Last*	*Years*	r	*Mean*	*SD*
Ash	4	2	0	1619	1807	509	0.40	0.87	0.358
Butternut	1	1	0	1766	1799	34			
Tulip	2	1	0	1613	1710	166			
	1	1	0	1728	1803	76			
Total	8	5	0	1613	1807	785			

9.7 Conclusion

Chapter 9 reported the analysis of diverse cultural landscapes. It included African-American, Native American, and European American resources with fascinating histories. Dendrochronology was able to expand the knowledge and interpretation of these sites. We included examples of the challenges of preserving historic sites when results do not concur with the expectations of owners and managers. Chapter 10 reports the use of tree-ring analysis to document the evolution of landscapes over time.

9.8 Literature cited

American State Papers. 1832. *Documents, Legislative and Executive, of the Congress of the United States, Class VIII: Public Lands (1789 to 1815)*. Gales and Seaton, Washington, DC. 856 p.

Belle Grove Plantation. 2018. https://bellegrove.org/about/a_community_restoration_project. Accessed 25 September 2018.

Belson, J. 1999. *Letter Describing Sanders Cabin Archaeological Investigations at H38 (SER-OP)* October 8. Kentucky Heritage Council, Frankfort, KY.

Bennett, J.D. 1996. *National Historic Landmark Nomination: Eleutherian College Classroom and Chapel Building*. National Park Service, Washington, DC.

Clay Hill Forest. 2018. *Digital Historic Trail*. www.clayhillforest.org/signs/digitaltrail.html. Accessed 15 July 2018.

Department of Public Instruction. 1917. *State of Indiana: Twenty-Eighth Biennial Report of the State Superintendent of Public Instruction*. Fort Wayne Printing Company, Fort Wayne, IN. State of Indiana. 1,040 p.

Detroit Free Press. 1875. *Death of George DeBaptiste*. Detroit, MI. February 23.

Detroit Historical Society. 2018. *DeBaptiste, George*. https://detroithistorical.org/learn/encyclopedia-of-detroit/debaptiste-george. Accessed 11 October 2018.

DeWeese, G.G., W. Bishop, H. Grissino-Mayer, B.K. Parrish, and S.M. Edwards. 2012. Dendrochronological dating of the Chief John Ross House, Rossville, Georgia. *Southeastern Archaeology* 31: 221–230.

Dufour, J.J. 1826. *The American Vine-dresser's Guide: Being a Treatise on the Cultivation of the Vine and the Process of Winemaking*. S.J. Browne, Cincinnati, OH. 314 p.

Dumas, A.A., N. Mooney, V. Moore, and C. Sly. 2017. Cabins as far as the eyes can see: An introduction to the black belt slave housing survey. *Alabama Review* 70: 22–49.

Egan, D. 2003. Defining cultural and ethnographic landscapes. *Ecological Restoration* 21: 258–260.

Evans, M.J, A. Roberts, and P. Nelson. 2001. Ethnographic landscapes. *CRM* 24: 53–56.

Frost, J. 1854. *Thrilling Adventures among the Indians*. J.W. Bradley, Philadelphia, PA. 512 p.

Furnish, M.A. 2014. *A Rosetta Stone on Slavery's Doorstep: Eleutherian College and the Lost Antislavery History of Jefferson County, Indiana*. West Lafayette, PhD Dissertation. Purdue University.

Furnish, M.A. 2016. Black Hoosiers and the formation of an antislavery stronghold in the Central Ohio Valley. *Ohio Valley History* 16: 6–27.

Hardesty, D.L. 2000. Ethnographic landscapes: Turning nature into culture. In: *Preserving Cultural Landscapes in America*. (A.R. Alanen and R.Z. Melnick, eds.). The Johns Hopkins University Press, Baltimore, MD. 250 p.

Historic Landmarks. 2006. *Switzerland County Interim Report. Indiana Historic Sites and Structures Inventory*. Historic Landmarks Foundation of Indiana, Indianapolis, IN. 115 p.

Indiana Department of Natural Resources. 2018. *The Story of Georgetown District in Madison, Indiana*. www.in.gov/dnr/historic/files/georgetown.pdf. Accessed 11 October 2018.

Indiana Historical Bureau. 2018. *Eleutherian College*. www.in.gov/history/markers/5.htm. Accessed 11 October 2018.

Indiana Historical Society Library. 2018. www.indianahistorylibrary.worldcat.org/title/jacob-weaver-letters-1814–1848/oclc/19771674&referer=brief_results. Accessed 10 October 2018.

Indianapolis News. 1922. *Probably the Oldest Brick House in Indiana Situated at Lamb*. Indianapolis, IN. September 9.

Kentucky Heritage Council. 1996. *Clay Hill Slave House TA-98* (Letter to Mark Whitt of Campbellsburg University). Frankfort, KY.

Kentucky Heritage Council. 2015. *Kentucky Individual Building Survey Form: Clay Hill Cabin (KHC 2014–1)*. Frankfort, KY.

Kniffen, F., and H. Glassie. 1966. Building in wood in the Eastern United States: A time-place perspective. *Geographical Review* 56: 40–66.

Lewis, D.B., W.L. Whitney, H.D. Grissino-Mayer, E.R. Cook, and R.D. Jones. 2009. Dendrochronological dating of Eastern Red Cedar (*Juniperus virginiana* L.) logs from Alfred's cabin, the hermitage, home of President Andrew Jackson. *Tree-Ring Research* 65: 47–55.

Madison Courier. 2011. *DeBaptiste Home on Endangered Landmark List*. https://madison-courier.com/Content/News/News/Article/DeBaptiste-home-on-endangered-landmark-list/178/961/62794. Accessed 5 October 2018.

Madison Republic. 1846. *The Flood*. Madison, IN. September 9.

Montell, W.L., and M.L. Morse. 1976. *Kentucky Folk Architecture*. University Press of Kentucky, Lexington, KY. 105 p.

Monticello. 2018. *Landscapes of Slavery*. www.monticello.org/mulberry-row. Accessed 23 June 2018.

Mount Vernon. 2018. *Slave Quarters*. www.mountvernon.org/library/digitalhistory/digital-encyclopedia/article/slave-quarters/. Accessed 13 August 2018.

Narayanan, B. 2003. *Lyman and Asenath Hoyt House National Register of Historic Places Registration Form*. National Park Service, Washington, DC.

Nash, S.E. 1999. *Time, Trees, and Prehistory: Tree-Ring Dating and the Development of North American Archaeology, 1914–1950*. University of Utah Press, Salt Lake City, UT. 294 p.

National Park Service. 2018. *Georgetown Neighborhood*. www.nps.gov/nr/travel/madison/Georgetown_Neighborhood.html. Accessed 5 April 2018.

Newtown History Center. 2018. http://newtownhistorycenter.org/stone-house-restoration-project/. Accessed 11 October 2018.

Project Reach. 2018. http://reach.blackrivertech.org/?page_id=211. Accessed 11 October 2018.

Roberts, W.E. 1996. *The Log Buildings of Southern Indiana*. Trickster Press, Bloomington, IN. 228 p.

Schelbert, L. 1996. Swiss. Pp 592–613. In: *Peopling Indiana: An Ethnic Experience*. (R.M. Taylor and C. McBirney, eds.). Indiana Historical Society, Indianapolis, IN. 703 p.

Schlarb, E.J. 2014. *An Archaeological Assessment of the 40.4 HA Bryan/Dewitt Addition to the Clay Hill Memorial Forest, Taylor, County, Kentucky* (KAS Report No. 262). Kentucky Archaeological Survey, Lexington, KY.

Stachowiak, L.A., E.A. Schneider, M.L. Rochner, S.A. Collins, and H.D. Grissino-Mayer. 2016. Dendrochronological dating of historic log cabins on the Belle Meade Plantation, Davidson County, Tennessee, USA. *Vegetation History and Archaeobotany* 25: 105–115.

Strezewski, M., and D. Rubino. 2017. *Dendrochronological Dating of a Burned Native American Structure at Fort Ouiatenon, Indiana*. Presented at The 81st Annual Meeting of the Society for American Archaeology. Vancouver, British Columbia, Canada.

Thompson, W.C. 1923. Eleutherian institute: A sketch of a unique step in the educational history of Indiana. *Indiana Magazine of History* 19: 109–131.

Turner, G. 1836. *Traits of Indian Character: Volume 1*. Key & Biddle, Philadelphia, PA. 207 p.

Weakley, Harraman & Co. 1885. *History of Dearborn, Ohio, and Switzerland Counties, Indiana*. Weakley, Harraman & Co., Chicago, IL. 1,386 p.

Weaver, D. 2011. *The Letters of Jacob Weaver: Vevay and Jacksonville 1813–1848*. Musée de Venoge, Vevay, IN. 49 p.

Weaver, T.R. 1996. *Venoge Farmstead. National Register of Historic Places Registration Form*. National Park Service, Washington, DC.

Williamsburg. 2018. *Slave Quarters at Carter's Grove*. www.history.org/Almanack/places/hb/hbslave.cfm. Accessed 5 October 2018.

Woollen, W.W. 1883. *Biographical and Historical Sketches of Early Indiana*. Hammond & Co., Indianapolis, IN. 568 p.

Chapter 10

Chronicling landscape evolution using tree-ring analysis

10.1 Introduction

In Chapter 10 we demonstrate how we use dendrochronology to chronicle the evolution of two landscapes over time (Figure 10.1). Technologic, economic, and political changes influence a site's purpose and use. Most all historic places have changed in some way through modifications to meet the evolving needs of their owners. Features were regularly altered, added, and removed. Land use and architectural spaces were often redefined in dramatically different ways. Obsolete landscapes and buildings risk abandonment and removal.

The first landscape we examine is our search for evidence of the late 18th- and early 19th-century Fort Wayne military outpost. Folklore says that timbers from the fortification are in the Dr. Merchant Huxford House, a residence constructed decades after the military abandonment of the region following the War of 1812. The second landscape is a rural farmstead established by Scottish immigrants. We demonstrate how tree-ring analysis can identify the construction dates for several domestic and agricultural buildings, and document a farmstead's evolution over time.

10.2 The fort in the doctor's house

Fort Wayne, Indiana, folklore claims the Dr. Merchant Huxford House contains timbers from the military outpost constructed under the command of General Anthony Wayne in 1794. Built on the outskirts of the young town, and within a quarter mile (0.4 kilometers) of the fort site, the doctor's house was believed to have been built in the 1850s. We had been seeking to expand the geographic range of our chronologies in northern Indiana and had contacted a preservation colleague regarding potential historic sites to analyze. She knew of the stories surrounding the house and thought it might be a great candidate for study. In our preliminary research we found that accounts contemporary to the house's construction describe how timbers from the decaying fort were recycled into new

Figure 10.1
Locations of Fort Wayne and the Robert Glenn Farmstead.
Source: Authors.

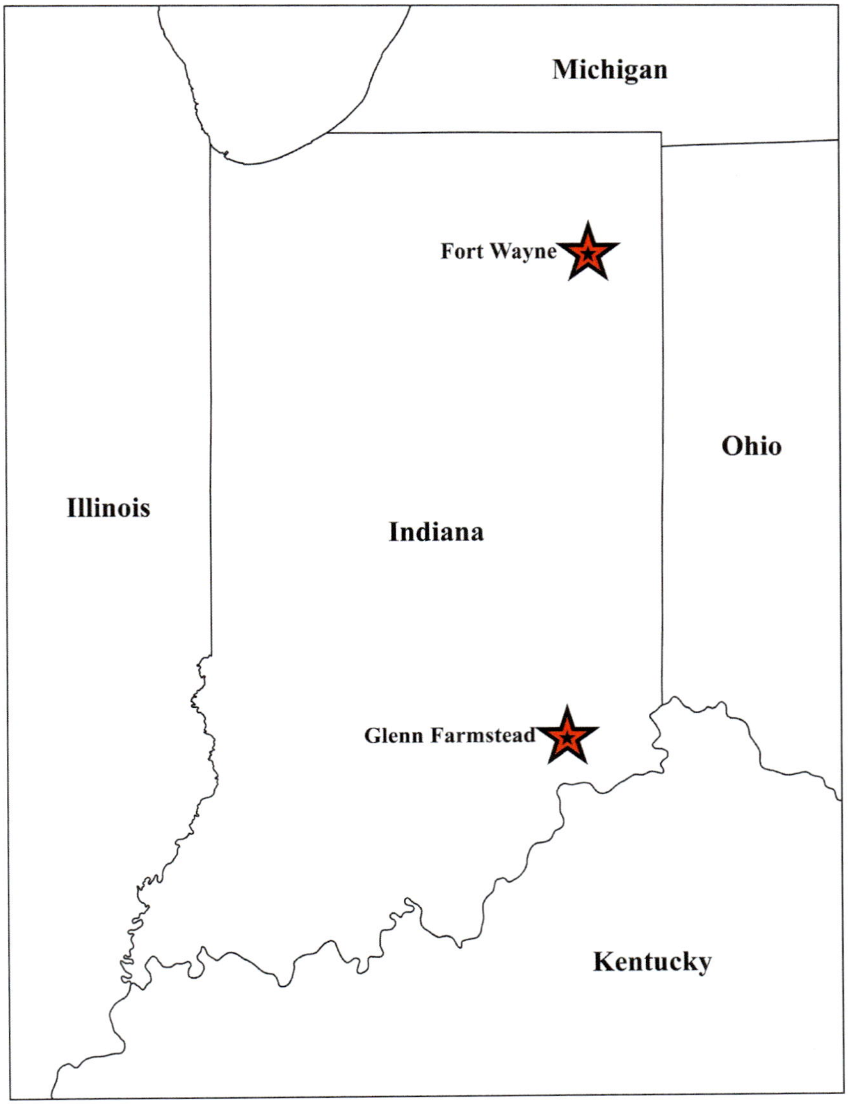

construction projects. We have not discovered how the lore became associated with the Huxford House, but stories are routinely reported in local newspapers and histories that the house was constructed about 1854 and that it might contain timbers from the historic garrison (News-Sentinel 2014; Kilbane 2016; ARCH 2017; Fort Wayne Community Development 2017). We approached tree-ring analysis with an effort to support or dispel these accounts by establishing construction dates for the doctor's residence. We also sought to identify and date timbers contemporary with Fort Wayne's 25-year military era of 1794 to 1819 (Baas et al. 2019).

10.3 Fort Wayne archival and secondary sources

We found a variety of unique sources to help us interpret the results of our analysis. Multiple 19th- and 20th-century scholarly histories have been written about the construction of the defensive fort and of the development of the city of Fort Wayne (Brice 1868; Brant and Fuller 1889; Griswold 1917; Poinsatte 1969; Woehrmann 1971). They describe how the outpost was located at the strategic confluence of the St. Joseph, St. Marys, and Maumee Rivers in northeast Indiana (Figure 10.2). In the late 18th century, North America's dense forests limited foot travel to established wilderness trails and the movement of goods to major streams. The ten-mile (16-kilometer) portage linking the St. Marys and Wabash Rivers made water travel possible from Quebec to New Orleans. Whoever controlled the confluence controlled the movement of a significant portion of goods from the North American interior to Europe (Woehrmann 1971).

The confluence was settled about 1700 by the Miami Tribe who established the village of Kekionga. As part of New France, Fort Miamis was constructed in 1715, was rebuilt in 1749, and lasted until 1763 when it was destroyed during Pontiac's Rebellion, a Native American uprising against the British following the French and Indian War (also considered part of the geographically broader Seven Years' War; Woehrmann 1971). Following the American Revolution, the regional tribes maintained allegiance to Britain. In the resulting Northwest Indian Wars, the United States sought to gain control of the Northwest Territory and to end British influence. Therefore, in the 1790s, the United States built a series of forts about a day's march apart throughout river systems south of the Great Lakes in a network designed to efficiently move and protect soldiers and supplies (Griswold 1917).

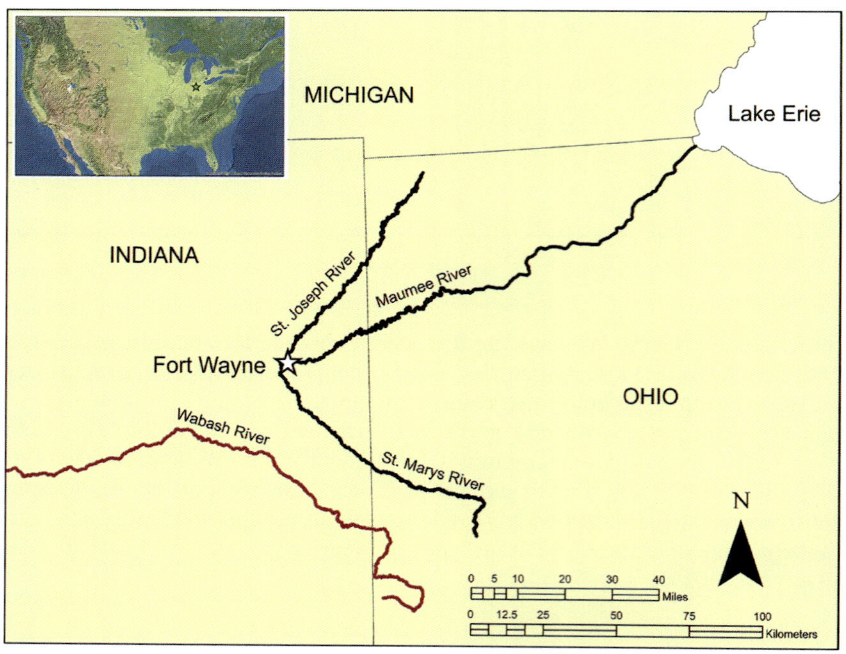

Figure 10.2
Fort Wayne was constructed at the confluence of the St. Joseph, St. Marys, and Maumee Rivers. The ten-mile portage between the St. Marys and Wabash Rivers allowed water travel from Lake Erie to the Gulf of Mexico.

Source: Authors.

General Anthony Wayne (1745–1796), appointed by President George Washington to lead the Western Army and conduct military campaigns against native tribes, constructed Fort Wayne in the fall of 1794. The fort was located on the high ground adjacent to the confluence of the three rivers and consisted of two timber blockhouses and a square timber palisade (Figure 10.3). The fort established a US military presence in the region and served as an outpost to Wayne's headquarters in Greenville, Ohio, 70 miles (113 kilometers) southeast. In 1800, Colonel John Hunt constructed a new and more substantial fort approximately 300 feet (91 meters) from the original defenses. Hunt's fort played a minor role in the War of 1812 when Native Americans laid siege in the fall of that year, and the fort's inhabitants required rescue (Woehrmann 1971). In 1815, the fort's commander Major John Whistler began a reconstruction of Hunt's fort. Whistler reconstructed Hunt's 1800 fort in the same location with new palisades, blockhouses, and support buildings. This final version of Fort Wayne was abandoned by the military in 1819, but it continued as an Indian Agency until 1827. Sections of the garrison were destroyed by the construction of the Wabash and Erie Canal in the 1830s, and the fort was completely removed in 1852 (Griswold 1917).

We learned from local histories that Dr. Merchant Huxford (1798–1878) was a native of Massachusetts who came to Fort Wayne in 1833. A trained pharmacist, he operated a drugstore and practiced medicine in the growing town. Huxford served as mayor from 1846–1849. The doctor owned a large tract of rural land between Spy Run and the St. Joseph River on the city's northern boundary where he built a substantial residence (Figure 10.4). The Greek Revival style, two-story house was constructed with a brick exterior and an

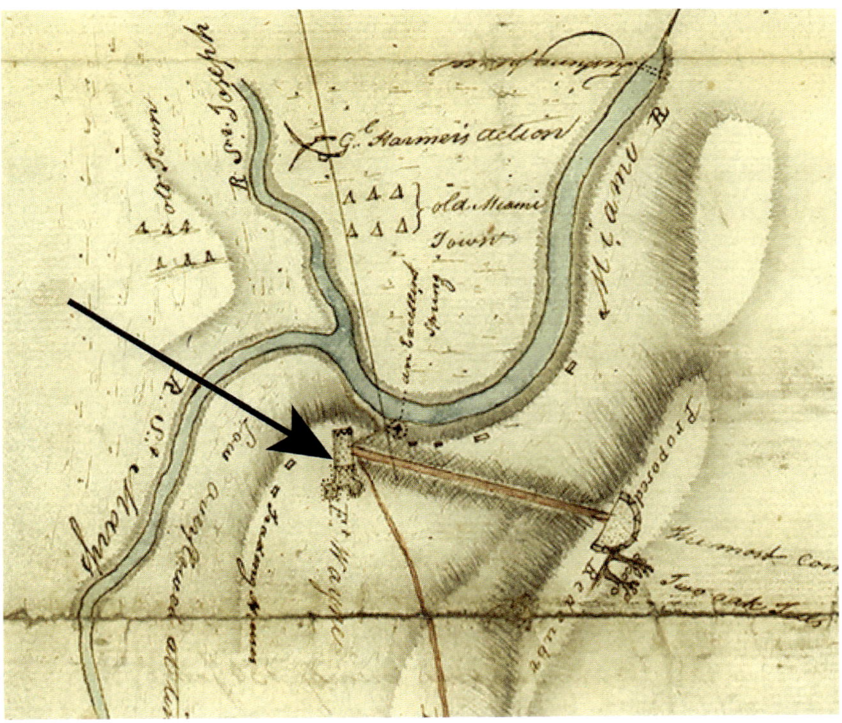

Figure 10.3
1795 illustration of the first version of Fort Wayne showing a timber structure on the high ground adjacent to the confluence. The arrow points to the timber structure.

Source: Neff, P. M. Indiana Fort Wayne: ms. map of Fort Wayne said to have been made on July 18, 1795, for General Anthony Wayne. 1795. Map. Retrieved from the Library of Congress, <www.loc.gov/item/2007629021/>.

interior structure of hewn timber beams and saw-cut floor joists and wall studs (Griswold 1917; Fort Wayne Community Development 2017). Over time, the city expanded beyond the house, and the residence became part of a gridded, late 19th-century urban neighborhood. In the late 20th century the house was divided into apartments, and in the 1990s the roof was completely replaced (Hall 1993). More recently, the house was purchased by a local preservation organization and is being restored for use as their office headquarters. Their rehabilitation of the house made tree-ring analysis possible by exposing interior timber beams, joists, and studs.

10.4 The archaeology of North American forts

We were curious if there were archaeological investigations of other North American forts that might guide our research, and if there were ever investigations of the fortifications in Fort Wayne. We found that the several texts reporting investigations were helpful for a broad understanding of fort locations and design, presenting case studies of archaeological investigations, and interpreting the changes in warfare and defenses over time (Whittaker 2009; Starbuck 2008, 2011; Burns et al. 2010; Babits and Gandulla 2013). Babson is effective in explaining how the blockhouse, a feature common to North American forts and all three versions of Fort Wayne, had "an overhang and monitored roof, which provided a watch or lookout, and the curtain made up of the buildings, protected by picketing, became after the Revolution the stock plan for the forts built by the United States on its western frontier" (Babson 1968).

The dense forests of eastern North America provided an abundant and accessible supply of timber. Therefore, 18th- and 19th-century French and Indian War–vintage forts and trading posts were nearly always wood, and they included both a blockhouse of horizontal timbers to protect the fort's inhabitants and a

Figure 10.4
West façade of the Dr. Merchant Huxford House.
Source: Authors.

picket wall made of vertical timber posts. French and American forts in the West were usually similar in layout, constructed of locally harvested green timber, and had a limited lifespan. The forts,

> were built on the same general plan, because of the most convenient material at hand [was wood], with lines of high pointed and contiguous pickets, and blockhouses at the corners, affording a means of defense generally effective against the Indian mode of warfare. Some were much more elaborate than others. The simplest had but one blockhouse surrounded with pickets; others were square stockades with block-houses at diagonal corners, and the most pretentious had blockhouses at all four corners. Constructed of green timbers cut in adjoining forests, the extreme durability of those defenses probably did not exceed 30 years.
>
> (Snyder 1913)

We found that the archaeological investigations of forts similar to Fort Wayne were often limited in scope, and did not provide our project much guidance. DeRegnaucourt and Hoelscher (1996) describe an investigation of Fort St. Marys constructed by Wayne in 1795 and, like the second Fort Wayne, was rebuilt with a stockade and two blockhouses in 1813 as a supply and defensive post for the War of 1812. The National Park Service describes the series of Wabash River forts constructed by French, British, and US armies between 1732 and 1813 at Vincennes, Indiana. We did learn that they believe the timbers from Fort Knox II, built in 1803 and abandoned in 1816, were likely scavenged by the residents of the town for new construction, and that current Vincennes structures might contain timbers from the garrison (National Park Service 2006). The only tree-ring analysis of North American forts available to us discovered structures, rumored to be part of Fort Armstrong, were not remains of the 1813 garrison constructed in central Alabama (Therrell et al. 2017).

Archaeological investigations of the city of Fort Wayne's multiple fort sites is challenged by two centuries of urban growth, changes in land use over time, and levee and floodwall construction. An in-depth investigation excavated General Wayne's presumed 1794 fort site (Andres 2008). While the excavation produced late 18th- and 19th-century artifacts, none could be directly tied to the garrison, and the remains of the fort's timber structures were not located. Unfortunately, no archaeological evidence of any of the fort's layout or construction materials that might add to our research exists at this time.

While these investigations were beneficial in describing the general design of defensive structures, and to a limited extent the type and use of building materials, they do not discuss construction techniques (e.g., timber hewing and notching), species identification of the timber used in 17th- and 18th-century frontier forts, or the application of timber analysis such as dendrochronology and dendroarchaeology.

10.5 Unique sources for documenting Fort Wayne

Our research of Fort Wayne and Dr. Huxford's house was aided by two unique historical documents. The first is Major John Whistler's measured plan drawing of the 1815 version of the fort showing the layout of the garrison's buildings and pickets and a surrounding landscape of buildings, soldier's gardens, river landing, and the US Indian Agency (Figure 10.5). The drawing is archived in the Allen County Public Library Digital Collections (2017). The second document is a mid-19th-century daguerreotype of what is reported to be an image of the last remaining building of the military outpost (Figure 10.6). It is archived at the Allen County Historical Society. We will explain how these sources fit into our research later in the chapter.

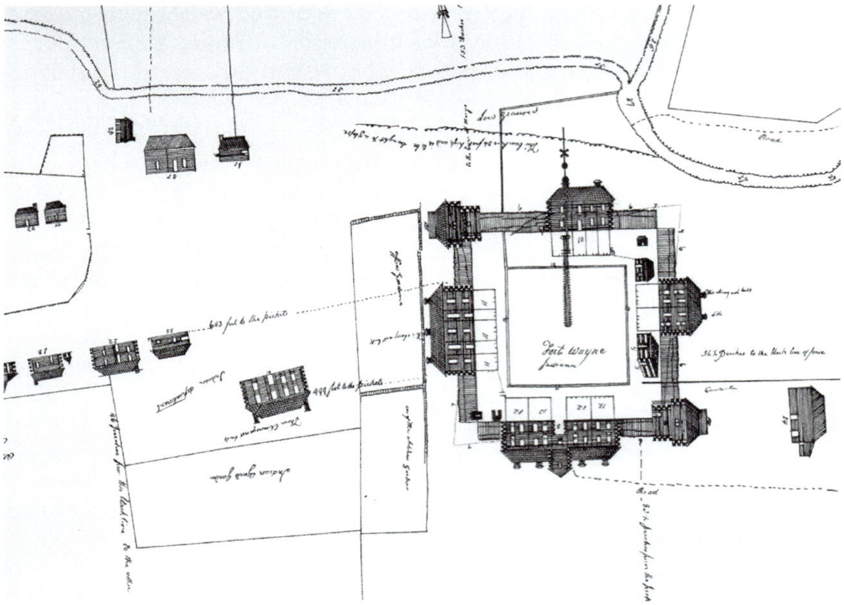

Figure 10.5
Major John Whistler's 1815 plan drawing of Fort Wayne. North is oriented to the top of the illustration.

Source: Allen County Public Library Digital Library Collection (http://contentdm.acpl.lib.in.us/digital/collection/p16089coll56/id/1140/).

Figure 10.6
1852 daguerreotype of the last structure historians have associated with Fort Wayne.

Source: Charles C. Stevens created the image (Allen County-Fort Wayne Historical Society).

10.6 Dendrochronology results

Since our goal was to determine the construction date of the Dr. Merchant Huxford House, we focused sampling on all bark- and wane-bearing timbers – timbers that, if accurately crossdated, would provide harvest or tree death dates. Sampling was performed in all stories of the house so that we would minimize the risk of only sampling renovations or repairs (Figure 10.7).

We successfully crossdated 34 samples from 28 timbers (Tables 10.1 and 10.2). White oak was the most common timber type (28 samples from 22 timbers), and we created a white oak chronology of 2,291 rings spanning from 1546 to 1850 (Table 10.2).

Tree-ring analysis identified a construction period for the Dr. Merchant Huxford House ranging from the late 1840s to the mid-1850s. Analysis also identified three timbers having an 1815 harvest or cutting date – the year of Whistler's reconstruction of the fort. These timbers were found in the house's cellar, and in timber stockpiled from the house's roof replacement. From these conclusions, we asked: Where could timbers with death dates from 1815 be obtained in the city of Fort Wayne in the mid-1850s? To answer this question, we sought to discover what was happening to the fort while the garrison was still active in the 1810s and while the young city was expanding in the 1850s.

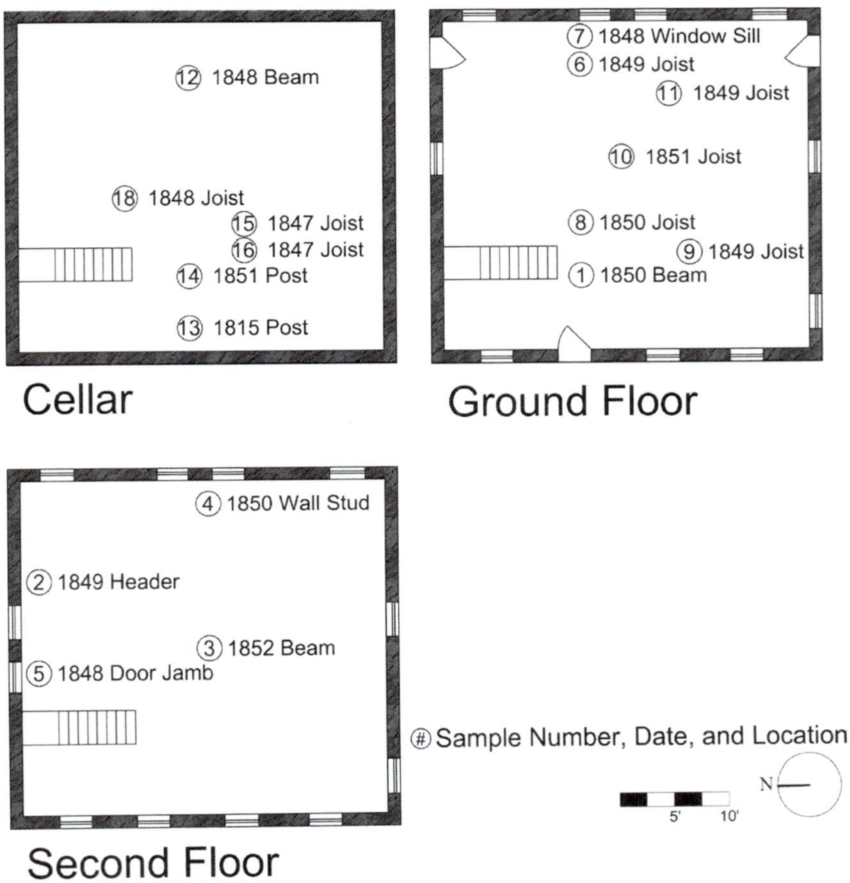

Figure 10.7
Dr. Merchant Huxford House plan showing sample numbers and locations. See Table 10.1 for a provenience of each sample.
Source: Authors.

Table 10.1 Samples collected from the Huxford House (Fort Wayne, Indiana). "First" and "last" refer to the first and last rings present in the sample. If no date is provided, the sample could not be reliably crossdated. "Outer ring" refers to the outermost ring of the sample: b = bark, w = wane, etc.; see Table 7.1 for a complete list of codes used for outer ring condition. "r" is the correlation coefficient obtained by correlating the ring widths of this sample with all other timbers of the same type from the building (series intercorrelation; Chapter 5). "Timber type" is the type of timber; see Table 3.1 for a list of possible tree species represented by the various timber types. "Provenience" is the timber's function in a structure; see Chapter 2 for a description of the various proveniences. Note: Timber type and provenience are only given once if more than one sample was obtained from an individual timber. An r-value is not reported if less than three timbers of a particular type were obtained from a structure. A full description of the data presented in the table is provided in Chapter 7.

Sample	*First*	*Last*	*Outer ring*	r	*Timber type*	*Provenience*
HUX01A	1733	1850	w	0.61	White oak	Beam; ground and 2nd floor
HUX01B	1789	1850	w	0.34		
HUX02A	1786	1849	w		Tulip	Header; 2nd floor wall
HUX03A	1780	1852	w	0.61	White oak	Beam; 2nd floor and attic
HUX04A	1776	1850	w	0.48	White oak	Stud; 2nd floor
HUX05A	1775	1848	w		Hard maple	Door jamb; 2nd floor
HUX06A	1797	1849	w		Beech	Floor joist; 2nd floor
HUX07A	1784	1848	w	0.30	White oak	Window sill; ground floor
HUX08A	1794	1850	w		Beech	Floor joist; 2nd floor
HUX09A	1780	1849	w	0.50	White oak	Floor joist; 2nd floor
HUX10A	1830	1851	w	0.70	White oak	Floor joist; 2nd floor
HUX11A	1791	1849	w		Red oak	Floor joist; 2nd floor
HUX12A	1742	1815		0.68	White oak	Beam; ground floor and cellar
HUX12B	1803	1848	w	0.72		
HUX13A	1668	1815	w	0.39	White oak	Post; cellar
HUX14A	1707	1835		0.36	White oak	Post; cellar
HUX14B	1757	1833		0.44		
HUX14C	1834	1851	w	0.72		
HUX15A	1817	1847	w	0.71	White oak	Floor joist; ground floor
HUX16A	1820	1847	w	0.60	White oak	Floor joist; ground floor
HUX17A	1725	1828		0.43	White oak	Floor joist; 2nd floor
HUX18A	1729	1848	w	0.59	White oak	Floor joist; 2nd floor
HUX25A	1544	1793		0.36	White oak	Loose 2nd floor; most likely joist; sawn
HUX50A	1791	1849	w	0.62	White oak	Loose 2nd floor; most likely joist; sawn
HUX51A	1740	1834		0.63	White oak	Loose 2nd floor; most likely joist; sawn
HUX52A	1683	1794		0.63	White oak	Loose 2nd floor; most likely joist; sawn
HUX53A	1740	1805		0.71	White oak	Loose 2nd floor; most likely joist; sawn
HUX55A	1740	1802		0.72	White oak	Loose 2nd floor; most likely joist; sawn

Sample	*First*	*Last*	*Outer ring*	r	*Timber type*	*Provenience*
HUX56A	1707	1790		0.64	White oak	Loose 2nd floor; most likely joist; sawn
HUX70A					Tulip	Loose 2nd floor; baseboard
HUX71A					Tulip	Loose 2nd floor; tongue and groove flooring
HUX72A					Tulip	Loose 2nd floor; plank
HUX73A					Tulip	Loose 2nd floor; plank
HUX74A					Tulip	Loose 2nd floor; plank
HUX75A					Tulip	Loose 2nd floor; plank
HUX88A	1689	1815	w	0.40	White oak	Loose 2nd floor; most likely post; hewn
HUX89A	1705	1814		0.65	White oak	Loose 2nd floor; most likely post; hewn
HUX89B	1771	1815		0.53		
HUX89C	1704	1815	w	0.52		
HUX96A					Bass	Loose 2nd floor; most likely roof sheathing
HUX97A					Sycamore	Loose 2nd floor; plank
HUX98A					Walnut	Loose 2nd floor; most likely brace
HUX99A	1770	1850	w		Sycamore	Loose 2nd floor; most likely roof sheathing

Table 10.2 Summary table of the timber type chronologies prepared from Huxford House (Fort Wayne, Indiana). "Series" and "Timbers" are the number of successfully crossdated series and timbers from the building. "Undated" is the number of timbers that could not be crossdated reliably. "First" and "last" refer to the first and last dates in the chronology. "Mean" and "SD" are the mean and standard deviation (mm) of the ring widths in the chronology. "r" is the mean series intercorrelation for each timber type (see Chapter 5). r is not provided if less than three series were obtained for a particular timber type. No total r, mean, or SD are provided since each chronology was prepared using only a single timber type. A full description of the data presented in the table is provided in Chapter 7.

Timber type	*Samples*	*Timbers*	*Undated*	*First*	*Last*	*Years*	r	*Mean*	*SD*
Basswood	0	0	1						
Beech	2	2	0	1795	1848	107		1.10	0.62
Hard maple	1	1	0	1776	1847	72		1.44	0.53
Red oak	1	1	0	1793	1848	56		1.39	0.30
Sycamore	1	1	1	1771	1849	79		0.76	0.42
Tulip poplar	1	1	6	1787	1848	62		1.40	0.60
Walnut	0	0	1						
White oak	28	22	0	1546	1850	2291	0.53	1.17	0.36
Total	34	28	9	1546	1850	2667			

Following the end of the War of 1812, Major Whistler reconstructed the fort in 1815–16. Whistler constructed six hewn timber buildings, including two blockhouses two stories in height. The timber was harvested over the autumn and winter months from a forest three miles (4.8 kilometers) from the fort site. The fort was complete by autumn of 1816, indicating they were promptly placed green, or unseasoned, into the new structure (Woehrmann 1971).

Whistler's plan drawing recorded the new fort's layout and surroundings with simple illustrations of the front façade of each of the fort's structures (Figure 10.5). The buildings from this third version of the fort were about 35 feet (10.7 meters) high and constructed of hewn timbers. The fort "enclosed an area about 150 feet (46 meters) square in pickets ten feet [3 meters] high, and set in the ground, with a block house at the southeast and northwest corners, two stories high. The [block house's] second floor projected and formed a bastion in each where the guns were rigged" to cover the exterior of all four sides of the enclosure. Additional two-story buildings helped form the walls. Their shed roofs angled toward the fort's interior to prevent them from being set on fire, and to protect the soldiers extinguishing the flames if they were (Griswold 1917).

An 1858 *Fort Wayne Times* newspaper article explains that much of the "season of 1815 was spent rebuilding the fort; and when completed, as with the first fort erected in 1794, was a most substantial affair. The timber from which it had been built, was obtained principally from what is now the east end" of the city. The timber pickets were 12.5 feet (3.8 meters) in length, "in sets, with cross pieces, two feet [0.6 meters] from the top, let in and spiked, and a trench dug, two and a half feet [0.8 m] deep, into which they were raised" resulting in a wall 10 feet (3 m) in height (Brice 1868). The area around the fort had been ordered cleared of timber during the War of 1812 to provide clear lines of site to any approaching army. In the area surrounding the fort there were soldier's gardens, a cemetery, roads, and a boat landing on the river (Brice 1868; Griswold 1917).

Whistler's 1816 drawing accounts for 14 timber houses and small buildings around the fort. Captain James Riley, an army surveyor preparing the fort's land for sale in 1819, mentions the presence of about 30 houses sheltering French and American families that would have been built after the 1812 siege when all structures outside the fort were burned to eliminate cover for attacking armies (Brice 1868). Therefore, it appears that the presence of timber construction dating to 1815 include Whistler's fort, several private houses, and structures connected with the Indian Agency.

Following the abandonment of the fort in 1819 and the close of Indian Agency activities in 1827, the fort slowly vanished. The blockhouse nearest the river was removed for the construction of the Wabash and Erie Canal (Griswold 1917). By 1838 only one building was still standing, and most all the pickets had been removed. While there were efforts to save the fort's remaining structure, it was torn down in 1852 because it harbored "undesirables" (Griswold 1917; Woehrmann 1971).

We wondered if there was evidence of the fort being used in local buildings, and we were able to locate several accounts. Historian Bert Griswold quotes Fort Wayne resident John W. Dawson's 1838 account that, except for one structure, the fort "had all been removed by the people for building purposes" (Griswold 1917). Logs from a house constructed in 1815 "within view of the fort" were used in Washington Hall, an early tavern and meeting house constructed about 1822 (Brice 1868). Timbers from the fort were reportedly used to construct the bell tower of First Presbyterian Church in 1845 (Fort Wayne Daily Sentinel 1882). Unfortunately, these buildings no longer exist to provide tree-ring samples for analysis.

We found that timber species and construction techniques distinguish the fort-era timbers from the rest of the house. Timber species used in the house are primarily white oak, but include red oak, sycamore, beech, tulip poplar, and hard maple. In January of 1816, Major Whistler informed the War Department that he was "constructing six buildings of barkless white oak" (Woehrmann 1971). All three samples cotemporary with the fort are white oak. The house's timbers are both hewn and sawn, but the hewn timbers we found to be contemporary with the fort are distinct from those of the house. The three white oak timbers (HUX13, HUX88, HUX89) contemporary to the fort are from split logs (i.e., half a tree) as opposed to the white oak beams (HUX01, HUX03, HUX12) hewn in the 1840s and 1850s, which are made from whole logs that were squared (Figure 10.8).

Tree-ring analysis demonstrated that timber death dates for the construction of the doctor's house have a five-year range of 1847 to 1852. We also know that fort-era timbers are located in the cellar and the attic, and that samples throughout the house are both hewn and saw cut. The construction of the house would have proceeded in phases: brick exterior, interior first floor, interior second floor, attic floor, and roof. We found that timber death dates parallel the progress of construction as you move upward through the structure. The first floor beam and joist platform dates to 1847–48 (HUX12, 15, 16, 18). The second floor beam and joist platform dates to 1849–52 (HUX06, 08, 09, 10, 11). One attic floor sample dates to 1852 (HUX03). Two roof timbers stockpiled in the house and a support post in the cellar date to 1815 (HUX13, 88, 89).

Based on our analysis, we believe the construction of the house had started by 1847, that the first floor was completed around 1848, and that the second floor was completed around 1852. In addition, about 1852, fort-era timbers were used to construct the roof and were installed in the cellar as additional structural support. By the time the interior was completed (plaster, windows, etc.), the finished construction date of 1853 or 1854 suggested in local literature appears feasible.

Dendrochronology cannot conclusively confirm that the timbers from Fort Wayne are in the Dr. Merchant Huxford's House. However, dendrochronology presents evidence that it is plausible. Tree-ring data identify timbers contemporary with the fort. Accounts of the reconstruction of the fort describe the use of oak, the species of fort-era timbers found in the Huxford House. The Huxford House is located close to the fort, and the expense, convenience, and use of recycled timbers was probably attractive to the house's builders. Tree-ring studies have demonstrated that the use of recycled timbers from past structures in new buildings was common (Grissino-Mayer and van de Gevel 2007; Baas et al. 2017; Thiebaud Hay Press Barn in Chapter 12). The timbers dating to the fort era are from split logs, and the timbers dating to the construction of the house are from hewn whole logs or commercially saw-cut dimensional lumber, indicating clear differences in preparation methods. There are reports of timbers from Whistler's Fort Wayne being recycled into new buildings. Tree-ring data identified a Huxford House construction date contemporary with descriptions of the 1852 demolition of the fort.

The accounts we found describe how the last remaining structure of the fort was removed in 1852, and tree-ring analysis shows that the timber construction

Figure 10.8
A comparison of split log and whole log hewn timbers. The top timber (HUX88) has an 1815 death date and is hewn from half, or a split, oak log. For comparison, the bottom timber is hewn from a whole log (the bottom timber is from a log house in Jefferson County, Indiana).

Source: Authors.

of the Huxford House's upper stories was active in 1852 thus supporting the local lore that timbers from the fort made it into the house. We reviewed Charles C. Stevens' 1852 daguerreotype (Figure 10.6) of what is reported to be one of the fort's blockhouses. Stevens convinced friends to pose on the remains of the building's walls and the loose timbers scattered on the ground around the structure. The building in the image is two stories in height and constructed of hewn timbers with V-notch joinery. The building has a central entry and hallway that is flanked by rooms, and suggests the vernacular I-house form (a two-story structure with a central passage flanked by two rooms). Central door openings are represented on the front façade's ground and second stories. Whistler's 1815 drawing of the fort depicts the garrison's structures, and the surrounding buildings beyond the fort in a surprising level of detail (Figure 10.5). For instance, the drawing illustrates the use of horizontal timber construction, and the locations for doors and windows. We could not identify a structure representative of the defensive portions of the fort – blockhouses, barracks, or other support building – like the building in Steven's daguerreotype. Rather, the building best represents one of the structures located near the fort's Indian Agency. Whistler shows a two-story log structure with central entry door openings on both the first and second stories and single windows in the rooms flanking the hall (Figure 10.9). It is clearly not the council house, but more likely a residence – perhaps for the US agent responsible for the department's operations. Therefore, we suggest that if timbers from the historic fort are in Huxford's house, the map and photographic sources tend to support that the timbers are from one of the garrison's support buildings associated with the Indian Agency.

10.7 Huxford House conclusions

Tree-ring analysis suggests construction of the Dr. Merchant Huxford House started about 1847 and finished about 1853. We cannot dismiss the possibility that the timber interior of the house was constructed all at once from a stockpile of

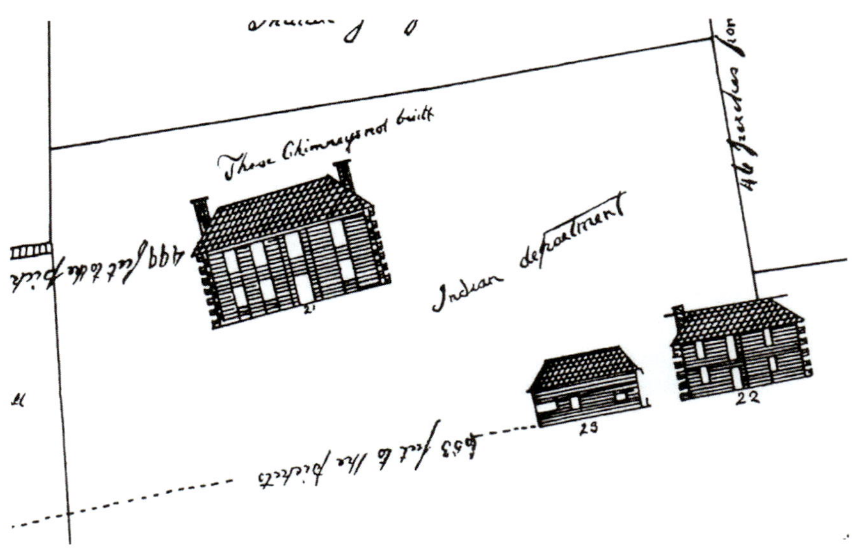

Figure 10.9
Detail from Major John Whistler's 1815 plan drawing of Fort Wayne (Figure 10.5). Building No. 22 best represents the structure in Charles Stevens' 1852 daguerreotype shown in Figure 10.6.

Source: Allen County Public Library Digital Library Collection (http://contentdm.acpl.lib.in.us/digital/collection/p16089coll56/id/1140/).

wood building materials (such as a lumberyard). In this scenario, the completion date is still about 1853. The house contains at least three hewn timbers with 1815 death dates. These timbers are prepared significantly differently from the hewn timbers dating to the 1840s and 1850s in that they are from split logs rather than whole logs. Written, illustrative, and photographic evidence suggests that timbers with 1815 death dates available as building materials in the city of Fort Wayne about 1850 was the structure in Charles C. Stevens' daguerreotype. We believe this structure is part of the fort's Indian Agency located outside the garrison's defensive pickets, and not part of one of the fort's blockhouses as had been reported. Therefore, tree-ring analysis supports, but does not prove, that the doctor constructed his house using timbers from the old fort. A decisive connection, likely something written in an undiscovered military report, personal journal, history, or newspaper account is needed to directly connect Huxford to the fort. Until then, tree rings are our best line of evidence that the fort is in the doctor's house.

10.8 Dendrochronological analysis of a 19th-century farmstead: introduction

When we were invited to analyze a 19th-century barn at the Robert Glenn Farmstead in rural Jefferson County, Indiana, we quickly realized the potential of sampling all the farmstead's buildings. Farmstead landscapes evolve over time in response to economic and technological changes and a farm family's finances and needs. We wondered if tree-ring analysis could provide construction dates for the buildings that, with archival sources, would demonstrate these changes from a post-frontier Indiana to a modern, apparently successful, late 19th-century farmstead.

We approached the Glenn Farmstead as a "rural historic landscape" – a cluster of buildings purposefully arranged to facilitate the efficient operations of a farm. Rural historic landscapes include buildings for storing and processing crops, raising farm animals, and housing the family. Rural historic landscapes often have "developed and evolved in response to both the forces of nature and the pragmatic need to make a living" (McClelland 1999).

10.9 Archival and scholarly sources for the Glenn Farmstead

We found that archival and secondary scholarly sources for the Glenn Farmstead were limited to local histories and the record of the family in both census data and a 20th-century plat map. Robert Glenn (1799–1881) was a Scottish immigrant who came to the Indiana Territory in 1815 and purchased property in Shelby Township, Jefferson County, in 1825 (Knox 1940; Scott 2018). Glenn and his family settled among fellow Scottish immigrants in an area known as New Caledonia, a rural enclave centered on the Caledonia Presbyterian Church and Cemetery (Danner 1945). The domestic and agricultural buildings we analyzed are located on a knoll along Tate Ridge Road. Historically, the property was wooded, but over time it was cleared for pasture, and the production of grasses and grains. From the US Agricultural Census, we know the farmstead was 240 acres (97 hectares) in size. By 1880, the Agricultural Census reports that only 40 acres (16 hectares) remained wooded, while 100 acres (40 hectares) were tilled, and 100 acres were meadow. It appears the family focused on the production of grain (wheat, corn, and oats), apples, swine, and sheep. They were not significant participants in the region's hay and tobacco economies.

The 1840 US Population Census records Glenn living on the farmstead with his wife and five children. By the 1870 census period, Glenn is listed as a retired farmer, and we assume that his son William Glenn (1848–1935) was operating the farm. William inherited the farmstead following his father's death in 1881, and his name appears on a 1900 plat map as the property's owner (Cosby 1900).

The archival resources and local histories did provide us clues for when the farmstead structures were built. Glenn was the first owner of the property, buying it from the US government in 1825. Therefore, we assumed the buildings would be constructed after this date. A history of the Robert Glenn House (the second house constructed on the property) identifies its construction as "about 1857" (Knox 1940). The population census suggests that buildings constructed after 1870 were built by William Glenn after his father's retirement.

10.10 Farmstead structures

Six of the farmstead's buildings, all dating to the 19th century, were analyzed. The cluster of buildings encompasses about two acres (0.8 hectares) of the farmstead, and is divided between housing for the Glenn family and buildings supporting the operations of the farm (Figure 10.10). There are three primary buildings (the Apple House, Glenn House, and English Barn) and three outbuildings (a woodshed, corncrib, and sheep barn).

10.10.1 Apple House

The Apple House is believed to be the farmstead's original house, and was later used to store the produce from the farm's orchards (Figure 10.11). The one and a half story stone building measures 23 × 34 feet, is oriented east/west, and the

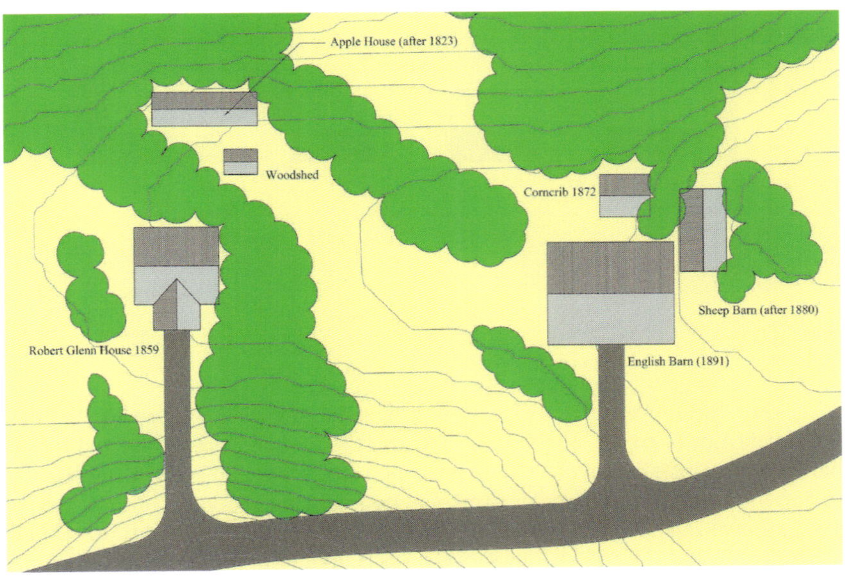

Figure 10.10
Illustration of the Robert Glenn Farmstead in 2017.
Source: Authors.

Figure 10.11
South façade of the Apple House.

Source: Authors.

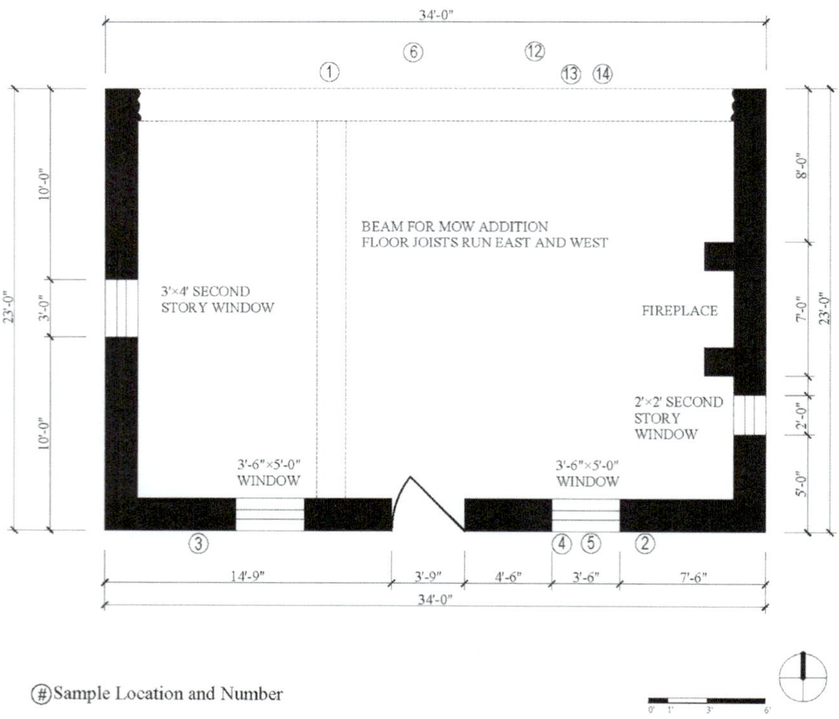

Figure 10.12
Apple House plan. See Table 10.3 for provenience of each sample.

entry is on the south wall that faces onto the adjoining road (Figure 10.12). The interior layout of rooms is unknown because the building was converted into a barn with the installation of a mow for hay and stabling for animals. It was likely a variation of a central passage or two-room house form (McAlester and McAlester 1988). Since the house is structurally unstable, the owner (properly) prohibited us from entering. Therefore, our sampling was limited to timber accessible from outside the building.

Fourteen samples were collected from the mow, roof rafters, and window frames (Table 10.3). They represented two timber types: tulip poplar and basswood. Unfortunately, only one sample provided a death date (RAH10), a rafter dating to 1870. Therefore, tree-ring analysis could not confirm that the Apple House is the Glenn's first residence, or that it is the farmstead's oldest remaining building. If we assume that the window frames are original, we can deduce that the house was constructed sometime shortly after 1823, the last ring we could crossdate (Figure 3.9). It appears a completely new timber roof, including the timber plate on top of the stone wall, was added in the latter half of the 19th century, but this conclusion is based on only one sample.

10.10.2 Robert Glenn House

The Robert Glenn House is a 40 × 45-foot, two-story, stone house (Figure 10.13). Family lore says that a local castle, or more likely a manor house, Glenn witnessed as a child in Edinburgh inspired the house's construction (Knox 1940). The house is, essentially, a two-story center gabled pavilion on the front of a central hall, double pile house (a house plan two rooms deep). Although the house form has Georgian, and therefore British roots, it was common throughout the United States (Peat 1962; McAlester and McAlester 1988). While it is debatable if the house form is Scottish, American, or a hybrid, Scots are renowned for building in stone (Bruce 2009), and the region is famous for its mortarless dry-laid stone walls.

Sampling was limited to the exposed and loose floor joists in the cellar, and the roof's plate (the bearing timber located on top of the house's stone wall and accessible from an upstairs closet). Fieldwork produced 12 series from 11 samples, and 1,052 rings were crossdated. All of the samples were tulip poplar, and four provided harvest dates (Table 10.4). A cellar floor joist dated to 1858, a section of roof sheathing dated to 1858, and the bearing plate produced a sample dating to 1859 (RGH01). Knox's (1940) description that the house's foundation was built

Table 10.3 Samples collected from the Glenn Apple House (Jefferson County, Indiana). See Table 10.1 for a description of the table contents; a full description of the data presented is provided in Chapter 7.

Sample	*First*	*Last*	*Outer ring*	r	*Timber type*	*Provenience*
RAH01A					Basswood	Beam supporting mow
RAH02A	1777	1845		0.71	Basswood	S plate
RAH03A	1774	1858		0.75	Basswood	S plate
RAH04A	1629	1746		0.67	Tulip	Window frame vertical
RAH05A	1700	1823		0.54	Tulip	Window frame vertical
RAH06A	1743	1839		0.81	Tulip	Rafter
RAH07A	1713	1792		0.73	Tulip	Rafter
RAH08A	1781	1843		0.71	Tulip	Rafter
RAH09A	1781	1842		0.69	Tulip	Rafter
RAH10A	1784	1870	w/e	0.52	Tulip	Rafter
RAH11A	1635	1679		0.63	Tulip	Rafter
RAH12A	1726	1823	sap	0.81	Tulip	Rafter
RAH13A	1771	1825		0.52	Tulip	Roof sheathing
RAH14A	1659	1740		0.72	Tulip	Roof sheathing

Figure 10.13
South façade of the Robert Glenn House.
Source: Authors.

Table 10.4 Samples collected from the Robert Glenn House (Jefferson County, Indiana). See Table 10.1 for a description of the table contents; a full description of the data presented is provided in Chapter 7.

Sample	First	Last	Outer ring	r	Timber type	Provenience
RGH01A	1799	1859	w	0.59	Tulip	N plate
RGH02A	1837	1858	w/e	0.76	Tulip	Roof sheathing
RGH03A	1745	1823		0.70	Tulip	Floor joist; ground floor
RGH04A	1640	1766		0.72	Tulip	Floor joist; ground floor
RGH05A	1783	1858	b	0.71	Tulip	Floor joist; ground floor
RGH05B	1786	1858	b	0.70	Tulip	
RGH06A	1651	1783		0.70	Tulip	Floor joist; ground floor
RGH07A	1746	1832		0.43	Tulip	Floor joist; ground floor
RGH08A	1691	1837		0.59	Tulip	Floor joist; ground floor
RGH09A	1753	1833		0.47	Tulip	Floor joist; ground floor
RGH10A	1773	1839		0.75	Tulip	Floor joist; ground floor
RGH11A	1636	1765		0.70	Tulip	Floor joist; ground floor

in 1857 and allowed to settle for two years before the remainder of the house was completed supports the 1859 construction date.

10.10.3 English barn

The English barn in the United States is also known as a New England, Yankee, Connecticut, or three-bay threshing barn (Noble and Seymour 1982; Noble and Cleek 1997). The barn's simple form is composed of a central bay containing an aisle (a driveway through the barn) that is flanked by bays for storing grain and hay or for stabling animals. The barn type is described in more detail in Chapter 12 (Figure 2.2).

The Glenns built a five-bay English barn that measures 48 × 60 feet (Figure 10.14). The center bay is an aisle that is flanked by two 14-foot wide bays to the east and two 10-foot wide bays to the west (Figure 10.15). The aisle and the adjacent bay to the east contain a threshing floor. The bent between these two bays contains a swing beam spanning 24 feet. A swing beam creates a larger area for the farmer to thresh wheat with a team of horses by eliminating a central support post (compare the position of posts on the west side of the aisle).

Figure 10.14
South façade of the Glenn English barn.
Source: Authors.

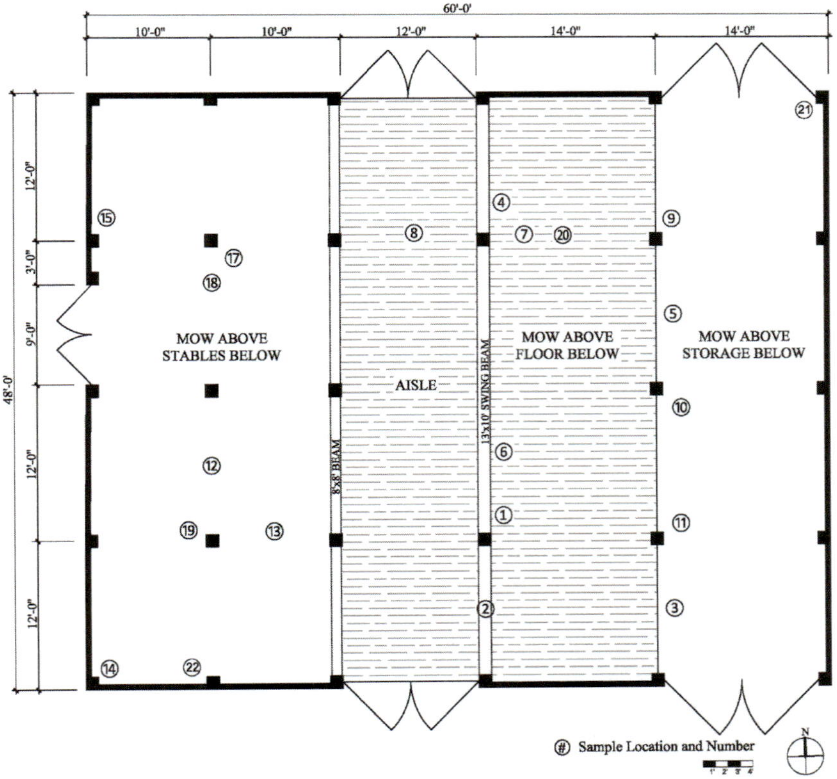

Figure 10.15
Glenn English barn plan. See Table 10.5 for a provenience of each sample.
Source: Authors.

We successfully crossdated 30 series from 22 timbers (Table 10.5). Samples were obtained from six different timber types. Tree-ring results suggest that the barn was constructed from timbers harvested over more than one growing season, because death dates (rings adjacent to wane or bark) of 1890 and 1891 were obtained. Nearly all of the primary structural timbers with discernable death dates had an outer ring of 1890. Samples with an 1891 death date were all floor joists between the ground level and the mow (RGB10, 11, and 13). Therefore, we concluded an 1891 construction date. We did find two white oak timbers with death dates of 1844 (RGB08 and 15), and wonder if they are recycled from one of the farmstead's past structures.

Table 10.5 Samples collected from the Glenn English Barn (Jefferson County, Indiana). See Table 10.1 for a description of the table contents; a full description of the data presented is provided in Chapter 7.

Sample	*First*	*Last*	*Outer ring*	r	*Timber type*	*Provenience*
RGB01A	1832	1890	w	0.80	Beech	Post; aisle
RGB02A	1848	1890	w	0.68	Red elm	Beam; ground/2nd floor; aisle
RGB03A	1876	1890	w	0.59	Beech	Beam; ground/2nd floor
RGB03B	1881	1890	w	0.54	Beech	
RGB04A	1869	1890	w	0.73	Red elm	Beam; aisle
RGB04B	1839	1890	w	0.72	Red elm	
RGB05A	1871	1890	w	0.73	Beech	Beam; ground/2nd floor
RGB05B	1814	1890	w	0.75	Beech	
RGB06A	1798	1890	w	0.59	Beech	Swing beam
RGB07A	1797	1890	w	0.74	Beech	Post; aisle
RGB08A	1797	1844	w	0.47	White oak	Beam; ground/2nd floor; aisle
RGB08B	1713	1844	w	0.40	White oak	
RGB09A	1859	1890	w	0.59	Basswood	Post
RGB10A	1813	1891	w	0.51	Beech	Floor joist; haymow
RGB11A	1829	1891	w	0.49	Beech	Floor joist; haymow
RGB12A	1714	1890	w		Hickory	Beam; W bay
RGB13A	1836	1891	b		White elm	Floor joist; haymow
RGB14A	1824	1890	w	0.55	Red elm	NW post
RGB15A	1799	1844	w	0.63	White oak	W sill
RGB15B	1742	1844	w	0.59	White oak	
RGB16A	1706	1859		0.55	White oak	Loose timber; likely S sill
RBG16B	1707	1846		0.55	White oak	
RGB17A	1865	1890	w	0.79	Basswood	Beam; top of mow between bents
RGB17B	1858	1890	w	0.73	Basswood	Beam; top of mow between bents
RGB18A	1790	1890	w	0.66	Beech	Beam; ground/2nd floor
RGB19A	1766	1890	w	0.53	Beech	Post
RGB20A	1863	1890	w	0.76	Basswood	Roof beam
RGB20B	1856	1890	w	0.69	Basswood	
RGB21A	1805	1890	w	0.55	Red elm	NE post
RGB22A	1771	1875	w12	0.43	White oak	S sill
RGB23A					White oak	Joist; aisle floor

10.10.4 Woodshed

The woodshed is a 22 × 26-foot gable roof, timber frame structure. When we collected samples, the structure had been collapsed for several years, so we used aerial photographs to estimate the dimensions (Figure 6.6). Therefore, the samples are from loose timbers, and their provenience could not be reliably determined (although four of the samples we recognized as roof rafters). Eleven tree-ring series were produced from 14 samples, representing a mixture of red elm, tulip poplar, basswood, white oak, and ash (Table 10.6).

The construction date for the woodshed was indeterminate. Four timbers produced a death date of 1841 (RWS01, RWS02, RWS05, and RWS07), and one timber (RWS03) has an outer ring, not a cutting or death date, of 1878. The results suggest two conflicting scenarios. The first is that the structure was built in the 1840s and post-1878 timbers were added later. The second suggests that, like other structures on the farmstead, that this is a post-1878 building with timbers that are recycled from 1840s buildings. We stress, again, that tree rings cannot always provide conclusive results. In the case of the woodshed, with the low sample size and the deteriorated condition of the building timbers, dates for structures were not obtainable.

10.10.5 Corncrib

The corncrib is a 20 × 24-foot gable roof, timber frame structure. The building does not represent typical corncrib or grain storage vernacular forms, although the current property owner and a past architectural survey identify it as such (IHSSI 1989; Noble and Cleek 1997; Reynolds 2017). However, the raised floor (built to avoid ground moisture from damaging the building's contents) suggests that the building was constructed to store grain or silage. Nineteen series were

Table 10.6 Samples collected from the Glenn Woodshed (Jefferson County, Indiana). See Table 10.1 for a description of the table contents; a full description of the data presented is provided in Chapter 7.

Sample	*First*	*Last*	*Outer ring*	r	*Timber type*	*Provenience*
RWS01A	1765	1841	w	0.65	Ash	Loose timber
RWS02A	1732	1834		0.50	Ash	Loose timber
RWS02B	1794	1841	w	0.63	Ash	
RWS03A	1738	1835		0.41	White oak	Loose timber
RWS03B	1791	1878		0.46	White oak	
RWS04A	1744	1859		0.46	White oak	loose timber
RWS05A	1786	1838		0.55	Ash	Loose timber
RWS05B	1796	1841	w	0.63	Ash	
RWS06A					Red elm	Loose timber
RWS07A	1770	1841	w	0.63	Ash	Loose timber
RWS08A	1710	1803		0.75	Tulip	Rafter
RWS09A					Basswood	Rafter
RWS10A	1743	1822		0.77	Tulip	Rafter
RWS11A					Basswood	Rafter

produced from 20 samples representing a mixture of tulip poplar, basswood, white oak, hickory, American beech, and ash (Table 10.7). Harvest dates could be determined for five of the timbers. Four timbers from the building's frame have an 1872 death date (GCC01–03 and GCC08).

10.10.6 Sheep barn

The sheep barn was a 22 × 39-foot gable roof, timber frame structure. The building collapsed before it was sampled, so its size was estimated by measuring the *in situ* stone footings. Our description of the sheep barn is based on oral history from the current owner (Reynolds 2017). The structure was oriented with the long walls north and south. Doors on the east and west façades opened onto a central aisle that provided the farmer access to interior stalls. Lofts were located over the stalls.

Twelve tree-ring series were produced from 14 samples representing a mixture of white oak, black walnut, and ash (Table 10.8). Only one timber (RSB09), an ash post or beam, provided an 1843 death date. However, other samples indicate that the barn was constructed sometime after 1880, based on the outer ring of the white oak sample RSB08. The building most likely contained recycled material, explaining the presence of the 1843 timber.

Table 10.7 Samples collected from the Glenn Corncrib (Jefferson County, Indiana). See Table 10.1 for a description of the table contents; a full description of the data presented is provided in Chapter 7.

Sample	*First*	*Last*	*Outer ring*	r	*Timber type*	*Provenience*
GCC01A	1678	1872	w	0.46	Ash	SW Post
GCC02A	1836	1872	w	0.51	Basswood	N Plate
GCC03A	1815	1872	w	0.47	Ash	NE Post
GCC04A	1689	1851			Hickory	Beam; ground floor
GCC05A	1767	1860		0.55	Beech	E sill?
GCC05B	1743	1867		0.53	Beech	
GCC06A	1727	1853		0.56	White oak	Loose timber
GCC07A	1749	1846		0.63	White oak	Loose timber
GCC08A	1792	1872	w	0.44	Basswood	S plate
GCC09A					Basswood	Loose timber
GCC10A	1697	1772		0.71	Tulip	Rafter
GCC11A	1739	1792		0.79	Tulip	Rafter
GCC12A	1732	1838		0.75	Tulip	Rafter
GCC13A	1678	1742		0.74	Tulip	Rafter
GCC14A	1770	1821		0.64	Tulip	Rafter
GCC15A	1771	1817		0.65	Tulip	Rafter
GCC16A	1689	1769		0.65	Tulip	Rafter
GCC17A	1756	1870	w	0.54	Tulip	Siding
GCC18A	1815	1860		0.67	Basswood	Loose beam
GCC19A	1803	1862		0.57	Basswood	Loose beam

Table 10.8 Samples collected from the Glenn Sheep Barn (Jefferson County, Indiana). See Table 10.1 for a description of the table contents; a full description of the data presented is provided in Chapter 7.

Sample	*First*	*Last*	*Outer ring*	r	*Timber type*	*Provenience*
RSB01A	1720	1816		0.54	White oak	S sill; S pen
RSB01B	1727	1818		0.53	White oak	
RSB02A	1713	1829		0.59	White oak	Mow beam; S pen
RSB03A					White oak	Loose timber
RSB04A	1676	1872		0.35	White oak	Mow beam or joist; N pen
RSB05A	1679	1831		0.53	White oak	Mow beam or joist; N pen
RSB06A	1705	1828		0.52	White oak	Post or beam; N pen
RSB07A					Walnut	Post or beam; N pen
RSB08A	1705	1880		0.65	White oak	Post or beam; N pen
RSB09A	1734	1843	w	0.45	Ash	Post or beam; N pen
RSB09B	1746	1843	w	0.45	Ash	
RSB10A	1694	1826		0.57	White oak	Post or beam; N pen
RSB11A	1700	1817		0.61	White oak	Post or beam; N pen
RSB11B	1706	1817		0.69	White oak	

10.11 Glenn Farmstead conclusions

Our objective in analyzing the Glenn Farmstead was to identify the construction dates for six structures and to identify periods of significant change on the property. The study resulted in the collection of 97 series from 81 timber samples, representing eight timber types: tulip poplar, white oak, American beech, basswood, ash, red elm, hickory, and white elm (in order from most samples to least) (Tables 10.9 and 10.10). We successfully crossdated 8,042 rings to create timber type chronologies spanning from 1630 to 1890. Construction dates were established for three of the six structures: Robert Glen House (1859), corncrib (1872), and the English Barn (1891). Minimum construction dates were established for the Apple House (after 1823) and the sheep barn (after 1880). No construction date was established for the woodshed. It appears the farmstead went through at least three distinct phases during the 19th century. The first was the early settlement and development of the farmstead. However, dendrochronology was not able to confirm the Apple House is from this period, but did identify the likely presence of structures from the 1840s that were recycled into farmstead structures built decades later. The second phase is the construction of the house in the late 1850s, demonstrating Glenn's financial success as a farmer. The final phase is the construction of the English barn toward the end of the century, likely a modernization of the farmstead and another major investment into the property by William Glenn, Robert's son.

Although the vernacular form, presence of a fireplace, and stone construction suggest the Apple House was the Glenn family's first residence built sometime after their 1825 purchase of the property, its age could not be confirmed through dendrochronology. If we use the 1870 death date of the roof rafter as the construction date for the house, it suggests the unlikely scenario that the Glenn's constructed a second smaller stone house a mere decade following the construction of

Table 10.9 Summary table of the timber type chronologies prepared from the Glenn Farmstead (Jefferson County, Indiana). See Table 10.2 for a description of table contents; a full description of the data presented is provided in Chapter 7.

Timber type	*Samples*	*Timbers*	*Undated*	*First*	*Last*	*Years*	r	*Mean*	*SD*
Ash	10	7	0	1679	1871	818	0.52	1.08	0.45
Basswood	11	9	4	1775	1889	506	0.64	1.41	0.59
Beech	13	10	0	1749	1890	902	0.61	1.22	0.58
Hickory	2	2	0	1690	1889	334		0.85	0.40
Red elm	5	4	1	1806	1889	259	0.62	1.53	0.52
Tulip poplar	33	32	0	1630	1869	2687	0.67	1.24	0.55
Walnut	0	0	1						
White elm	1	1	0	1839	1890	52		1.53	0.61
White oak	22	16	2	1678	1879	2484	0.53	0.85	0.37
Total	97	81	8	1630	1890	8042			

Table 10.10 Summary of tree-ring analysis of the six structures found at the Glenn Farmstead. "First" and "last" are the first and last tree rings present in all of the timber type chronologies prepared from each structure. "Years" is the total number of rings in all of the chronologies for each building. "Timber types" is the number of timber types from which a chronology was constructed for each building.

Structure	*Timbers*	*Samples*	*Undated*	*First*	*Last*	*Years*	*Timber types*
Apple house	13	13	1	1630	1869	1035	2
Barn	22	30	1	1712	1890	2032	6
Corncrib	18	19	1	1679	1871	1617	6
House	11	12	0	1637	1858	1052	1
Sheep barn	9	12	2	1678	1879	1473	3
Woodshed	8	11	3	1711	1877	833	3
Total	81	97	8	1630	1890	8042	9

their grand house in 1859. Furthermore, we cannot account for a house the family would have occupied between their 1825 property purchase and 1859 construction of the residence. Therefore, dendrochronology and archival sources suggest, but do not confirm, the current Apple House roof is a replacement to an earlier Glenn family residence.

Construction dates established links to primary and secondary sources (census data, atlas maps, and county histories) that helped to understand the evolution of the farmstead. Robert Glenn operated the farm from his 1825 property purchase until his retirement in the 1870s. His son William appears to have operated the farm following his father's retirement, and he inherited the farm upon Robert's death. Therefore, we directly link one of the structures to Robert Glenn (house), and two of the structures to his son William (English barn and corncrib).

We cannot account for barns predating the 1891 English barn, but it is almost certain that at least one existed. We also know that recycled timbers from the early 1840s made their way into the English barn, sheep barn, and woodshed structures. We propose the idea that the current farmstead layout is more of a reflection of William's tenure on the property than his father's, and that significant alterations to farmsteads tend to occur when property passes from one generation to the next (Groover 2008). Tree-ring analysis suggests that in the early 1870s William made modifications to the Apple House and constructed the corncrib. Tree-ring analysis also suggests that about the time the English barn was constructed in 1891, 1840s-era timber structures were removed and reused in new structures. Therefore, we wonder if the construction of the sheep barn and woodshed, which have recycled timbers from the 1840s, might also date to the 1890s. As stated in multiple chapters of this book, the Glenn Farmstead study affirms the historic practice of recycling timbers from old buildings. In the analysis of regional timber buildings, we have found that it was common to use recycled timbers from past structures in new buildings (Roberts 1996; Grissino-Mayer and van de Gevel 2007; Baas et al. 2017). This study, like others we have performed, reinforces the importance of collecting samples from throughout a building or landscape. Limited samples, especially from buildings with recycled timbers, can result in false construction dates (Baas et al. 2017).

The tree-ring analysis of the Glenn Farmstead advanced the understanding of the region's vernacular landscape by establishing the presence of the vernacular English barn form in the region late into the 19th century. We have also dated the construction of a nearby English barn to 1824 (Chapter 12), demonstrating the form's use and versatility to regional farmers for nearly 70 years despite changes in agriculture, technology (i.e., mechanization), economy, and transportation.

Finally, dendrochronology highlighted the use of tree species that were not typically considered the first choice of 19th-century builders: beech, red elm, white elm, hickory, and basswood. This is especially true for the English barn which includes nine samples representing these species. While we cannot draw direct correlations between the barn and the surrounding landscape, sampling suggests that the forests surrounding the barn in the late 19th century may have been harvested of desired species (e.g., ash, oak, tulip poplar) and that second-choice species were all that remained to construct the barn. We explore the concept of timber species bias in Chapter 13.

10.12 Literature cited

Allen County Public Library Digital Collections. 2017. *Drawing of Fort Wayne, IN, John Whistler*. http://contentdm.acpl.lib.in.us/digital/collection/p16089coll56/id/1140/rec/1. Accessed 2 September 2018.

Andres, C.R., D. McCullough, and M. Strezewski. 2008. *Intensive Survey of the Forts of Fort Wayne, Allen County, Indiana: Volume 1*. IPFW Archaeological Survey. Fort Wayne, IN. 233 p.

ARCH (Fort Wayne Architecture and Community Heritage). 2017. *Dr. Merchant W. Huxford House*. https://archfw.org/portfolio-items/drmerchant-w-huxford- house/. Accessed 20 June 2017.

Baas, C., T.N. Davis, and D.L. Rubino. 2019. The Fort in the Doctor's House: Using Tree-Ring Growth Patterns to Discover Historic Fort Wayne, IN, USA. *Journal of Conflict Archaeology* 13: 97–116.

Baas, C., P. Jiang, and D.L. Rubino. 2017. Using tree-ring growth patterns to identify construction dates of two nineteenth century timber structures in Craig Township, Switzerland County, Indiana. *Indiana Archaeology* 12: 12–41.

Babits, L.E., and S. Gandulla. 2013. *The Archaeology of French and Indian Frontier Forts*. University of Florida Press, Gainesville, FL. 324 p.

Babson, J.F. 1968. The architecture of early Illinois forts. *Journal of the Illinois State Historical Society* 61: 9–40.

Brant and Fuller. 1889. *Valley of the Upper Maumee River: Volume 1*. Brant & Fuller, Madison, WI. 498 p.

Brice, W.A. 1868. *History of Fort Wayne from the Earliest Known Accounts*. D. W. Jones & Sons, Fort Wayne, IN. 324 p.

Bruce, I. 2009. Viewpoint: The balloon frame, George Snow, Augustine Taylor, and all that. A view from abroad. *Buildings & Landscapes: Journal of the Vernacular Architecture Forum* 16: 1–8.

Burns, J.A., G.J. Drobnock, and J.M. Smith. 2010. Croghan at Aughwick: History, maps, and archaeology collide in the search for Fort Shirley. *PAST: Pioneer America Society Transactions* 33: 14–31.

Cosby, C.R. 1900. *Map of Jefferson Co., Indiana*. Madison, IN.

Danner, E. 1945. The Caledonia Church and cemetery. *Indiana Magazine of History* 41: 407–408.

DeRegnaucourt, T., and P. Hoelscher. 1996. *The Archaeology of Fort St. Marys: Major General Anthony Wayne's Fort 1795 to 1796*. Upper Miami Valley Archaeological Research Museum, Arcanum, OH. 128 p.

Fort Wayne Community Development. 2017. *Dr. Merchant W. Huxford House Local Historic District*. www.fwcommunitydevelopment.org/images/historic_preservation/docs/Merchant_Huxford_House_brief_history_no_photos.pdf. Accessed 2 September 2018.

Fort Wayne Daily Sentinel. 1882. *The First Presbyterian Church Destroyed by Fire*. Fort Wayne, IN. December 18.

Grissino-Mayer, H.D., and S.L. van de Gevel. 2007. Tell-tale trees: Historical dendroarchaeology of log structures at Rocky Mount, Piney Flats, Tennessee. *Historical Archaeology* 41: 32–49.

Griswold, B.J. 1917. *The Pictorial History of Fort Wayne, Indiana*. Robert O. Law Co., Chicago, IL. 735 p.

Groover, Mark D. 2008. *The Archaeology of North American Farmsteads*. University Press of Florida, Gainesville, FL. 150 p.

Hall, Glenn. 1993. Old home steeped in history. *Fort Wayne Journal-Gazette*. Fort Wayne, IN. July 18.

IHSSI. 1989. *Jefferson County Interim Report*. Indiana Historic Sites and Structures Inventory, Historic Landmarks Foundation of Indiana, Indianapolis, IN. 125 p.

Kilbane, K. 2016. Work progressing well on historic Huxford House. *Fort Wayne News-Sentinel*. Fort Wayne, IN. February 12.

Knox, J.L. 1940. Robert Glenn's Scottish home in Indiana. *Indiana Magazine of History* 36: 334–336.

McAlester, V., and L. McAlester. 1988. *A Field Guide to American Houses*. Alfred A. Knopf. New York, NY. 525 p.

McClelland, L.F. 1999. *Guidelines for Evaluating and Documenting Rural Historic Landscapes: National Register Bulletin No. 30*. National Park Service, Washington, DC. 33 p.

National Park Service. 2006. *George Rogers Clark: Fort Sackville*. George Rogers Clark National Historic Park visitor brochure.

News-Sentinel. 2014. Fort Wayne city council gives preliminary approval to designating circa 1854 home as local historic district. *Fort Wayne News- Sentinel*. Fort Wayne, IN. July 16.

Noble, A.G. and G.A. Seymour. 1982. Distribution of Barn Types in Northeastern United States. *Geographical Review* 72: 155–170.

Noble, A.G., and R.K. Cleek. 1997. *The Old Barn Book*. Rutgers University Press, New Brunswick, NJ. 222 p.

Peat, W.D. 1962. *Indiana Houses of the Nineteenth Century*. Indiana Historical Society, Indianapolis, IN. 195 p.

Poinsatte, C.R. 1969. *Fort Wayne during the Canal Era: 1828–1855*. Indiana Historical Bureau, Indianapolis, IN.

Reynolds, C. 2017. *Oral History Collected from Current Property Owner During Dendrochronology Fieldwork*.

Roberts, W.E. 1996. *The Log Buildings of Southern Indiana*. Trickster Press, Bloomington, IN. 228 p.

Scott, R.W. 2018. *Scottish Immigrants in the Caledonia Settlement*. www.ingenweb.org/inswitzer/census/scottishImmigr.html. Accessed 3 September 2018.

Snyder, J.F. 1913. Fort Kaskaskia. *Journal of the Illinois State Historical Society* 6: 58–71.

Starbuck, D.R. 2008. The "massacre" at Fort William Henry: History, archaeology, and re-enactment. *Expedition* 50: 17–25.

Starbuck, D.R. 2011. *The Archaeology of Forts and Battlefields*. University of Florida Press, Gainesville, FL. 128 p.

Therrell, M.D., B.S. Mozayen, and M.D. Gage. 2017. The search for Fort Armstrong: Dendroarchaeology of the Williamson "Snow Hill" plantation, Cherokee County, Alabama, U.S.A. *Dendrochronologia* 43: 59–65.

Whittaker, W.E. 2009. *Frontier Forts of Iowa: Indian Traders and Soldiers, 1682–1862*. University of Iowa Press, Iowa City, Iowa. 258 p.

Woehrmann, P. 1971. *At the Headwaters of the Maumee: A History of the Forts of Fort Wayne*. Indiana Historical Society, Indianapolis, IN. 306 p.

Chapter 11

New Harmony, Indiana

Tree-ring analysis of a communal and utopian landscape

11.1 Introduction

In 1814, Father George Rapp and approximately 1,000 followers established the town of New Harmony, Indiana, in the frontier wilderness of the Wabash River Valley (Figure 11.1). The small village where Rapp and his followers resided for ten years was the second of three American towns the group constructed after emigrating from the German Duchy of Württemberg in 1803. The millenarian group was known for communal living, hard work, and celibacy. They were collectively recognized as members of the Harmonist Society, or simply as Harmonists.

When the group outgrew the town of Harmony, Pennsylvania, which they had created 1804, they migrated to the seclusion of the Indiana Territory in 1814 to await Jesus Christ's Second Coming (Taylor 1987). Desiring more space and greater isolation, Rapp chose the Wabash River location for its access to New Orleans markets via the Ohio and Mississippi River systems. Over the following decade, the Harmonists created a landscape of gridded streets lined with houses, churches, and several substantial industrial and agricultural structures that included a granary, brewery, and rope-making factory. Agricultural fields and vineyards surrounded the village. The society became wealthy selling their goods to local settlers and exporting goods to distant markets.

When the region filled with non-Harmonist settlers, the group chose to isolate itself again by returning to a new location in western Pennsylvania where they created the town of Economy in 1825. Rapp sold the Indiana town to Welsh industrialist Robert Owen. Owen, in partnership with geologist William Maclure, sought to establish an industrial and social utopian community. Owen was well known in Europe and America as an industrialist who enacted social reforms at his wool mill in New Lanark, Scotland. Owen promoted a utopian philosophy endorsing freedom of religion, the communal raising of children, and the end of the institution of marriage because of the control and limitations it placed on women. Although the experiment in creating an industrial utopia in America brought the town international attention, it failed after only a few years. The town was eventually sold to individual property owners, and it has evolved as a typical rural Hoosier community (Arndt 1965; Taylor 1987; Hoyt 1996; Center for Historic Preservation 2007; Baas and Rubino 2013). New Harmony, Indiana, is currently a mixture of historic sites and museums commemorating Harmonist and Owen history.

Figure 11.1
State of Indiana showing the location of New Harmony.
Source: Authors.

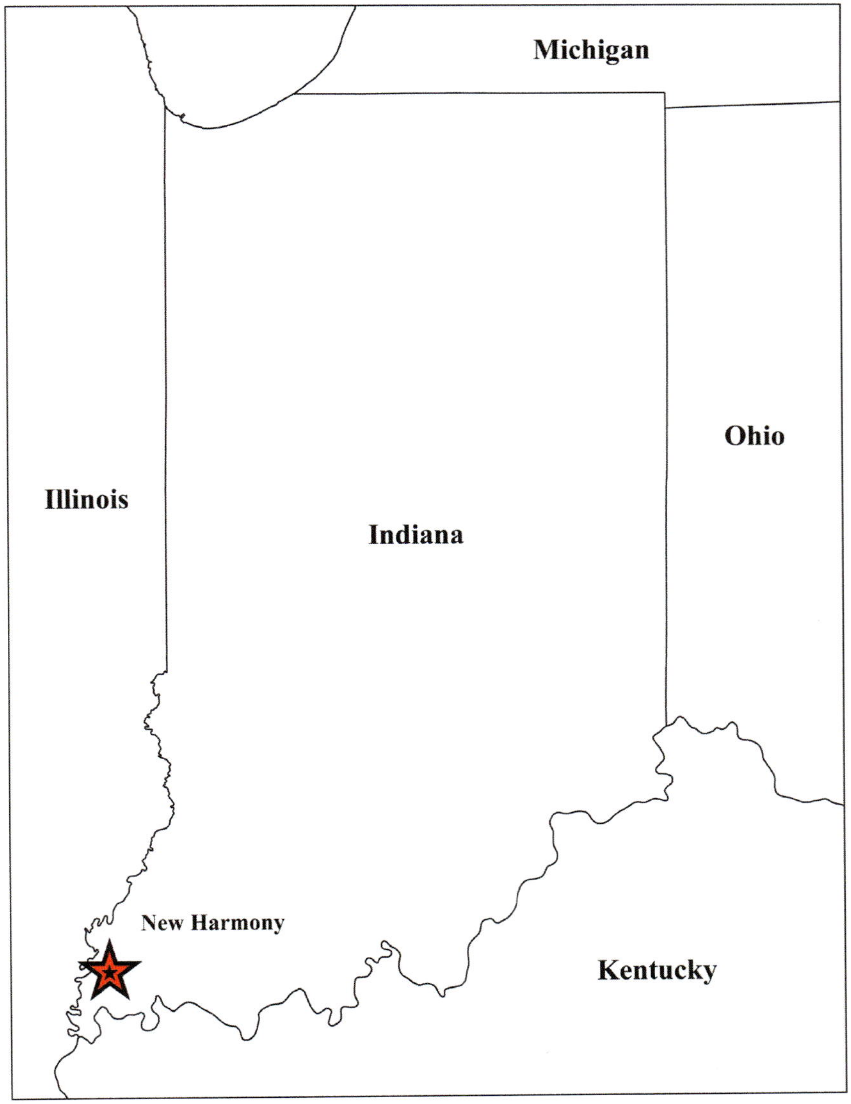

In 2012, we were invited by the Indiana State Historic Sites and Museums, Inc. to sample the Fauntleroy House while it was undergoing restoration. We returned to New Harmony several more times to sample the Scholle House and the two community houses (dormitories). Our analysis of Community House No. 2 and the Grayson House resulted in journal articles in *Indiana Archaeology* and *Proceedings of the Indiana Academy of Science*, parts of which are included in this chapter (Baas and Rubino 2013; Rubino and Baas 2014). We have found that New Harmony's history, social experimentation, and preservation make it a very special place, and we treasure collaborating with the town's dedicated managers and preservationists who are using the results of tree-ring analysis to advise their preservation and interpretive choices. We have found that scholars have been reasonably accurate in dating the construction of Harmonist structures without the use of dendrochronology. Our analysis has been most effective in identifying post-Harmonist and post-Owen changes as seen in both dormitories and the Fauntleroy House. It is here that we feel tree-ring analysis contributes new and meaningful knowledge of this fascinating landscape.

11.2 Harmonist housing

Housing was the principal challenge facing the Harmonists following their move to the Indiana frontier. Log houses were constructed upon the group's arrival in 1814 and served many families throughout their ten-year presence. Beginning about 1820, the log houses were steadily replaced with timber frame and brick residences and dormitories known as community houses. Archaeologist Michael Strezewski (2015) used population census data and period maps of the town to describe how between 1814 and 1820 Harmonists lived in nuclear family groups of adult men, women, and children. A household typically consisted of four to five individuals. After 1820, living arrangements underwent changes based on the replacement of the original log structures with brick and frame houses and communal dormitories. Brick houses appear to have been preferentially provided to heads of households involved in the operations of the society and a family's geographic proximity to their employment.

The Harmonists constructed four dormitories in New Harmony. The first was built about 1817, and the remaining three after 1820. Dormitory construction concluded by 1824, with Community House No. 4. Only Community House No. 2 and No. 4 still exist. Early historians incorrectly concluded that the dormitories were constructed for the Society's unmarried members. Each of the four dormitories were several stories in height, constructed of brick, and had exterior kitchens, gardens, and stables. They were distributed throughout the town. Recent scholarship has established that as many as 40 Harmonist families were assigned to reside in the buildings (Blair 1964; Strezewski 2015). It is not clear why Rapp chose to move toward dormitory housing of Harmonists. It may have been to provide an independent residence for adult children that could now fend for themselves, to move toward a more literal communal living embraced by other religious societies, or he may simply have wanted more control over certain Society members (Strezewski 2015).

The brick and frame Harmonist houses constructed to replace the settlement's original log buildings were a standardized design to accommodate mass production. The two-story, gable roof structures have a cellar and an attic. The interiors are composed of six rooms on two floors. The ground floor contains an entry hall, kitchen, and living room. The second story contains two bedrooms and the stairway hall. Frame houses are of *fachwerk* construction, consisting of a timber frame with low-fired brick infill (Figure 11.2). *Fachwerk* construction is an Old World building technique brought to America by German and other European immigrants (Kniffen 1965; Glassie 1968; Stanton 1985; Tishler 1986). In New Harmony, the exterior walls of brick houses are approximately 14 inches (36 centimeters) thick, and interior walls are *fachwerk*. Roofs and walls are insulated with Dutch biscuits, a wood plank wrapped in mud and straw and inserted into slots in the house's timber frame (their name comes from local slang; Figure 11.3).

Houses were located on small plots that included the residence, a small structure for livestock, and a fenced yard for a vegetable garden and fruit trees. Harmonist houses were entered from within the fenced yard, as opposed to the street. This landscape arrangement contrasts with the typical 19th-century American urban residence where entry doors face onto the street. Perhaps a house entrance unapproachable from the public street indicates that interaction between

Figure 11.2
Harmonists used *fachwerk* construction methods to build many of their buildings. The method consists of a timber frame with an infill of low-fired brick. It is an Old World building technique brought to America by German and other European immigrants. This example is from Community House No. 2.

Source: Authors.

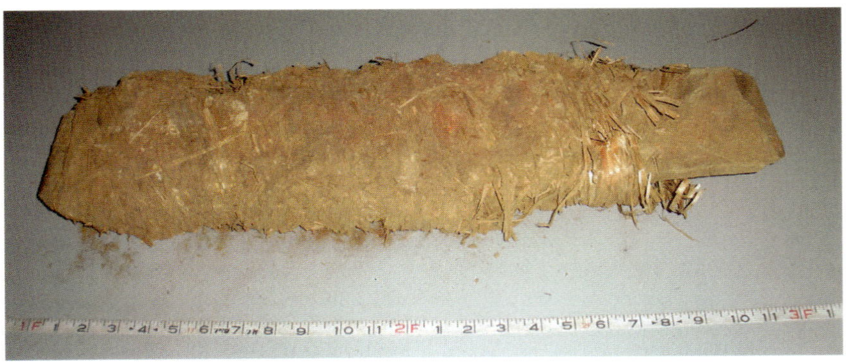

Figure 11.3
Dutch biscuits are short wood planks wrapped in mud and straw. They are inserted into grooves made in walls and roofs and provide insulation.

Source: Authors.

Harmonists should take place in communally held spaces such as the market, workplace, or church, but not at the family residence.

This chapter reports our use of tree rings to date the construction and modification of four Harmonist buildings – two dormitories and two houses (Figure 11.4). We sought to establish when a building was constructed or modified and to determine what parts of the buildings are Harmonist, Owenite, or from a period following the communal and utopian periods.

11.3 Archival and scholarly sources for New Harmony buildings

We found a variety of archival and scholarly sources to interpret dendrochronological results. Our study of New Harmony structures benefited from a series of maps documenting the town's layout. William Pickering, a local surveyor, assembled a measured plan of the town in 1824 to help Rapp sell the community. The illustra-

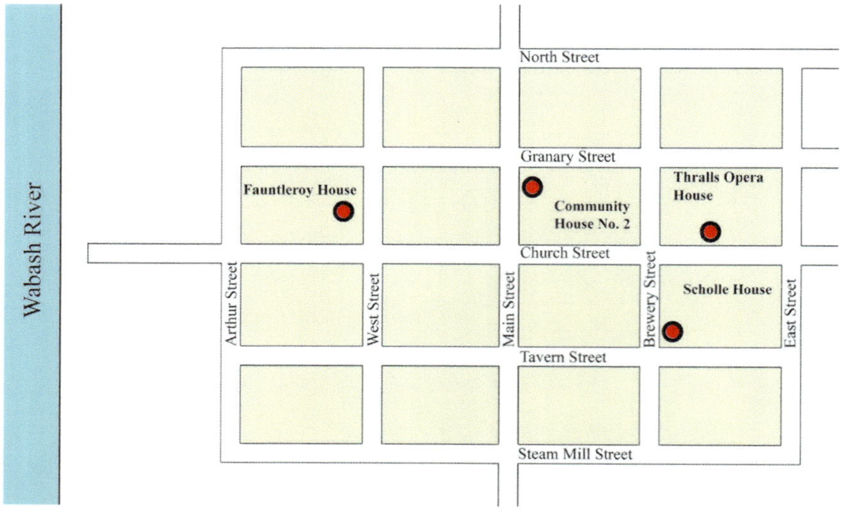

Figure 11.4
New Harmony's historic town grid with case study locations.
Source: Authors.

tion was completed just a few months before the society returned to Pennsylvania, and it shows the location of roads, buildings, and landscape features such as drainage ways, garden spaces, and agricultural fields. A second map was created from memory in 1832 by Harmonist Wallrath Weingärtner, eight years after the community had left Indiana. The map, *folk* in quality, depicts the town in plan and the town's buildings in two dimensions. The names of the head of household are listed for each house, so it is Weingärtner's map that allows a Harmonists' age, place of birth, and occupation to be tracked in the US Population Census.

Sanborn Insurance Company maps document the late 19th-century arrangement of the structures and illustrate additions to the original Harmonist building forms (Chapter 6). The years of map publications and updates for New Harmony are 1888, 1893, 1899, and 1907. While the maps do not provide construction dates, they do provide key snapshots in time that are effective in documenting changes to the buildings and their properties. Unfortunately, for the most part, 1950s and 60s restoration decisions called for the removal of building additions not associated with the Harmonists, so they are not available to sample.

George Rapp's letters to family, Society members, and business associates document many aspects of the town's development (Arndt 1978). Likewise, publications of the written accounts of New Harmony residents and visitors provide first-person accounts of life in the village (Lindley 1916; Brown 1972; Indiana Historical Commission 1975).

Late 19th- and early 20th-century photographs provided historic images of the buildings and their alterations. The photographs are located in historic archives such as the Indiana State Historic Sites and Museums, Inc., the University of Southern Indiana, and the US Library of Congress.

A multitude of historians and archaeologists have studied New Harmony. Dr. Karl Arndt's (1972 and 1974) seminal works on the Harmonist Society describes all three of the group's communities: Harmony, Pennsylvania; New Harmony, Indiana; and Economy, Pennsylvania. Anne Taylor's (1987) scholarship builds on Arndt's work extending it through Owen's experiment. Local historian Don Blair authored the most detailed examination of Harmonist construction (1964). Archaeologist Michael Strezewski has reported the excavation of significant sites within the town, as well as describing Indiana's Harmonists and the layout of their town (2015). There is National Park Service documentation through the National Register and the National Historic Landmark programs (Hurdis 2003), and all four buildings we analyzed have a Historic Structures Report (HSR) documenting construction and alterations over time (although without dendrochronology). Historic American Building Survey (HABS) documentation for Community House No. 2 (completed in 1934) and No. 4 (completed in 1940) is available from the Library of Congress.

We were challenged by time gaps in historical sources. Very little is known of the residents of the Owenite period unless they were notable participants connected with the utopian experiment. The Owen community was created and dissolved between the 1820 and 1830 US Population Census periods so the participants are not clearly recorded in the population census. It is not until the Sanborn Map Company began publishing measured illustrations of the town, along with mid- and late-century period photographs, do we get illustrative documentation of mid- and later 19th-century New Harmony. It is this window of time in which the village was transitioning into a typical Indiana town that dendrochronology is potentially the most useful for providing new information.

11.4 Community House No. 2

Community House No. 2 is a rectangular, three-story dwelling with a mansard roof, which Rapp called a "French" roof (Arndt 1978; Figure 11.5). It measures 82 × 45 feet, and is oriented north and south, with the long facade of the building facing west onto Main Street (Figure 11.6). The Harmonists constructed the dwelling with the east façade as the primary entrance, which opened on a garden, water well, and stables. However, the Main Street façade became the building's public face later in the 19th century when the structure evolved into a commercial building serving customers from the adjacent street.

Figure 11.5
HABS photograph of Community House No. 2 in 1940.

Source: Historic American Building Survey, Library of Congress, Prints & Photographs Division.

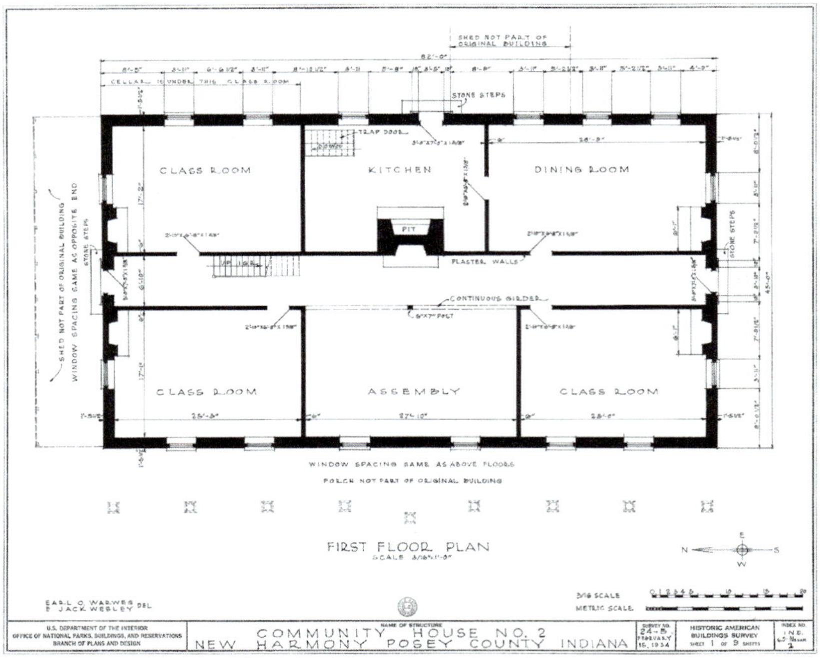

Figure 11.6

HABS documentation drawing of the first floor of Community House No. 2, completed in 1934.

Source: Historic American Building Survey, Library of Congress, Prints & Photographs Division.

The dwelling demonstrates two types of construction. The exterior walls are brick and are instrumental in bearing the weight of the roof. The interior walls and the third-story exterior walls are *fachwerk* (Blair 1964; Center for Historic Preservation 2007; Baas and Rubino 2013).

Community House No. 2 housed 50 Harmonists on the first two stories. Rapp chose not to finish the third story in bedrooms because he believed the kitchen could not efficiently serve the residents. Therefore, Community Houses No. 3 and No. 4 were constructed with only two stories (Arndt 1978). Following the Harmonists' departure for Economy, Pennsylvania, in 1824, Community House No. 2 became the property of Robert Owen. Owen used the structure as a school, a Masonic Hall, and apartments. William Maclure became owner of the dormitory in 1828 as part of the settlement of the Owen community's debt following the failure of the utopian experiment. Beginning in 1856, under a variety of different owners, the building housed such commercial uses as a tavern, grocery store, newspaper office, and apartments (Center for Historic Preservation 2007). The building was sold to the state of Indiana in 1940 and underwent a major restoration in the 1950s and several minor restorations over the ensuing decades. Recent renovations added elevator access to all three stories, office space, and a new HVAC system.

Estimates of the dormitory's construction date vary among sources, however George Rapp's May 1822 letter is the best indication for when the construction of the dormitory was completed. He writes that in the "large brick-and-stone house there are 40 people; these all eat at one table and live together in a brotherly manner, and there are many of our people who want to live in this manner" (Arndt 1978). An April 1823 letter further confirms the dormitory was in operation and explains the reason for not housing Harmonists on the third story. The "upper floor with its beautiful rooms cannot be occupied, because 50 persons have very good and comfortable room in the 2 lower stories, and if one wants to put more people into it, there will be too much activity with cooking, baking, and washing" (Arndt 1978).

The dorm's presence on the Pickering Map demonstrates it was built prior to 1824. The 1888 Sanborn Map Company maps illustrate that by the late 19th century the building had been divided into commercial uses, that a porch had been added to the Main Street façade, and timber additions were added to the north and east walls of the structure. HABS documentation, National Register, and National Historic Landmark documentation all list an 1822 construction date (Roy et al. 2000).

11.4.1 Tree-ring analysis of Community House No. 2

Tree-ring analysis was performed in the dormitory's third story and attic spaces. The third story is within the community house's wood-framed mansard roof, and its timbers were exposed as part of the building's restoration. The roof is built of timbers angled to give the Mansard roof its characteristic angle, so the third floor's exterior walls angle into its interior space. Dutch biscuits were installed between the timbers for insulation and likely add additional stability to the roof.

The unfinished attic is located over the third floor. It contains two timber trusses paralleling the linear spaces on the third floor below. The roof framing, including the trusses, is documented in the 1934 HABS drawings (Figure 11.7). The measured drawings illustrate the building's complex construction, including the attic timber truss, so we knew they existed in 1934. When we collected samples, it was unclear if the trusses were an original part of the Harmonist building, an Owenite addition, a modification of the building to support 19th-century commercial use, or part of the 20th-century remodeling of the dormitory.

11.4.2 Dendrochronology results for Community House No. 2

We were able to successfully crossdate 20 tulip poplar samples from 14 timbers and create a chronology consisting of 1,573 rings ranging from 1686 to 1858. We also dated two red oak timbers (Tables 11.1 and 11.2). Dendrochronology

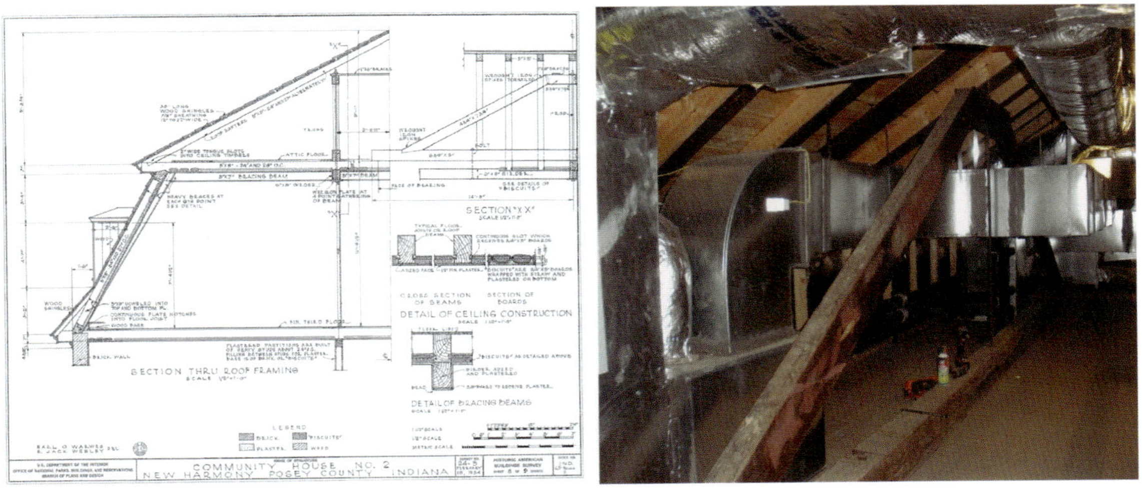

Figure 11.7
Left: HABS documentation drawing of roof and attic truss, Community House No. 2 completed in 1934. Right: Photograph of the attic roof truss at the time of tree-ring analysis.

Source: Left: Historic American Building Survey, Library of Congress, Prints & Photographs Division. Right: Authors.

Table 11.1 Samples collected from Community House No. 2 (New Harmony, Indiana). "First" and "last" refer to the first and last rings present in the sample. If no date is provided, the sample could not be reliably crossdated. "Outer ring" refers to the outermost ring of the sample: b = bark, w = wane, etc.; see Table 7.1 for a complete list of codes used for outer ring condition. "r" is the correlation coefficient obtained by correlating the ring widths of this sample with all other timbers of the same type from the building (series intercorrelation; Chapter 5). "Timber type" is the type of timber; see Table 3.1 for a list of possible tree species represented by the various timber types. "Provenience" is the timber's function in a structure; see Chapter 2 for a description of the various proveniences. Note: Timber type and provenience are only given once if more than one sample was obtained from an individual timber. An r-value is not reported if less than three timbers of a particular type were obtained from a structure. A full description of the data presented in the table is provided in Chapter 7.

Sample	*First*	*Last*	*Outer ring*	r	*Timber type*	*Provenience*
CHS01A	1743	1821	w	0.65	Tulip	Post; 3rd floor
CHS01B	1731	1821	w	0.69		
CHS02A	1731	1821	w	0.54	Tulip	Beam; 3rd floor
CHS03A	1755	1821	w		Red oak	Post/door frame; 3rd floor
CHS04A	1726	1820	w		Red oak	Post; 3rd floor
CHS05A	1728	1821	w	0.66	Tulip	Floor joist; attic
CHS05B	1798	1821	w	0.54		
CHS06A	1803	1822	w	0.75	Tulip	Post; 3rd floor
CHS06B	1728	1822	w	0.80		
CHS07A	1798	1822	w	0.56	Tulip	Post; 3rd floor
CHS07B	1722	1821		0.76		
CHS08A	1723	1822	w	0.78	Tulip	Post; 3rd floor
CHS08B	1682	1822	w	0.68		
CHS09A					Sweet gum	Post; 3rd floor
CHS10A	1741	1859	w	0.75	Tulip	Diagonal truss; attic
CHS11A	1766	1859	w	0.78	Tulip	Horizontal truss; attic
CHS12A	1739	1859	w	0.68	Tulip	Diagonal truss; attic
CHS13A	1770	1859	w	0.81	Tulip	Diagonal truss; attic
CHS14A	1803	1822	b	0.64	Tulip	Rafter
CHS14B	1782	1820		0.77		
CHS15A	1785	1822	b	0.86	Tulip	Roof joist/rafter
CHS16A					Sweet gum	Rafter
CHS17A	1695	1820		0.74	Tulip	Unknown timber from storage
CHS18A	1703	1821	w	0.64	Tulip	Unknown timber from storage

confirms that Community House No. 2 was completed during the Harmonist period in Indiana (1814–1824), and that it was most likely finished by the end of 1822. Sampling indicates that the third story was likely constructed during 1821 and 1822 (CHS01–08). Attic roof rafters have 1822 death dates that

Table 11.2 Summary table of the timber type chronologies prepared from New Harmony, Indiana. "Samples" and "timbers" are the number of successfully crossdated samples and timbers from a building. "Undated" is the number of timbers that could not be crossdated reliably. "First" and "last" refer to the first and last dates in the chronology. "Mean" and "SD" are the mean and standard deviation (mm) of the ring widths in the chronology. "r" is the mean series intercorrelation for each timber type (see Chapter 5). No total r, mean, or SD are provided since each chronology was prepared using only a single timber type. A full description of the data presented in the table is provided in Chapter 7.

Timber type	*Samples*	*Timbers*	*Undated*	*First*	*Last*	*Years*	r	*Mean*	*SD*
				Community House No. 2					
Sweet gum	0	0	2						
Red oak	2	2	0	1727	1820	158		1.22	0.20
Tulip	20	14	0	1686	1858	1573	0.71	1.29	0.61
Total	22	16	2	1686	1858	1731			
				Opera House					
Hard elm	0	0	1						
Soft maple	0	0	1						
Tulip	37	20	2	1704	1885	2112	0.64	1.13	0.44
Total	37	20	4	1704	1885	2112			
				Scholle					
Red oak	4	3	1	1712	1819	208	0.59	1.66	0.35
Tulip	4	4	1	1702	1853	311	0.33	1.01	0.43
White oak	11	8	1	1641	1819	944	0.53	1.17	0.33
Total	19	15	3	1641	1853	1463			
				Fauntleroy					
Red oak	4	3	0	1723	1818	273	0.54	1.64	0.41
Tulip	15	11	2	1699	1855	994	0.59	1.11	0.49
White oak	3	3	2	1710	1819	223	0.42	1.32	0.25
Total	22	17	4	1699	1855	1490			

indicate it is original and not a replacement from earlier renovations (CHS14 and 15).

Analysis also revealed the origins of the attic truss. Previous studies described it as Harmonist and suggested that the scale of the mansard roof required the additional support (Center for Historic Preservation 2007). However, dendrochronology indicates the trusses were added to the building about 1859, so they are neither from the Harmonist nor Owen periods (CHS10–13). Achille Emery Fretageot who had purchased the building in the mid-1850s likely added the trusses (Center for Historic Preservation 2007). Since the trusses span the locations where there were originally third-floor interior posts supporting the ceiling, they were likely added for structural support following the removal of interior partitions.

While the results of this study were successful in establishing a completion date for the dwelling's construction, and the date of attic's modification, it was not successful in establishing the starting date for construction. Since the samples are from the upper stories of the dormitory, they tell more about the conclusion of the construction process rather than the beginning. We have concluded that the timbers were harvested and added to the building in the year of 1822, because the entire 1821 growth ring is present, and at least part of the 1822 ring was found. These logs were most likely harvested after mid-growing season (about June of 1822). Since the sample locations were from the top stories of the structure, a date for the beginning of the house's construction could not be identified. Sample locations in the lower stories are unavailable because the exterior walls are brick, and the interior timber walls have plaster surfaces. Also, site managers believe that a significant amount of historic timber material in the lower stories was likely removed during the mid-century restorations when the interior was "gutted" (Evansville Courier 1980).

11.5 Community House No. 4/Thrall's Opera House

Community House No. 4 was the final dormitory built by the Harmonists. The structure was constructed of brick and has a gable roof (as opposed to the mansard roof of Community House No. 2). As we discussed, the structure is a story smaller than the previously constructed dormitories because Rapp believed too many residents would overtask the kitchen. It was believed to have been constructed about 1824, and it is unclear if it was ever used by Harmonists since it was so close to their move to Pennsylvania in the winter and spring 1825. Following Robert Owen's departure from New Harmony in 1827, the building was used as a grocery store and a school. The Thespian Society, a local theater group, purchased the building in 1856 and named it Union Hall. The society removed all the interior walls and the northern two-thirds of the second floor, leaving the remaining third as a balcony. A stage was constructed at the north end of the first floor. In 1867, a group of men that included Eugene Thrall bought the venue. In 1888, Thrall purchased the structure from his partners and added a two-story Romanesque Revival entry foyer to the Church Street façade, and a rectangular addition on the rear of the building. The new front entrance displayed a prominent pediment, bracketed cornice, and arched windows (Figure 11.8). Thrall passed away during the renovations, and his wife completed the expansion and named the theater The Thrall Opera House in his honor (Center for Historic Preservation 2008a). The theater had a capacity of 350 people, and its movable seats could be stored beneath the stage to open the room for events such as dances (Figure 11.9). In 1913, Ida Hyne opened an automotive garage in the opera house by expanding the front entry, removing the stage, and installing a concrete floor.

The opera house's roof contains a large wooden pulley. Site managers are unclear if it was installed by Harmonists to assist in the construction of the dormitory, or if it was installed during the building's theater period to move curtains and stage sets. Author DLR was asked to examine the wheel to see if it was a candidate for tree-ring analysis. The pulley, essentially a spoked wheel, is approximately 36 inches (1 meter) in diameter, and the outer edge is grooved to house a rope (Figure 11.10). Analysis identified that it is constructed from a variety of wood species (tulip, white oak, and basswood). No bark or wane was present so

Figure 11.8
Photograph of Thrall's Opera House (formerly Community House No. 4), showing the 1888 entry addition to the Harmonist structure.

Source: Authors.

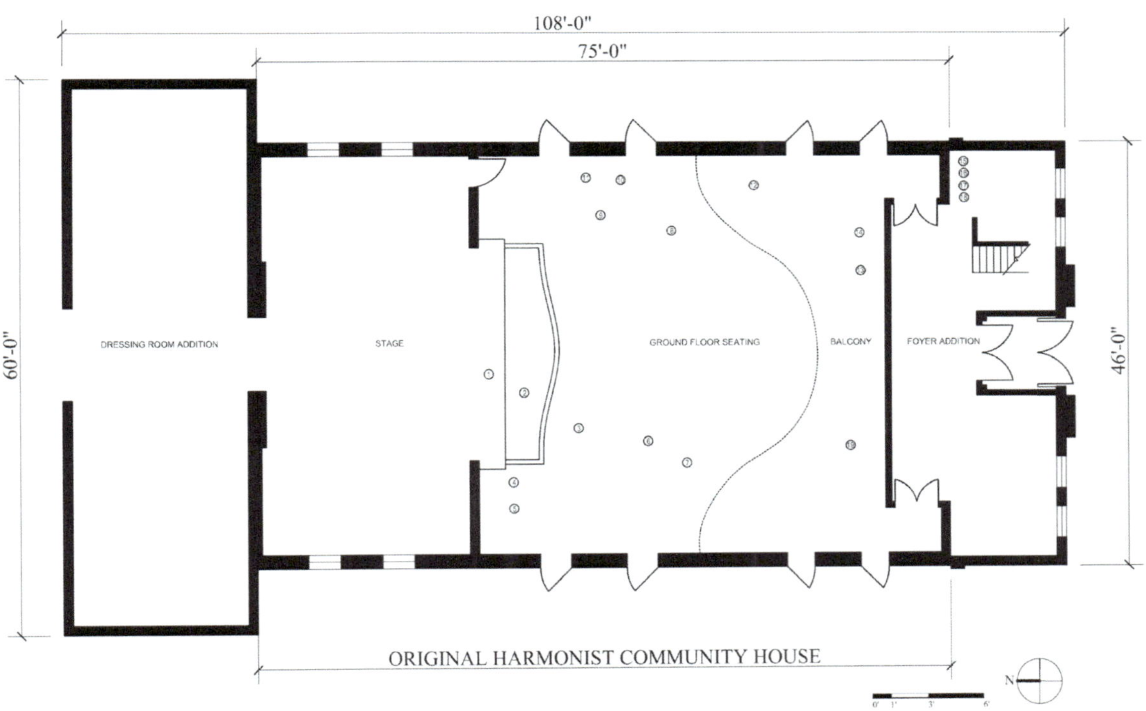

Figure 11.9
Plan drawing and sample locations of Thrall's Opera House (formerly Community House No. 4).

Source: Authors.

sampling was not performed. However, several aspects of its construction suggest it is not Harmonist. Some timbers contain circular saw marks indicating it was likely produced later in the 19th century. Its haphazard construction (for instance the use of a recycled floorboard to build the wheel's support) is not representative of Harmonist artisanship. Finally, the use of such a pulley for the construction of major buildings is not mentioned in archival sources, and none of the remaining Harmonist structures shows evidence of having pulleys.

Figure 11.10
Photo of pulley located in Thrall's Opera House (View is looking up).

Source: From the collection of the Indiana State Museum and Historic Sites.

11.5.1 Community House No. 4/Thralls Opera House archival and scholarly sources

Most primary and secondary sources place Community House No. 4 construction at 1823 or 1824. It is included on both the Pickering and Weingärtner maps, indicating the structure was present in 1824. Sanborn maps effectively illustrate the building's evolution (Figure 11.11). The 1888 map shows a plan view of Union Hall. The 1893 map identifies the building as the Opera House and shows the foyer and dressing room additions.

We found that secondary sources proposed a range of build dates. HABS provides photos of the building from 1940 when it was a garage, but the documentation did not include dates for construction or the alterations. Newspaper reports covering the structure's restoration identified the dormitory's construction as 1823. (Indianapolis Star 1968; Strupp 1968). The National Register application reports an 1818 construction date, with 1888 alterations. National Historic Landmark documents and the Historic Structures Report identify an 1824 construction date, with 1856 and 1888 alterations.

11.5.2 Community House No. 4/Thralls Opera House dendrochronology results

Thirty-seven samples from 20 timbers produced a tulip poplar chronology of 2,112 rings ranging from 1704–1885 (Tables 11.2 and 11.3). Sampling locations were limited to the unfinished attic space containing exposed roof and wall timbers. Tree-ring analysis supported previously reported dates for the construction of, and alterations to, the building. It appears that the roof structure for the Harmonist dormitory was constructed in 1823, so an 1824 finish date of construction appears correct (TOP01, TOP03–04, TOP06–12, and TOP19). The attic timber dating 1818, presumably installed toward the end of the construction period

Figure 11.11
Sanborn Map Company images of Union Hall in 1888 (top), and Thrall's Opera House in 1893 (bottom) showing the addition of the entry foyer and dressing rooms.

Source: Indiana Spatial Data Portal, Indiana University.

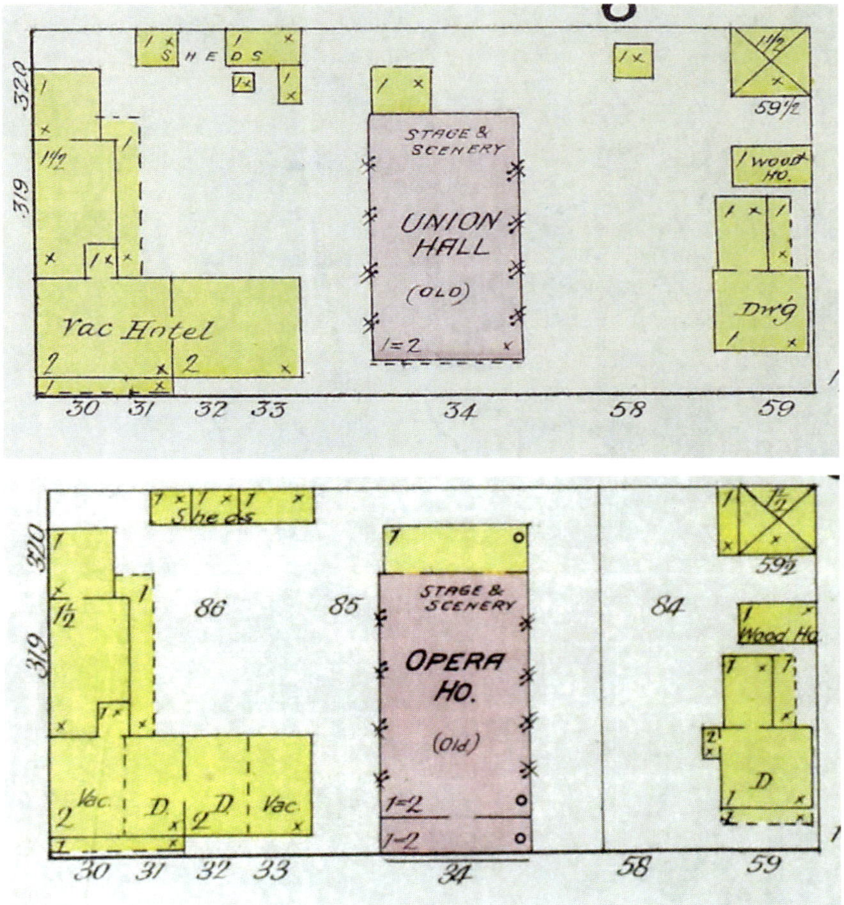

Table 11.3 Samples collected from Community House No. 4/Thrall's Opera House (New Harmony, Indiana). See Table 11.1 for a description of the table contents; a full description of the data presented is provided in Chapter 7.

Sample	*First*	*Last*	*Outer ring*	r	*Timber type*	*Provenience*
TOP01A	1731	1822	w/e	0.77	Tulip	Floor Joist, attic
TOP01B	1782	1818		0.71		
TOP02A	1828	1847		0.89	Tulip	Horizontal brace attached to floor joists
TOP02B	1718	1851		0.59		
TOP02C	1838	1856	b	0.71		
TOP02D	1775	1826		0.48		
TOP02E	1716	1769		0.58		
TOP03A	1798	1823	w/e	0.81	Tulip	Diagonal roof truss
TOP03B	1730	1823	w/e	0.73		

(Continued)

Table 11.3 Continued

Sample	*First*	*Last*	*Outer ring*	r	*Timber type*	*Provenience*
TOP03C	1741	1787		0.71		
TOP04A	1749	1822	w	0.55	Tulip	Floor Joist, attic
TOP04B	1771	1822	w/e	0.56		
TOP05A	1789	1818	w	0.75	Tulip	Diagonal roof truss
TOP05B	1786	1818	w	0.77		
TOP06A	1780	1823	w/e	0.73	Tulip	Vertical post supporting north ridge pole
TOP06B	1704	1772		0.37		
TOP07A	1808	1823	w/e	0.68	Tulip	Diagonal roof truss
TOP07B	1808	1822	w	0.57		
TOP08A	1732	1822	w	0.77	Tulip	Floor Joist, attic
TOP08B	1800	1822	b	0.77		
TOP09A	1778	1822	w	0.61	Tulip	Post supporting S ridge pole
TOP10A	1792	1823	w/e	0.64	Tulip	Diagonal roof truss
TOP10B	1751	1780		0.62		
TOP11A	1768	1822	w	0.76	Tulip	Diagonal roof truss
TOP11B	1707	1822	w	0.56		
TOP12A	1729	1822	w	0.53	Tulip	Diagonal roof truss
TOP12B	1783	1822	w	0.38		
TOP12C	1730	1783		0.66		
TOP13A	1781	1886	w	0.72	Tulip	Stud, façade addition
TOP14A	1756	1886	w	0.71	Tulip	Stud, façade addition
TOP15A	1703	1791		0.56	Tulip	Loose timber; likely floor joist in façade addition
TOP16A	1818	1882		0.36	Tulip	Floor joist; façade addition
TOP17A					Soft maple	Floor joist; façade addition
TOP18A					Hard elm	Diagonal roof truss
TOP19A	1749	1822	w	0.70	Tulip	Post supporting N ridge pole
TOP20A	1793	1857	w	0.72	Tulip	Loose
TOP20B	1795	1857	w	0.74		
TOP21A	1791	1857	b	0.74	Tulip	Loose
TOP22A						Loose
TOP23A	1798	1857	w	0.71	Tulip	Loose
TOP24A					Tulip	Loose

might suggest an earlier start to construction, but more likely represents timber from the Society's lumber mill stock (TOP05). Analysis also confirmed the Thespian Society's mid-1850s alterations (TOP02, 20, 21, and 23) and Thrall's 1888 Romanesque Revival foyer addition (TOP13–14 and 16).

11.6 Mattias Scholle House

In 1808, at the age of 12, Mattias Scholle immigrated with his family from Germany to the United States. As followers of Father Rapp they joined the Harmonists at their first community of Harmony, Pennsylvania. He then traveled with the Society to the Indiana Territory in 1814. The 1820 US Population Census lists Scholle as head of household, and sharing the residence with three brothers and three sisters (Center for Historic Preservation 2008b). Mattias was a shoemaker, and he produced boots and shoes that supplied local markets as well as shops in St. Louis, New Orleans, Pittsburgh, and Boston (Figure 6.5). It is unclear what role he took in the Society to warrant a brick home, but his name appears in Society letters connecting him to roles in the distribution of Harmonist-made shoes, and as Rapp's representative charged with assessing the status of Harmonist crops (Arndt 1978). The house is representative of the standardized Harmonist form of three rooms on each story (Figure 11.12). Over time, the house was expanded with small structural additions, most notably a Victorian era entry porch and other small additions (Broz 1979).

11.6.1 Archival and scholarly sources for the Scholle House

The archival and scholarly sources we used in our research of the Scholle House were the Pickering and Weingärtner maps, Sanborn Map Company maps, photographs, and letters. Secondary sources available to us were the Blair, Arndt, and Strezewsi's Harmonist histories, the Historic Structures Report, and National Register and Historic National Landmark documents.

Figure 11.12
Mattias Scholle House showing the south and east façades.
Source: Authors.

Prior to tree-ring analysis, the house was thought to have been constructed sometime between 1816 and 1824 (Indiana Bell 1967; Center for Historic Preservation 2008b). The Weingärtner map identified Mattias Scholle as the head of household (Figure 11.13), and the US Population Census provided information regarding his age and occupation. We knew the house was constructed by 1824 because it was illustrated on the Weingärtner and Pickering maps. Arndt's Harmonist history includes an inventory from January of 1819 of the Society's buildings that identifies a single brick house, presumably Rapp's, and 150 log houses. This suggests that brick house construction had not yet started, and the source narrows the assumed construction date to between 1819 and 1824. The National Register nomination reports that the house was constructed circa 1823, and National Historic Landmark documents report 1822. However, neither document provides sources.

Sanborn Map Company maps and photographic primary sources document the late 19th and early 20th-century additions to the structure (Figure 11.14). The maps show that an entry porch was added between 1888 and 1899, and then enlarged by 1907. Between 1899 and 1907, a small one-story addition to the

Figure 11.13
Detail of the Weingärtner map showing a depiction of the Mattias Scholle House. The house is a typical two-story Harmonist brick structure with three rooms on each floor.

Source: Pennsylvania State Archives, Old Economy Village.

Figure 11.14
1893 Sanborn Map Company image of the Mattias Scholle House. The image shows the additions to the original Harmonist structure (pink). The additions were removed in the 1960s restoration.

Source: Indiana Spatial Data Portal, Indiana University.

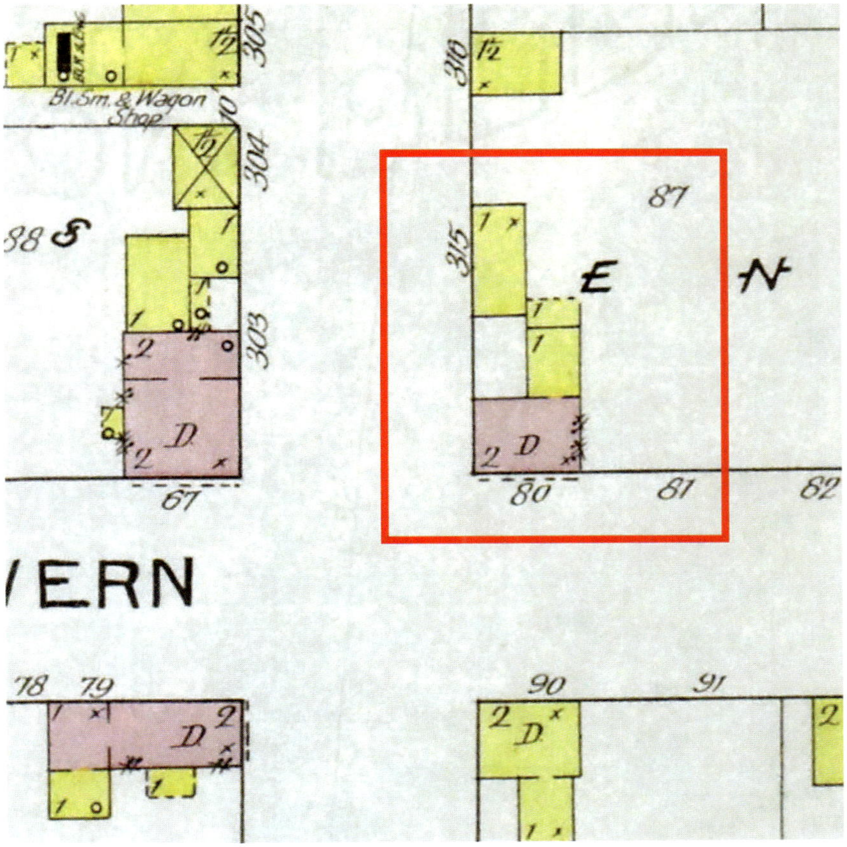

house was removed, and an outbuilding was replaced with a smaller structure. All of these additions were removed during building restoration.

11.6.2 Dendrochronology results for the Scholle House

We successfully crossdated 19 samples from 15 timbers to produce chronologies spanning from 1641 to 1853 (Tables 11.2 and 11.4; Figure 11.15). Twelve timbers had wane or bark, and provided harvest dates. Tree-ring analysis supports a construction date of 1820 based on the death dates of two of the roof rafters (SCH02 and SCH05). The death dates of 1818 and 1819 for other rafters (SCH01, 03, 04, 06, and 07) and a beam (SCH16) suggest the house may have been constructed over the course of a few years, or that the timbers, especially rafters, are from Harmonist lumber stock cut in previous years. Attempts to sample cellar timbers that might indicate the start of construction did not produce dates. The plank used in a Dutch biscuit dates 1816, four years prior to the house's construction. This suggests they were mass produced and stored until installed into a structure, but this conclusion is based on only one sample (SCH14A). Conversely, a dead tree could have been used to construct the biscuits.

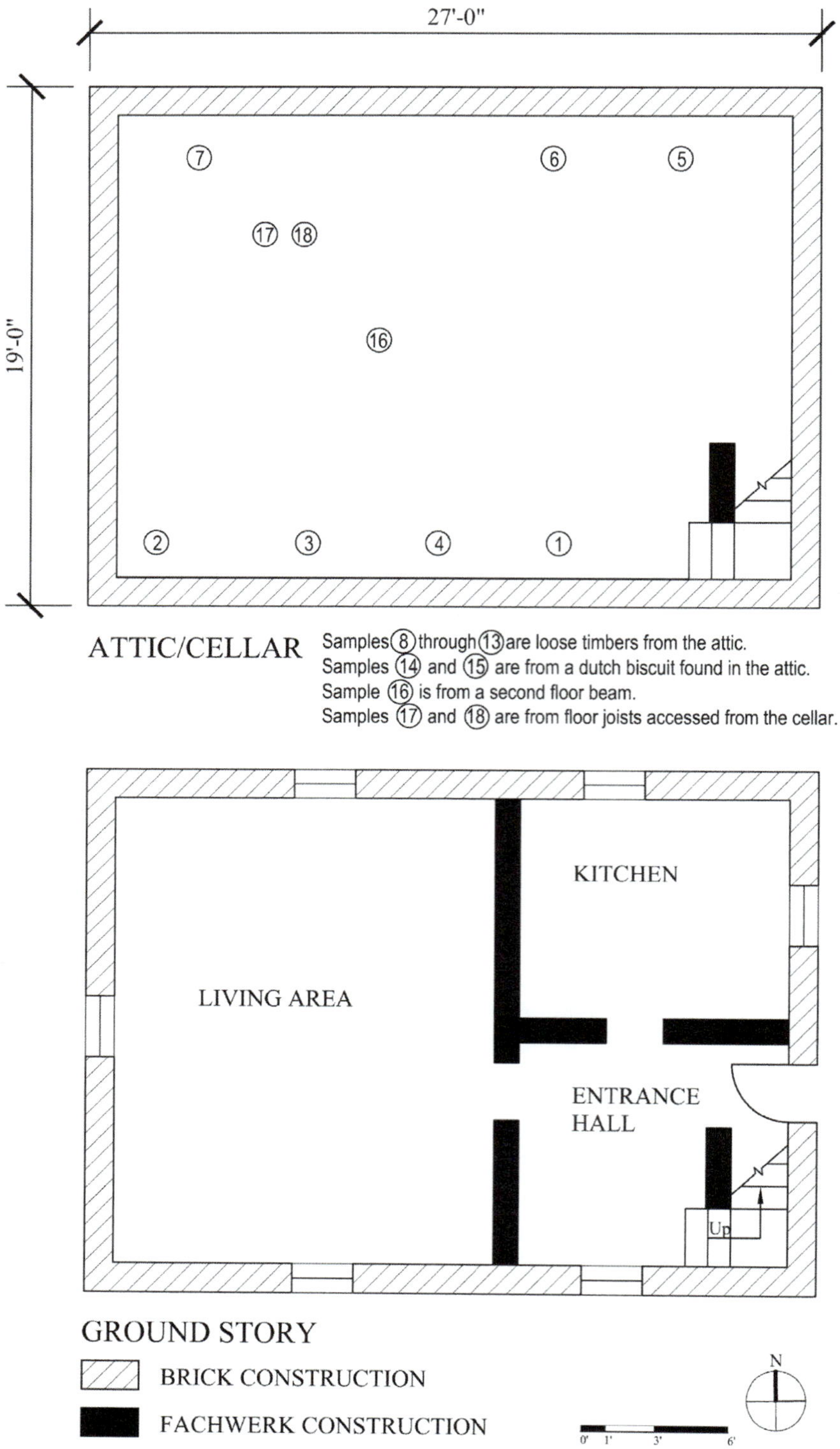

Figure 11.15
Plan drawing of the Mattias Scholle House showing sample locations.

Source: Authors.

Table 11.4 Samples collected from Mattias Scholle House (New Harmony, Indiana). See Table 11.1 for a description of the table contents; a full description of the data presented is provided in Chapter 7.

Sample	*First*	*Last*	*Outer ring*	r	*Timber type*	*Provenience*
SCH01A	1780	1818	w	0.65	Red oak	Rafter
SCH01B	1747	1780		0.55		
SCH02A	1708	1820	w/e	0.44	White oak	Rafter
SCH03A	1790	1819	w/e	0.72	White oak	Rafter
SCH03B	1738	1819	w/e	0.62		
SCH04A	1787	1819	w/e	0.75	White oak	Rafter
SCH04B	1760	1819	w/e	0.50		
SCH05A	1699	1820	w	0.55	White oak	Rafter
SCH06A	1669	1819	w/e	0.33	White oak	Rafter
SCH07A	1718	1811		0.60	White oak	Rafter
SCH07B	1806	1818	w	0.76		
SCH08A	1711	1820	w/e	0.65	Red oak	Loose timber in attic
SCH09A	1701	1811		0.50	Tulip	Loose timber in attic
SCH10A	1800	1854	w	0.12	Tulip	Loose timber in attic
SCH11A	1688	1818	w	0.61	White oak	Loose timber in attic
SCH12A	1640	1788		0.51	White oak	Loose timber in attic
SCH13A	1739	1805		0.25	Tulip	Loose timber in attic
SCH14A	1780	1816	w	0.42	Red oak	Loose Dutch biscuit
SCH15A					Red oak	Loose Dutch biscuit
SCH16A	1720	1819	w	0.29	Tulip	Beam; 2nd floor and attic
SCH17A					Tulip	Joist; ground floor
SCH18A					White oak	Floor beam in cellar

11.7 Fauntleroy House

The Fauntleroy House is a wood-framed version of the Harmonist house form (Figure 11.16). Franz Pfiel, a tanner in the Harmonist Society, originally occupied the residence. Robert Henry Fauntleroy (1806–1849) and Jane Dale Owen (1806–1861) purchased the house in 1841 (Center for Historic Preservation 2007a). Fauntleroy was a scientist and engineer from Virginia who had come to New Harmony about 1830. Jane Dale was Robert Owen's daughter who moved to New Harmony in 1833. The Fauntleroy's transformed the original Harmonist house by constructing a one and a half story addition to the rear of the house. The addition, called a hall, became the main entrance into the house via an entry door and porch on the south façade. The hall contains a staircase to the upstairs bedrooms. The original Harmonist structure was also raised approximately 18 inches, resulting in taller lower story rooms. Robert died in 1849, and ownership of the house was transferred to Jane. Their daughter Constance founded the Minerva Society, one of the earliest American women's literary clubs, in the house in 1859 (Walker 1913). Fauntleroy family descendants sold the house to the state of Indiana in 1939.

Figure 11.16
Photograph of the north façade of the Fauntleroy House. The original Harmonist house is on the left, the hall addition is the shed roof in the center, and a later addition is to the right.

Source: Authors.

11.7.1 Archival and scholarly sources for the Fauntleroy House

Archival and scholarly sources for the Fauntleroy House are very similar to the Scholle House, and include the Pickering and Weingärtner maps, Sanborn Map Company maps, and photographs. In addition to Harmonist histories by Blair, Arndt, and Strezewski we reviewed works by Carmony and Elliott (1980) and Lockridge (1939). A Historic Structures Report has been completed for the house, and it is included in Historic National Landmark documents. A construction date range of 1815 to 1824 has been attributed to the house (Center for Historic Preservation 2007a). An early National Register documentation of the house reported it to be built in 1815 and that it was constructed of walnut, oak, and hickory (Mendinghall 1977). A later, more comprehensive National Register listing identified an 1822 construction date with 1840 modifications. Historic National Landmark documents also reported an 1822 date.

Historian Ross Lockridge (1939) wrote that the house's additions were constructed in the 1830s prior to the Fauntleroy's purchase, but he does not provide sources. Sanborn Map Company maps show that the two additions to the west side of the Harmonist structure were constructed prior to 1893 (Figure 11.17). Between 1893 and 1899 the property was subdivided and a neighboring house constructed; outbuildings were removed, a shed addition was added to extend the house to the west property line, and the south entry porch was enlarged.

11.7.2 Dendrochronology results for the Fauntleroy House

Crossdating permitted us to accurately date 22 samples from 17 timbers (tulip poplar, red oak, and white oak timbers were dated). We were able to produce chronologies spanning from 1699 to 1855 (Tables 11.2 and 11.5; Figure 11.18).

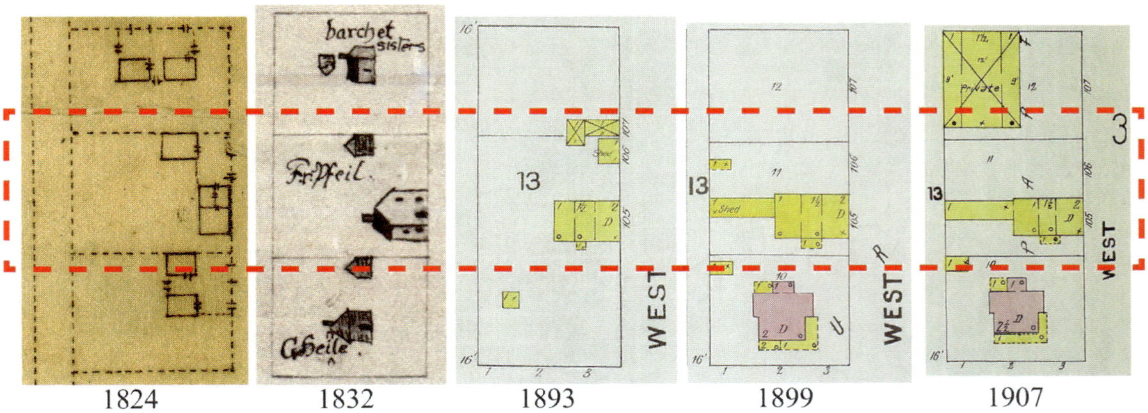

Figure 11.17
A series of maps illustrating the evolution of the Fauntleroy House and property: 1824 Pickering map; 1832 Weingärtner map; and the 1893, 1899, and 1907 Sanborn maps.

Source: From the collection of the Indiana State Museum and Historic Sites; Pennsylvania State Archives, Old Economy Village; Indiana Spatial Data Portal, Indiana University.

The structure was sampled during restoration, and there was an exhaustive search for wane and bark-bearing timbers. An 1820 construction date was assigned to the residence based on the death dates of sampled timbers (FNT08 and FNT13). We were able to establish an 1841 date for raising the house based upon the consistent death dates of the blocks inserted in the structure (Figure 11.19; FNT03–06). One sample in the hall addition dated to 1840 (FNT01), but other samples ranged from 1818 to 1856. Likewise, the rear room additions resulted in scattered dates from the 1810s and the 1830s, suggesting the use of recycled timbers. Two timbers were identified that possibly relate to the 1830s alteration mentioned by Lockridge (FNT09 and FNT20). Therefore, dendrochronology was unable to confirm a build date for the hall or the rear addition. It could not confirm that the hall was constructed at the time the house was raised in 1841. For the house additions, dendrochronology actually raised more questions than answered and shows the limitations of the science.

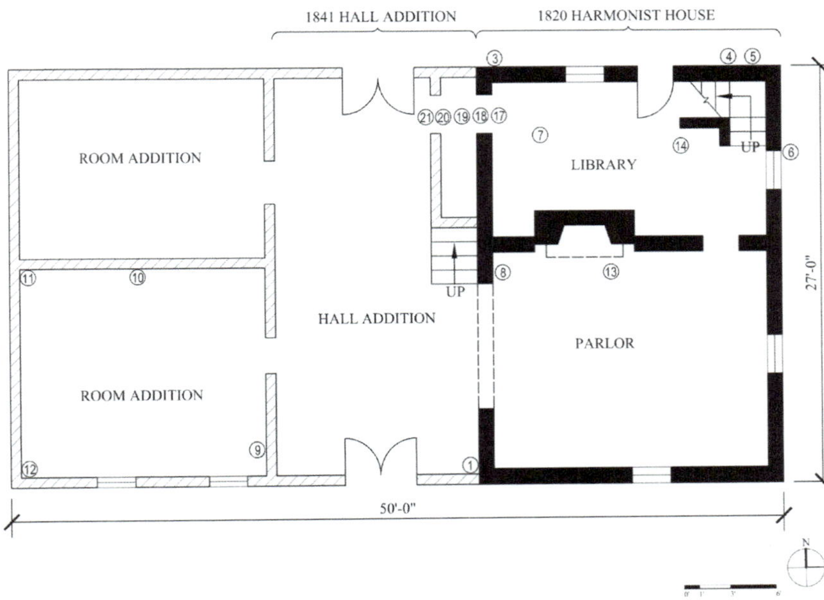

Figure 11.18
Plan drawing of the Fauntleroy House showing sample locations. The walls of the original Harmonist house are the solid hatch.

Source: Authors.

Table 11.5 Samples collected from Fauntleroy House (New Harmony, Indiana). See Table 11.1 for a description of the table contents; a full description of the data presented is provided in Chapter 7.

Sample	*First*	*Last*	*Outer ring*	*r*	*Timber type*	*Provenience*
FNT01A	1746	1840	w	0.67	Tulip	Stud; hall
FNT01B	1756	1840	w	0.76		
FNT02A					Tulip	Loose
FNT03A	1765	1828		0.50		Riser block; original house
FNT03B	1823	1841	w/e	0.85	Tulip	
FNT04A	1766	1841	w/e	0.76	Tulip	Riser block; original house
FNT05A	1810	1841	w/e	0.77	Tulip	Riser block; original house
FNT06A	1765	1841	w/e	0.75	Tulip	Riser block; original house
FNT07A	1785	1819	w	0.59	Red oak	Ceiling joist; original house
FNT07B	1738	1813		0.65		
FNT08A	1699	1820	w/e	0.49	White oak	Stud; original house
FNT09A	1763	1824		0.58	Tulip	Stud; addition
FNT09B	1781	1832	w	0.49		
FNT10A	1722	1819	w/e	0.40	Red oak	Stud; addition
FNT11A					Tulip	Post; addition
FNT12A	1698	1765		0.25	Tulip	Post; addition
FNT13A	1749	1820	w/e	0.40	White oak	Joist; 2nd floor; original house
FNT14A	1745	1819	w	0.55	Red oak	Joist; 2nd floor; original house
FNT15A					White oak	Loose; attic
FNT16A					White oak	Loose; attic
FNT17A	1780	1856	b	0.51	Tulip	Lath nailer; attic; addition
FNT18A	1786	1846		0.77	Tulip	Lath nailer; attic; addition
FNT18B	1786	1856	w	0.49		
FNT19A	1773	1818	w	0.33	White oak	Lath nailer; attic; addition
FNT20A	1723	1834	b	0.42	Tulip	Rafter; addition
FNT21A	1767	1856	w	0.45	Tulip	Lath nailer; attic; addition

Figure 11.19
Photograph of timber blocks inserted into the Fauntleroy House in the 1840s to add approximately 18 inches of vertical space to the lower story. The blocks rest on a new sill plate, and the original Harmonist timbers are fastened together with metal angles.

Source: From the collection of the Indiana State Museum and Historic Sites.

11.8 Summary of the tree-ring analysis of Harmonist buildings

The tree-ring analysis of the four New Harmony buildings resulted in 100 samples from 68 timbers. We measured 6,796 rings spanning 1641 to 1885.

Historian Don Blair describes a Harmonist preference for tulip poplar (*Liriodendron tulipifera*) over other timber species (Blair 1964). The tree is abundant throughout the American Middle West, but is not a species present in the Harmonists' German homeland. According to Blair, Harmonist carpenters found that it had ample structural strength, was easy to shape, and handled the humid and moist riverside climate of southern Indiana. Our analysis tends to support a Harmonist bias for tulip poplar, which was overwhelmingly the most common species we sampled. However, we are tentative to call this definitive since other species were documented. The Scholle and Fauntleroy Houses contain several red and white oak timbers, and maple, elm, and sweet gum are also present. Also, we have not completed an entire timber type identification for an entire Harmonist structure like we have with other buildings. More information about species bias is included in Chapter 13.

11.9 Conclusion

Dendrochronology of buildings in New Harmony, Indiana, was able to identify build and alteration dates for several Harmonist dwellings. The results of the analysis also raised many questions that dendrochronology could not answer, such as construction dates for the additions to the Fauntleroy House. Whether conclusive or not, the results added new and unique knowledge to the documentation of the historic town. In Chapter 12, we look at how tree-ring analysis is used to interpret a regional agricultural economy.

11.10 Literature cited

Arndt, K.J.R. 1965. *George Rapp's Harmony Society: 1785–1847*. University of Pennsylvania Press, Philadelphia, PA. 713 p.

Arndt, K.J.R., ed. 1978. *A Documentary History of the Indiana Decade of the Harmony Society 1814–1824: Volume I 1814–1819*. Indiana Historical Society, Indianapolis, IN. 978 p.

Baas, C., and D.L. Rubino. 2013. Using tree-ring growth patterns to identify the date of construction of community house no. 2, New Harmony, Indiana. *Indiana Archaeology* 8: 11–31.

Blair, D. 1964. *Harmonist Construction*. Indiana Historical Society, Indianapolis, IN. 82 p.

Brown, P. 1972. *Twelve Months in New Harmony*. Porcupine Press, Philadelphia, PA. 128 p.

Broz, R. 1979. *Mattias Scholle House: National Register of Historic Places Inventory – Nomination Form*. National Park Service, Washington, DC.

Carmony, D.F., and J.M. Elliott. 1980. New harmony, Indiana: Robert Owen's seedbed for utopia. *Indiana Magazine of History* 76: 161–261.

Center for Historic Preservation. 2007a. *Historic Structure Report for the Fauntleroy House: New Harmony Historic Site*. The Center for Historic Preservation, Department of Architecture, Ball State University, Muncie, IN. 195 p.

Center for Historic Preservation. 2007b. *Historic Structure Report for Community House Number Two, New Harmony State Historic Site, New Harmony, Indiana*. The Center for Historic Preservation, Department of Architecture, Ball State University, Muncie, IN. 233 p.

Center for Historic Preservation. 2008a. *Historic Structure Report for Thrall's Opera House: New Harmony Historic Site, New Harmony, Indiana*. The Center for Historic Preservation, Department of Architecture, Ball State University, Muncie, IN. 146 p.

Center for Historic Preservation. 2008b. *Historic Structure Report for the Scholle House: New Harmony Historic Site, New Harmony, Indiana*. The Center for Historic Preservation, Department of Architecture, Ball State University, Muncie, IN. 124 p.

Evansville Courier. 1980. Remember when. *The Evansville Courier*. Evansville, IN. January 18.

Glassie, H. 1968. *Pattern in the Material Folk Culture of the Eastern United States*. University of Pennsylvania Press, Philadelphia, PA. 316 p.

Hoyt, G.R. 1996. Germans. Pp. 146–181. In: *Peopling Indiana: An Ethnic Experience*. (R.M. Taylor and C. McBirney, eds.). Indiana Historical Society, Indianapolis, IN. 703 p.

Hurdis, F. 2003. *New Harmony National Historic Landmark Boundary Increase*. National Historic Landmarks Survey, National Park Service, Washington, DC.

Indiana Bell. 1967. Indiana Bell gives historic new harmony home to state. *Indiana Bell News*. January–February.

Indiana Historical Commission. 1975. *New Harmony as Seen by Participants and Travelers*. Porcupine Press, Philadelphia, PA. 150 p.

Indianapolis Star. 1968. New harmony beholds restored opera house. *Indianapolis Star*. Indianapolis, IN. November 20.

Kniffen, F. 1965. Folk housing: Key to diffusion. *Annals of the Association of American Geographers* 55: 549–577.

Lindley, H. 1916. *Indiana as Seen by Early Travelers*. Indiana Historical Commission, Indianapolis, IN. 596 p.

Lockridge, R.F. 1939. *The Old Fauntleroy Home*. New Harmony Memorial Commission, New Harmony, IN. 256 p.

Mendinghall, J.S. 1977. *New Harmony Historic District*. National Register of Historic Place Nomination Form, National Park Service, Washington, DC.

Roy, J., J. Eickhoff, and S. Branigin. 2000. *New Harmony Historic District*. National Register of Historic Places Nomination Form, National Park Service, Washington, DC.

Rubino, D., and C. Baas. 2014. Using tree-ring growth patterns to date the construction of a nineteenth century dogtrot house in Posey County, Indiana. *Proceedings of the Indiana Academy of Science* 123: 14–22.

Stanton, G.W. 1985. *Brought, Borrowed, or Bought: Sources and Utilization Patterns of the Material Culture of German Immigrants in Southeastern Indiana, 1833–1860*. PhD. Dissertation, Department of Folklore, Indiana University, Bloomington, IN. 444 p.

Strezewski, M. 2015. Harmonist demography and town planning in New Harmony, Indiana. *Communal Societies* 35: 1–27.

Strupp, S.D. 1968. No title. *Mount Vernon Democrat*. Mount Vernon, IN. October 26.

Taylor, A. 1987. *Visions of Harmony: A Study of Nineteenth-Century Millenarianism*. Clarendon Press, Oxford, UK. 285 p.

Tishler, W.H. 1986. Fachwerk construction in the German settlements of Wisconsin. *Winterthur Portfolio* 21: 275–287.

Walker, C.M. 1913. Concerning the Hoosier: An appreciation. *Indiana Magazine of History* 9: 23–46.

Chapter 12

Innovation to obsolescence

Tree-ring analysis of a regional barn type

12.1 Early commercial production and shipping of hay

Throughout the 19th century, the growth of urban horse populations in US cities created a steadily increasing need for hay (McShane and Tarr 2007). To meet this demand, mid-Ohio Valley farmers supplied distant markets with timothy hay (*Phleum pratense* L.) via the Ohio and Mississippi Rivers. The New Orleans market was their primary, downstream destination for hay that was subsequently exported to East Coast cities.

Switzerland County, Indiana, is recognized as the heart of this commercial system (Brown 1858; Figure 12.1). It was in the county's rural village of Allensville that Samuel Hewitt invented and patented the beater hay press in 1843 (Figure 12.2). Like a guillotine, Hewitt's automated, animal-powered machine compacted timothy by dropping a massive wooden block from a height of multiple stories into a hay-filled box. Horses in the basement of the barn turned a sweep that, through a set of pulleys, lifted the beater to the top of the barn where it was released to fall between a set of timber rails onto the grass. This sequence took place multiple times until the box was full. The invention was first referred to as a "beater press." However, Hewitt was a well-known Mormon, so it became commonly known as the "Mormon Hay Press," or "Mormon Beater Hay Press."

Compressed hay is easier to store, ship to market, and sell than loose hay. New techniques were developed to reduce its bulk properties through pressing and baling. Hewitt's machine was unique from other inventions because it pressed 400-pound (180-kilogram) bales that simplified handling, storage, and transportation (Figure 12.3). The machine was appealing to regional farmers because it used horsepower to maximize the pressing strength of a metal screw and incorporated a wooden beater to replace the time-consuming process of compressing hay using human weight. Most appealing was the machine's ability to produce bales that made it possible to load twice as much hay onto boats than bales made with other presses (Baas and Rubino 2013, 2014).

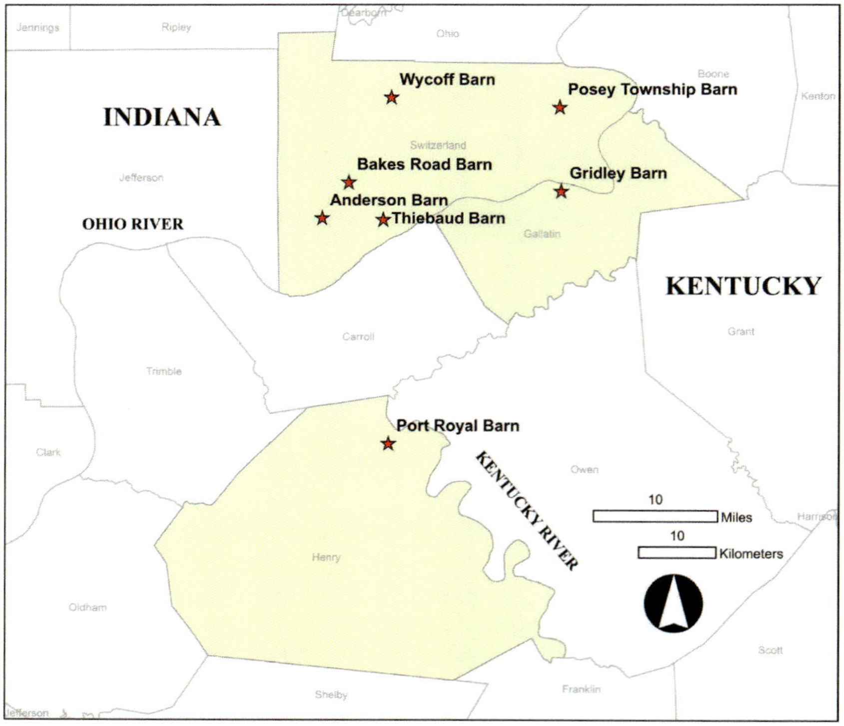

Figure 12.1
Barn and farmstead locations analyzed in Chapter 12.
Source: Authors.

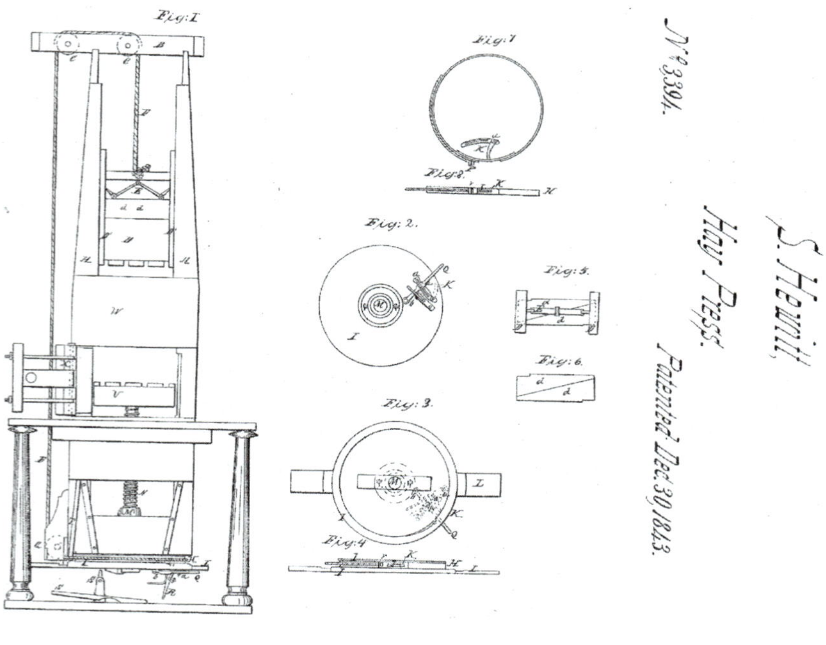

Figure 12.2
Samuel Hewitt's 1843 patent for a beater hay press. US Patent No. 3,394, issued December 30, 1843.
Source: US Patent Office.

Figure 12.3
The upper two stories of the Mormon beater hay press in the Denver Markland Hay Press Barn. The hay is placed in the upper box, and the bale is removed from the lower. The bottom portion of the press (the screw, sweep, and pulley for horse) is located in the cellar. The photo also includes a portion of the barn bent that carries the weight of the machine. The press and barn are restored and operational, and are located at the Switzerland County Agricultural Museum.

Source: Authors.

12.2 Defining, interpreting, and preserving a regional vernacular barn type

The hay press was housed in a vernacular barn form with spaces for curing, baling, and moving loose and baled hay. While historical documents describe the press and its operations in detail, very little is recorded about press barns. Recently, several barns have been restored with operating presses. The co-author Christopher Baas was involved with the rehabilitation of the Leavenworth-Lang-Cole press and barn now located at O'Bannon Woods State Park in southern Indiana. He also developed the masterplan for the Switzerland County Agriculture Museum that now has two press barns and one operating press. The public is now able to visit these presses and barns and observe the press in action.

About 2010 Baas contacted co-author Darrin L. Rubino regarding the tree-ring analysis of surviving press barns. Since that time, we have analyzed 16 barns. We are working with private and public barn owners, public agencies, historical societies, and museums to define, interpret, and preserve this rare vernacular barn type that is the icon for the historic regional economy. This chapter reports the tree-ring analysis of 16 press barns, our attempts to identify a pattern in the evolution in their form, the use of recycled timbers in barn construction, and our search for a press barn precedent that was modified to house a press and its operations. We analyze the barns within the context of the historic farmstead landscape, therefore additional buildings associated with the press barns are included in our analyses.

12.3 Archival sources and scholarship of Mormon beater hay press barns

To interpret hay press barns we rely upon local histories, newspapers, agricultural magazines, maps, atlases, census data, and patents. Documentation of hay presses, press barns, and the region's hay culture is primarily located in local histories celebrating Hewitt's invention (Dufour 1925), and the "Hay King" Ulysses Schenck's production and shipping enterprise that made his family one of the wealthiest in the region (Beach 1987). Folklorist Warren E. Roberts (1993) described a Crawford County, Indiana, barn and press using fieldwork and oral history (this is the press barn now on display at O'Bannon Woods State Park). The authors have published articles about the materials used to construct hay presses (Baas and Rubino 2013), and the tree-ring analysis of several hay press barns (Baas and Rubino 2012, 2014; Baas et al. 2017).

Hay press technology evolved rapidly throughout the 19th century when several inventors proposed different methods for baling hay using human or animal power with various combinations of levers, beaters, and screws. We found the effectiveness of these inventions was debated in agriculture-based periodicals such as *Prairie Farmer*, *Ohio Cultivator*, and *The Country Gentleman*. Essential details of pressing hay are found in the news and advertisements published in the region's newspapers. The economic livelihood of hay production was so engrained in the local culture that advice for growing and harvesting timothy, market reports, advertisements for press-related products, and shipping news were routinely published. Press barns were listed in real estate advertisements, and their misfortunate loss to fire was news.

US patents provided us with dates for press inventions, describe how a hay press operates, and shows the evolution of pressing technology (US patents are viewable online). Patents described distinct variations in press forms ranging from portable pressing machines used in the field, to stationary designs, to machines like the Mormon beater press that required the supportive framework of a building to operate. Historic maps and atlases allowed us to connect farmers to specific press barns. Many map and atlas publishers identified the structures as landmarks in their illustrations (Baas and Rubino 2014; Baas et al. 2017). Additionally, agricultural data from the US Census allowed us to track a farmer's hay production over time.

12.4 Testing theories for an evolution in press barn forms

As a vernacular building type, the Mormon beater hay press barn can be categorized into the fundamental components of construction, use, and form (Glassie 1968; Baas and Rubino 2013). Press barns have gable roofs and are three stories in height, or can also be described as two stories over a cellar. They are constructed of timber frames and are enclosed in vertical siding. We have observed how much larger joists and bent timbers are in comparison to the non-press barns in the region. This demonstrates the additional strength designed into the structures to handle the stresses imposed by the press operations and the weight of the hay, horses, and wagons. Rock cellar walls or footings support the frames. Since the barns are located on farmsteads with both flat and rolling topography, they are either constructed into a hillside, like a bank barn, or are freestanding. The barn accommodates wagon traffic on the baling floor level, so hillside barns utilize topography for access. Barns on flat terrain use earthen ramps and wood bridges (Figure 12.4).

Press barns were constructed to house a beater hay press and to facilitate the operations of pressing and baling hay (as opposed to curing tobacco, threshing wheat, housing livestock, or dairying operations). The barn and press were constructed as one. The press was an integral part of the barn's structural framework, and the press bents were important to the structural integrity of the press. We have not seen evidence of, nor do archival sources support, presses being added to existing barns. The baling floor was used for storing and pressing hay, and the cellar space housed the horse-related operations of the press. A three-story-high space was allotted for the storage and curing of the season's hay harvest. Ramps, bridges, and aisles accommodated wagon traffic into and through the barn.

There are three variations of press barn plans we have observed in our ten years of press barn research. Variation 1, the *platform press barn*, represents the structure in its simplest form. This variation includes a basic timber frame with a baling floor platform. This affordable barn did not have spaces to store hay. A ramp accessed the baling floor, but it is unclear if it accommodated wagon traffic. The sole documented example of this variation is located on a Kentucky River farmstead near Port Royal, Kentucky (Figure 12.5). The barn was expanded for curing tobacco, so it is unclear how its original roof was constructed or if the barn was always enclosed with plank siding (Baas and Rubino 2014).

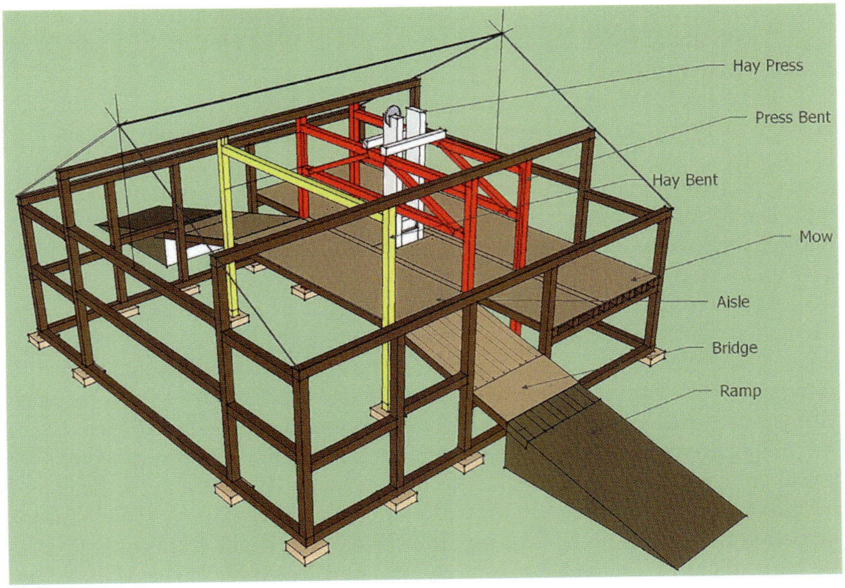

Figure 12.4
The timber framing of a single-aisle press barn. The press bents that carry the press are illustrated in red. The hay bent, in yellow, allows for the movement of loose and bailed hay from storage to the press. The barn represented here is located near East Enterprise, Indiana.

Source: Authors.

Figure 12.5
Image of the Port Royal, Kentucky, platform press barn. The original barn was constructed in 1862, and it was later expanded for curing tobacco.
Source: Authors.

The Port Royal Hay Press Barn is significant because it demonstrates the minimum structure required to operate the press. John K. Harris, an agent for the press inventor, promoted this minimal form (Harris 1851). Therefore, the barn lacks several character-defining features seen in other press barn forms, such as wagon aisles and storage spaces. The barn was constructed in a 35 × 35-foot square that was later expanded to 75 × 75 feet by constructing shed additions to all four sides (Figure 12.6).

We were invited by local preservationists to analyze the barn. Analysis generated 22 dated samples from 12 American beech and tulip poplar timbers, resulting in 1,646 measured rings ranging from 1623 to 1861 (Tables 12.1 and 12.2; Baas and Rubino 2014).

Variation 2, the *single-aisle press barn*, has an interior wagon aisle running through the barn's center. The aisle runs perpendicular to the gable roof and has double doors at each end. The space between bents is called a bay, and this variation typically contains four or five bays. Three of these bays contain the baling floor. The forth, and often a fifth, bay is open from the cellar to the roof, and is used for hay curing and storage. This storage space often consumes half of the barn, and would hold a season's harvest.

The Wycoff Hay Press Barn, formerly located in Allensville, Indiana, represents the typical single-aisle beater form (Figure 12.7; Baas and Rubino 2012). Since we sampled the barn, it has been purchased, disassembled, and placed in storage. John N. Wycoff, a merchant rather than a farmer, constructed the barn within the town of Allensville. Wycoff's barn measured 50 × 54 feet (Figure 12.8). Our analysis generated 20 crossdated samples from 12 timbers, resulting in 1,957 measured rings ranging from 1723 to 1865 (Tables 12.3 and 12.4; Baas and Rubino 2012). The barn is of special interest because the beater hay press inventor, Samuel Hewitt, identified Allensville as his residence, and it was thought that the barn might be a tangible link to the inventor. However, dendrochronology

Figure 12.6
Plan drawing of the Port Royal, Kentucky, platform hay press barn.

Source: Authors.

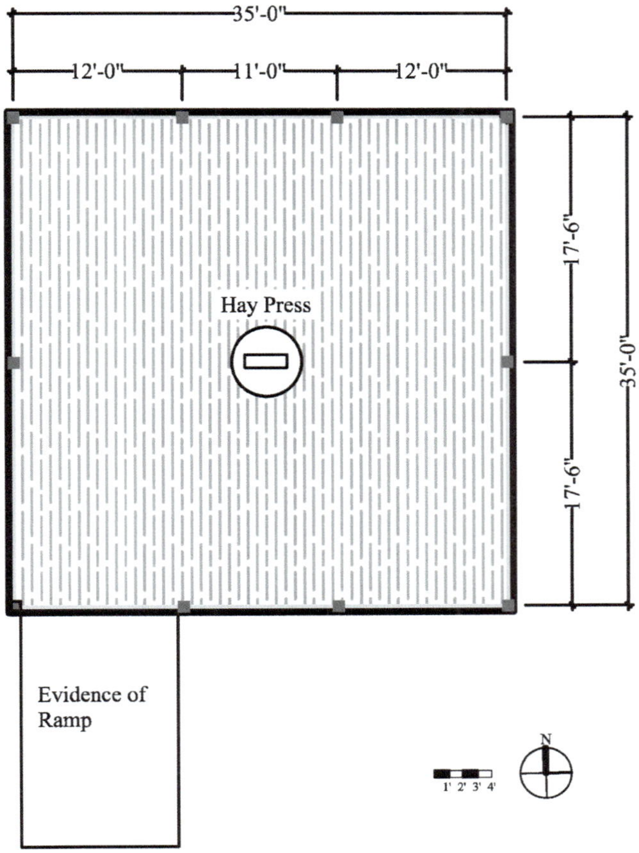

Table 12.1 Samples collected from the Port Royal, Kentucky, Hay Press Barn. "First" and "last" refer to the first and last rings present in the sample. If no date is provided, the sample could not be reliably crossdated. "Outer ring" refers to the outermost ring of the sample: b = bark, w = wane, etc.; see Table 7.1 for a complete list of codes used for outer ring condition. "r" is the correlation coefficient obtained by correlating the ring widths of this sample with all other timbers of the same type from the building (series intercorrelation; Chapter 5). "Timber type" is the type of timber; see Table 3.1 for a list of possible tree species represented by the various timber types. "Provenience" is the timber's function in a structure; see Chapter 2 for a description of the various proveniences. Note: Timber type and provenience are only given once if more than one sample was obtained from an individual timber. An r-value is not reported if less than three timbers of a particular type were obtained from a structure or if the sample did not overlap temporally with the other samples obtained for a particular timber type. A full description of the data presented in the table is provided in Chapter 7.

Sample	First	Last	Outer ring	r	Timber type	Provenience
CHU01A	1725	1861	w	0.69	Beech	Press platform timber
CHU02A	1816	1861	w	0.78	Beech	Press platform timber
CHU02B	1799	1861	w	0.57		
CHU02C	1704	1771		0.54		

(Continued)

Table 12.1 Continued

Sample	*First*	*Last*	*Outer ring*	r	*Timber type*	*Provenience*
CHU03A	1848	1862	w/e	0.75	Beech	Press platform timber
CHU03B	1846	1862	w/e	0.69		
CHU03C	1709	1862	w/e	0.54		
CHU04A	1708	1862	w/e	0.63	Beech	Press platform timber
CHU04B	1804	1862	w/e	0.70		
CHU05A	1711	1861	w	0.58	Beech	Press platform timber
CHU06A	1731	1861	w	0.54	Beech	Press platform timber
CHU07A					Tulip	Floorboard
CHU08A	1659	1719		0.59	Tulip	Floorboard
CHU09A	1706	1821	w/e	0.70	Tulip	Beam
CHU09B	1818	1862	w/e	0.52		
CHU10A	1790	1859		0.76	Tulip	Beam
CHU10B	1793	1858		0.77		
CHU10C	1827	1853		0.82		
CHU11A	1652	1736		0.85	Tulip	Beam
CHU11B	1622	1736		0.86		
CHU12A	1829	1857	w	0.52	Tulip	Floor joist
CHU12B	1829	1857	w	0.71		
CHU13A	1786	1857	w	0.47	Tulip	Floor joist

shows that the barn was constructed in 1866, 23 years after the press patent and after the inventor had moved away from Switzerland County.

Variation 3, the ***double-aisle press barn***, has two aisles that flank the press. The aisles allow additional traffic flow through the barn, and for interior U-turns by horses and wagons. The double-aisle Gridley Hay Press Barn (Gallatin County, Kentucky) measures 38 × 57 feet, and was later enlarged to 67 × 78 feet (Figures 12.9 and 12.10). Tree-ring analysis identified an 1864 construction date; we did not sample the additions, only the original press barn. We crossdated ten samples from eight timbers, resulting in 905 measured rings ranging from 1713 to 1863 (Tables 12.5 and 12.6; Baas and Rubino 2014).

Table 12.2 Summary table of the timber type chronologies prepared from the Port Royal, Kentucky, Hay Press Barn. "Samples" and "timbers" are the number of successfully crossdated samples and timbers from the building. "Undated" is the number of timbers that could not be crossdated reliably. "First" and "last" refer to the first and last dates in the chronology. "Mean" and "SD" are the mean and standard deviation (mm) of the ring widths in the chronology. "r" is the mean series intercorrelation for each timber type (see Chapter 5). r *is not provided if less than three timbers were obtained for a particular timber type. No total* r*, mean, or SD are provided since each chronology was prepared using only a single timber type. A full description of the data presented in the table is provided in Chapter 7.*

Timber type	*Samples*	*Timbers*	*Undated*	*First*	*Last*	*Years*	r	*Mean*	*SD*
Beech	11	6	0	1705	1861	972	0.61	1.18	0.57
Tulip	11	6	1	1623	1861	674	0.71	1.13	0.64
Total	22	12	1	1623	1861	1646			

Figure 12.7
Image of the single-aisle Wycoff Hay Press Barn formerly located in Allensville, Indiana. The barn was constructed in 1866.

Source: Authors.

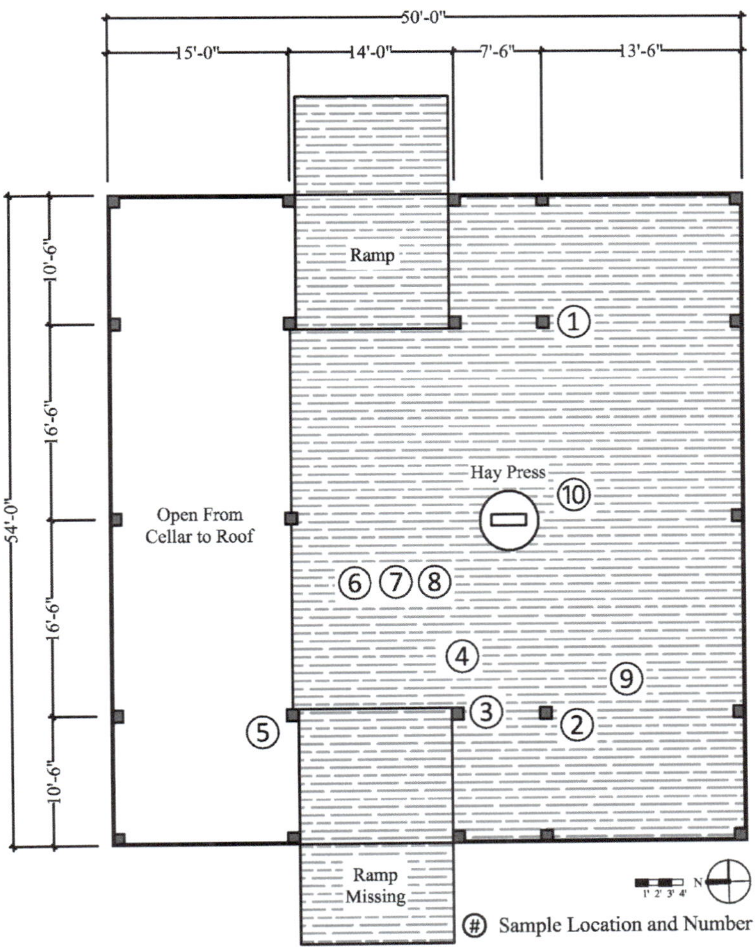

Figure 12.8
Plan drawing with sample locations of the single-aisle Wycoff Hay Press Barn formerly located in Allensville, Indiana.

Source: Authors.

Table 12.3 Samples collected from the Wycoff Hay Press Barn (Allensville, Indiana). See Table 12.1 for a description of the table contents; a full description of the data presented is provided in Chapter 7.

Sample	*First*	*Last*	*Outer ring*	r	*Timber type*	*Provenience*
ALA01A	1737	1865	w	0.81	Tulip	N post; E press bent
ALA01B	1807	1865	w	0.80		
ALA01C	1723	1806		0.80		
ALA02A	1724	1865	w	0.76	Tulip	S post; E press bent
ALA03A	1736	1865	w	0.65	Tulip	S post; W press bent
ALA04A	1778	1865	w	0.07	Tulip	Bottom chord; W press bent
ALA04B	1768	1865	w	0.73		
ALA05A	1751	1805		0.74	Tulip	SW post
ALA05B	1806	1844		0.67		
ALA06A	1734	1866	w	0.52	White oak	Floor joist; W mow
ALA07A	1734	1866	w	0.47	White oak	Floor joist; ground floor
ALA08A	1840	1865	w		Red oak	Floor joist; ground floor
ALA08B	1727	1782				
ALA09A	1771	1809		0.79	White oak	Floor joist; ground floor
ALA09B	1722	1864	w	0.43		
ALA10A	1775	1845		0.78	Hickory	Floor joist; ground floor
ALA10B	1679	1865	w	0.61		
ALA11A	1682	1865	w	0.59	Hickory	Floor joist; ground floor
ALA12A	1779	1862		0.74	Tulip	Post; cellar
ALA12B	1740	1865	w	0.71		

Table 12.4 Summary table of the timber type chronologies prepared from the Wycoff Hay Press Barn (Allensville, Indiana). See Table 12.2 for a description of table contents; a full description of the data presented is provided in Chapter 7.

Timber type	*Samples*	*Timbers*	*Undated*	*First*	*Last*	*Years*	r	*Mean*	*SD*
Hickory	3	2	0	1776	1864	433	0.63	0.99	0.43
Red oak	2	1	0	1728	1781	54		1.20	0.50
				1841	1864	24		1.63	0.40
Tulip	11	6	0	1724	1864	1009	0.74	1.37	0.79
White oak	4	3	0	1723	1865	437	0.50	1.05	0.38
Total	20	12	0	1723	1865	1957			

12.4.1 Results of the tree-ring analysis of hay press barns

Tree-ring analysis successfully identified construction dates for 16 Mormon beater press barns. This sample of barns included 1 single platform barn, 11 single-aisle barns, and 4 double-aisle barns. We had hypothesized that the beater press barn had evolved from the most basic form of the platform barn, to the single-aisle, and finally to the double-aisle form. Our results identified a 24-year window of construction, with the earliest barn being built in 1850, and the most recent constructed in 1874 (Table 12.7). However, tree-ring analysis did not support our

Figure 12.9
Image of the double-aisle Gridley Hay Press Barn located near Warsaw, Kentucky.

Source: Authors.

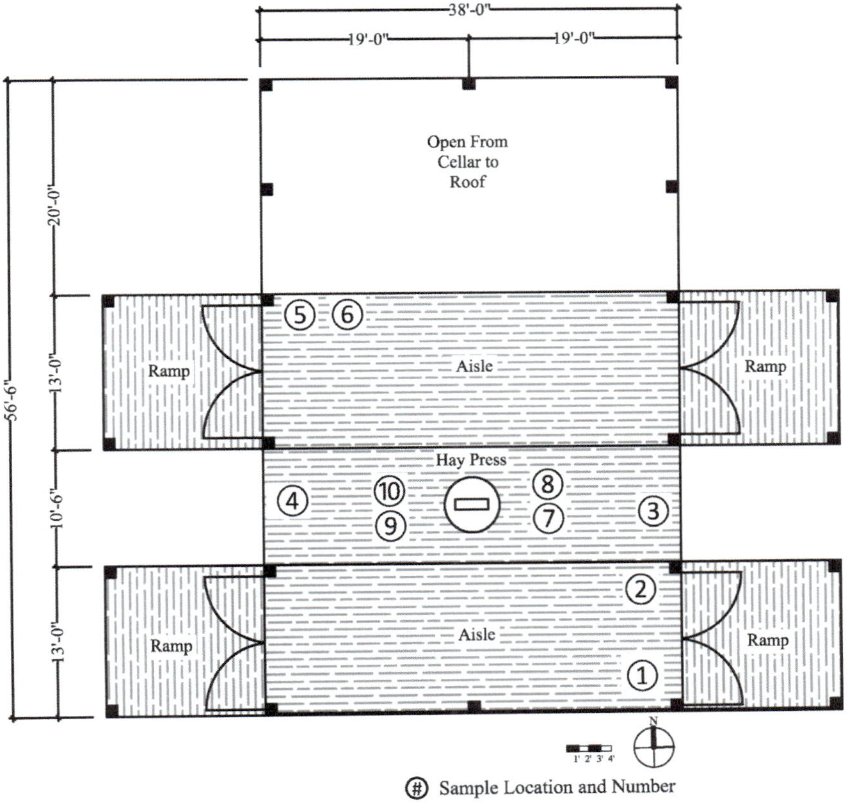

Figure 12.10
Plan drawing with sample locations of the double-aisle Gridley Hay Press Barn located near Warsaw, Kentucky. The barn was constructed in 1865.

Source: Authors.

Table 12.5 *Samples collected from the Gridley Hay Press Barn (Gallatin County, Kentucky). See Table 12.1 for a description of the table contents; a full description of the data presented is provided in Chapter 7.*

Sample	*First*	*Last*	*Outer ring*	r	*Timber type*	*Provenience*
GRI01A	1845	1864	w	0.58	Beech	Post
GRI01B	1848	1864	w	0.64		
GRI02A	1739	1854	+8	0.43	White oak	Post
GRI02B	1824	1855	+6	0.69		
GRI03A	1751	1863		0.66	Beech	Post
GRI04A					Beech	Post; press bent
GRI05A	1724	1854		0.34	White oak	Post
GRI06A					Hard maple	Remnant tenon in mortise of GRI05A
GRI07A	1712	1864	w	0.50	Beech	Floor Joist; ground floor
GRI08A	1738	1864	w	0.53	Beech	Floor Joist; ground floor
GRI09A	1758	1864	w	0.58	Beech	Floor Joist; ground floor
GRI10A	1749	1864	w	0.53	Beech	Floor Joist; ground floor

Table 12.6 *Summary table of the timber type chronologies prepared from the Gridley Hay Press Barn (Gallatin County, Kentucky). See Table 12.2 for a description of table contents; a full description of the data presented is provided in Chapter 7.*

Timber type	*Samples*	*Timbers*	*Undated*	*First*	*Last*	*Years*	r	*Mean*	*SD*
Beech	7	6	1	1713	1863	638	0.56	1.30	0.62
Hard maple	0	0	1						
White oak	3	2	0	1727	1854	267	0.42	1.03	0.52
Total	10	8	2	1713	1863	905			

hypothesis for an evolution of form over time, and suggests that a farmer chose to build a barn based on other factors.

We found that 7 of the 16 barns were constructed during the American Civil War (1861–1865) when the demand for hay to feed military horses dramatically inflated prices (Baas and Rubino 2013, 2014). The seven barns include the platform barn, four single-aisle barns, and two double-aisle barns. Perhaps the 1862 construction date of the Port Royal platform barn is an example of a farmer adopting a quick and affordable way to participate in the wartime hay economy.

Table 12.7 *Table of hay press barn types, number sampled, and range of construction dates.*

Barn form	*Number sampled*	*Range of construction dates*
Platform	1	1862
Single aisle	11	1850–1866
Double aisle	4	1857–1874
Total	16	1850–1874

12.5 Tree-ring analysis of the Justi and Mary Thiebaud Farmstead

The Thiebaud Farmstead is located in Craig Township, Switzerland County, Indiana. The family came to the Midwest as part of the small wave of settlers from the Swiss parish of Montreux that attempted to establish commercial wine production in the United States. In 1802 Swiss immigrant John James Dufour petitioned Congress to purchase land for establishing wine production, and was granted rights to purchase 2,500 acres in the Indiana Territory.

In 1817, Frederick and Harriet Thiebaud, with their eight children, purchased property along the Ohio River downstream from Vevay, the county seat. The property extended inland from the riverbanks to a bluff-top ridge, an elevation change of more than 400 feet (120 meters). Like many Swiss settlers, Frederick attempted viticulture by terracing the slopes oriented toward the river, but wine production was never regionally successful (Baas 2004).

In the mid-1840s, Frederick's son and daughter-in-law, Justi and Mary Thiebaud, inherited the farm at a time ripe for prosperity. Switzerland County had transitioned from a frontier to an agricultural economy. The river provided affordable transportation of produce and manufactured goods, and Vevay had become a successful and productive river town. Like most farmsteads passed between generations, Justi expanded and modernized his parents' property (Groover 2004). By 1860 the farmstead had expanded from 80 to 180 acres (32 to 73 hectares), and by 1870 they had purchased a second property for a total of 340 acres (138 hectares; Table 12.8). Justi and Mary were well positioned to take part in the hay culture. Their farm's ridge-top soils were well suited for hay (Cox 1872), and the farm's riverbank location saved them the expense of hauling the bales to the river, which was required of inland farmers.

Justi constructed a 29 × 60-foot press barn (Figure 12.11) that is two stories in height, and rests on a one-story stone cellar. It was later enlarged to 58 × 60 feet. The cellar is built into the slope so that its north end retains the hillside, and its southern end is freestanding (Figure 12.12). The barn varies from the typical single-aisle form in that the baling floor extends throughout the entire barn, and there is no three-story space for seasoning hay. We were curious if this was a later alteration from when the barn was modified to cure tobacco, or if this was how the barn was originally constructed.

From archival sources, we could generalize a broad period for construction of between Hewitt's 1843 patent date to the crash of the hay economy in the late

Figure 12.11
Image of the single-aisle Thiebaud Hay Press Barn located at the Switzerland County Agricultural Museum near Vevay, Indiana. The barn was constructed in 1850, and is the earliest press barn to be analyzed. Note the vertical joints in the gable end cellar walls where the barn was expanded.

Source: Authors.

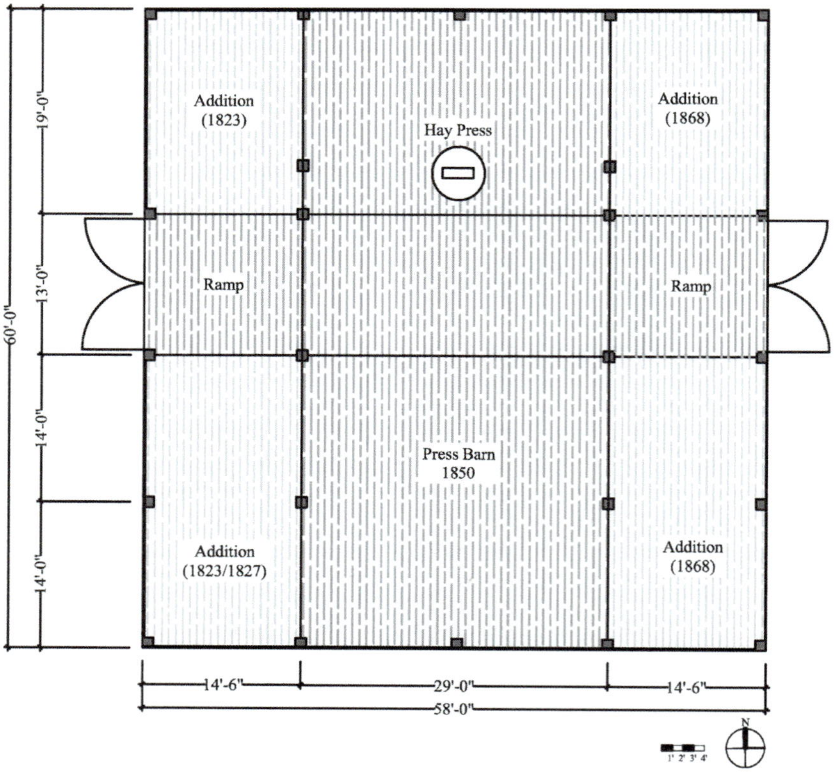

Figure 12.12
Plan drawing of the single-aisle Thiebaud Hay Press Barn located at the Switzerland County Agricultural Museum near Vevay, Indiana. The original barn was constructed in 1850, and tree-ring analysis suggests that it was later expanded in 1868 and/or at an undeterminable date.

Source: Authors.

1870s. The agricultural census provided more detail. The data illustrate a steady increase in the Thiebaud's farm size and hay production between 1850 and 1870. Hay production nearly disappeared by 1880 (Table 12.8). The jump in hay production from 6 to 30 tons (5,400 to 27,000 kilograms) between the 1850 federal and 1857 state census periods supported a construction date within this window of time.

Tree-ring analysis (not including the white pine floorboards) produced 83 crossdated samples obtained from 63 timbers representing six timber types (8,019 rings ranging from 1564 to 1879; Tables 12.9 and 12.10). Our analysis established a construction date of 1850 for the Thiebaud Hay Press Barn, which agrees

Table 12.8 Hay production and farm size for the Thiebaud Farmstead (Craig Township, Switzerland County, Indiana) for the years 1850 to 1880. The data are from federal and state agricultural censuses. The family's hay production steadily increases from 1850 to 1870. The dramatic rise between the 1850 and 1860 census periods helped interpret the 1850 construction date of the family's hay press barn. The drop in production between 1870 and 1880 parallels the drop in market prices for hay. A 400-pound bale = 180 kilograms; 1 acre = 0.4 hectares; 1 ton = 907 kilograms.

Year	*Farm size in acres*	*Tons of hay*	*Number of 400-lb bales*
1850	80	6	30
1857	80	30	150
1860	180	39	195
1870	340	80	400
1880	340	10	50

with the agricultural census data. The barn appears to have been expanded a few decades later when additions were placed on both of the eave sides of the barn, increasing it to its current 58 × 60 feet.

Although tree-ring analysis of the Thiebaud Hay Press Barn was successful in determining a build date, it cannot answer all of the questions regarding how it was modified. The barn's west addition contains timbers from trees harvested in 1823 and 1827. These timbers would have most likely been recycled from another building (Table 12.9). Strezewski (2004) states that many abandoned foundations

Table 12.9 Samples collected from the Thiebaud Hay Press Barn (Craig Township, Switzerland County, Indiana). See Table 12.1 for a description of the table contents; a full description of the data presented is provided in Chapter 7.

Hay press barn						
Sample	*First*	*Last*	*Outer ring*	r	*Timber Type*	*Provenience*
THE01A	1745	1850	w	0.72	White oak	Floor joist; ground floor; aisle
THE02A	1713	1848		0.62	White oak	Post; W elevation
THE02B	1826	1850	w	0.83		
THE03A	1713	1850	w	0.61	White oak	Floor joist; ground floor
THE03B	1767	1849		0.67		
THE04A	1700	1850	w	0.53	Ash	Floor joist; ground floor
THE04B	1726	1850	w	0.56		
THE05A	1712	1850	w	0.55	White oak	Post; W elevation
THE06A	1812	1850	w	0.68	White oak	Floor joist; ground floor
THE06B	1767	1850	w	0.74		
THE07A	1769	1850	w	0.75	White oak	Post; press bent
THE07B	1705	1768		0.58		
THE08A	1736	1850	w	0.61	Ash	Horizontal brace/nailer
THE09A	1742	1850	w	0.56	White oak	Post; E elevation
THE09B	1786	1850	w	0.48		
THE09C	1783	1850	w	0.58		
THE10A	1700	1850	w	0.54	White oak	Post; E elevation
THE11A	1831	1850	w	0.73	White oak	SE post
THE11B	1774	1828		0.59		
THE12A	1774	1842		0.62	Ash	Horizontal brace/nailer
THE13A	1649	1850	w	0.42	White oak	Floor joist; recycled press timber?
THE13B	1830	1850	w	0.61		
THE14A					Beech	Floor joist; ground floor
THE15A	1743	1850	w	0.69	Beech	Floor joist; ground floor
THE15B	1752	1850	b	0.69		
THE16A	1735	1850	w	0.67	Beech	Floor joist; ground floor
THE17A					White oak	Loose; cellar; brace?
THE30A	1829	1850	w	0.40	White oak	E sill
THE30B	1764	1850	w	0.58		
THE31A	1728	1850	w	0.73	White oak	Floor joist; ground floor
THE35A	1727	1819		0.62	White oak	NE post
THE36A	1719	1849	sap	0.56	White oak	Post; cellar

(Continued)

Table 12.9 Continued

East addition						
Sample	*First*	*Last*	*Outer ring*	r	*Timber Type*	*Provenience*
THE18A	1822	1868	w	0.83	Ash	Ceiling joist; tool room
THE18B	1739	1808		0.38		
THE19A	1717	1868	b	0.55	Ash	Ceiling joist; tool room
THE20A	1755	1868	w	0.57	Ash	Ceiling joist; tool room
THE32A	1736	1825	w	0.44	Ash	Ceiling joist; tool room
THE33A	1764	1851		0.66	Ash	Ceiling joist; tool room
TNB05A	1760	1868	b	0.74	Ash	Floor joist; ground floor
TNB06A	1749	1858	w	0.70	Ash	Floor joist; ground floor
TNB07A	1787	1868	b	0.67	Ash	Floor joist; ground floor
TNB07B	1749	1868	b	0.64		
TNB08A	1818	1867	w	0.46	Ash	Horizontal brace/nailer
TNB08B	1842	1867	w	0.33		
TNB09A	1789	1868	w	0.42	Ash	Horizontal brace/nailer
TNB10A	1762	1868	b	0.52	Ash	Ceiling joist; tool room
TNB11A	1801	1865		0.61	Ash	NE plate
TNB11B	1832	1865		0.79		
TNB12A	1770	1868	w	0.52	Ash	SE Plate
West addition						
Sample	*First*	*Last*	*Outer ring*	r	*Timber Type*	*Provenience*
THE21A					Red oak	Floor joist; ground floor
THE37A	1715	1823	b	0.55	Ash	Floor joist; ground floor
THE38A	1739	1823	w	0.42	Ash	Floor joist; ground floor
THE39A	1761	1823	w	0.66	Ash	Floor joist; ground floor
TNB01A	1723	1827	w	0.57	Ash	Floor joist; ground floor
TNB01B	1715	1819		0.68		
TNB01C	1793	1827	w	0.76		
TNB01D	1796	1827	w	0.76		
TNB02A	1695	1823	w	0.42	Ash	Floor joist; ground floor
TNB03A	1760	1823	w	0.51	Ash	Floor joist; ground floor
TNB04A	1770	1827	b	0.60	White oak	Floor joist; ground floor
North addition						
Sample	*First*	*Last*	*Outer ring*	r	*Timber Type*	*Provenience*
THE22A	1680	1841		0.39	White oak	NW post
THE23A					Ash	Horizontal brace
THE24A					White oak	N Sill
THE25A					White oak	Post; N elevation
THE26A	1732	1853	w	0.42	Ash	Horizontal brace
THE26B	1732	1853	w	0.35		
THE27A					White oak	Post; N elevation
THE28A	1773	1853	w	0.34	Ash	Horizontal brace

North addition						
Sample	*First*	*Last*	*Outer ring*	r	*Timber Type*	*Provenience*
THE29A					White oak	NE post
THE34A	1744	1817		0.25	Ash	S Sill
THE60A	1820	1867	w		Walnut	Loose timber
THE61A	1716	1830		0.53	Ash	Loose plank
THE62A	1666	1781			Walnut	Loose plank
TNA01A	1753	1868	b	0.40	Ash	Loose timber
TNA02A	1704	1818		0.52	Tulip	Loose timber
TNA03A	1772	1868	w	0.64	Ash	Loose timber
TNA04A	1762	1868	w	0.77	Ash	Loose timber
TNA05A	1727	1850	w	0.53	Ash	Loose timber
TNA06A					Red oak	Loose timber
TNA07A	1646	1731		0.52	Tulip	Loose timber
TNA08A	1734	1825	w	0.54	Ash	Loose timber
TNA09A	1795	1868	w		Red elm	Loose timber
TNA10A	1746	1880	w	0.38	Beech	Loose timber
TNA11A	1682	1825	w	0.51	Ash	Loose timber
TNA12A	1706	1825	w	0.47	Ash	Loose timber
TNA13A	1570	1772		0.40	Tulip	Loose timber
TNA14A					White pine	Loose timber
Floorboards						
Series	*First*	*Last*	*Outer*	r	*Type*	*Provenience*
THE63A	1562	1763		0.26	White oak	Loose
THE64A	1602	1775		0.61	Tulip	Loose
THE65A					Red pine	Loose
THE66A	1572	1767		0.59	Tulip	Loose
THE66B	1563	1715		0.73		
THE67A	1724	1813		0.58	White pine	Loose
THE67B	1723	1841		0.44		
THE68A					White oak	W addition
THE69A	1599	1747		0.47	Tulip	Press barn
THE70A					White pine	Loose
THE71A					White pine	Loose
THE72A	1697	1804		0.68	White pine	Loose
THE73A					White pine	Loose
THE74A	1719	1802		0.60	White pine	Loose
THE75A	1718	1779		0.78	White pine	Loose
THE76A					White oak	W addition
THE77A	1718	1817		0.63	White pine	W addition
THE77B	1718	1790		0.63		
THE78A	1678	1777		0.52	White pine	W addition
THE78B	1678	1773		0.61		

(Continued)

Table 12.9 Continued

			Floorboards			
Series	*First*	*Last*	*Outer*	r	*Type*	*Provenience*
THE79A					White pine	W addition
THE80A	1717	1827	w	0.55	White pine	Loose
THE81A					White pine	Loose
THE82A					White pine	Loose
THE83A					White pine	Loose
THE84A					White elm	Loose
THE85A	1675	1764		0.40	White pine	Subfloor; press barn
THE86A	1685	1777		0.42	White pine	Loose
THE87A					White pine	Loose
THE88A	1697	1793		0.41	White pine	Loose
THE89A	1749	1827	w	0.66	White pine	Loose
THE90A	1667	1840		0.31	White pine	Loose
THE91A					White pine	Loose
THE92A	1668	1778		0.63	White pine	Loose
THE93A	1736	1812		0.65	White pine	Loose
THE94A					White pine	Loose
THE95A	1694	1827	w	0.66	White pine	Loose
THE95B	1695	1827	w	0.66		
THE96A	1740	1848	w	0.57	Tulip	Subfloor; press barn
THE97A					White oak	W addition
THE98A					White oak	Loose
THE99A					White oak	W addition
THB01A	1722	1804		0.67	White pine	Loose
THB02A					White pine	Loose
THB03A					White pine	Loose
THB04A	1735	1829		0.60	White pine	Loose
THB05A	1612	1783		0.42	White pine	Loose
THB06A	1617	1772		0.31	White pine	Loose
THB07A	1726	1794		0.55	White pine	Loose
THB08A	1689	1827	b	0.55	White pine	Loose
THB09A	1691	1779		0.39	White pine	Loose
THB10A					White pine	Loose
THB11A	1751	1852		0.51	White pine	Loose
THB12A					White pine	Loose
THB13A					White pine	Loose
THB14A	1702	1816		0.78	White pine	Loose
THB15A					White pine	Loose

Table 12.10 Summary table of the timber type chronologies prepared from the Thiebaud Farmstead (Craig Township, Switzerland County, Indiana). See Table 12.2 for a description of table contents; a full description of the data presented is provided in Chapter 7.

	Samples	*Timbers*	*Undated*	*First*	*Last*	*Rings*	r	*mean*	*SD*
Barn									
Structural timbers									
Ash	42	33	1	1684	1867	3804	0.54	1.07	0.42
Beech	4	3	1	1738	1879	446	0.61	1.52	0.6
Red elm	1	1	0	1797	1867	71		1.85	0.51
Red oak	0	0	2						
Tulip	8	7	0	1564	1847	1151	0.55	1.10	0.51
Walnut	2	2	0	1667	1780	114		1.19	0.47
				1821	1866	46		1.05	0.42
White oak	26	17	11	1564	1849	2387	0.57	1.14	0.41
White pine	0	0	1						
Total	83	63	16	1564	1879	8019			
Conifer floorboards									
Red pine	0	0	1						
White pine	28	24	16	1613	1849	2853	0.54	1.17	0.53
Total	28	24	17	1613	1849	2853			
House									
Cottonwood	0	0	1						
Red elm	0	0	1						
Tulip	3	2	1	1783	1849	101	0.83	0.43	0.18
				1645	1781	137		1.35	0.77
White oak	9	7	0	1708	1855	870	0.55	1.10	0.36
Total	12	9	3	1645	1855	1108			
Corncrib									
Red elm	2	1	0	1793	1866	115		1.58	0.59
Total	125	97	36	1564	1879	12095			

are present on the farmstead; we hypothesize that these buildings could have furnished the timbers for the barn addition. Tree-ring analysis, however, cannot show when these timbers were incorporated into the press barn. The east addition to the barn contains many timbers with a death date of 1867 or 1868 (Table 12.9). Also, no outer ring on timbers without wane or bark had outer rings (not death dates) later than 1868. Were the trees used for timbers in the east addition felled in 1868 or were they salvaged from another structure that was built in 1868 and added later? The agricultural census data do support an 1868 barn expansion date. Between 1860 and 1870 the farm increased 160 acres (65 hectares) and hay production doubled, supporting the need for a larger barn.

The Thiebaud Hay Press Barn no longer has a press, and we believe (but cannot yet prove) that flooring was added over the open, three-story space we typically see in single- and double-aisle barns and that pieces of the press were used as joists. Sample THE13A, a floor joist, dates to the barn's construction but is in a location where one would not anticipate a floor joist on the ground-floor level (Table 12.9; this could be a missing timber from one of the press bents).

This is another example of how tree-ring analysis can help in lots of situations, but not when timbers of the same vintage are moved around within a structure. Careful sampling – we have been to this barn multiple times trying to resolve the patterns of its construction – and carefully reading of vernacular forms is essential to interpreting a resource. Another interesting aspect of the structure is that the planking used as siding and most of the floorboards are white pine, a species not found regionally. See Chapter 13 for a discussion of dating these timbers and how we determined their origin.

The north addition to the barn, which was used for dairy operations, contains timbers with numerous felling dates (Table 12.9). It has been removed since we collected samples. We were unable to determine the exact year in which this addition was built. Likely several different buildings provided recycled timbers for this addition. Tree-ring analysis is a very powerful technique to date the construction of a structure, but it does have limitations.

12.5.1 Thiebaud House and Corncrib

We also identified construction dates for the farmstead's house, and potentially the corncrib. Justi and Mary's house is representative of the "advanced center passage" house: one and a half stories in height, a symmetrical five-bay façade, a central corridor containing a stairwell and exterior doors, flanking parlors, and end chimneys (Figure 12.13). The small upper-story windows illuminate bedrooms. The vernacular house form has Greek Revival style features, such as a broad frieze and classically inspired portico (Knox 1948a; Noble 1984; Baas 2004). The house is approximately 44 × 20 feet in size. A roofed porch extends along the entire back wall of the residence, and connects the house to a stone summer kitchen.

Sampling the house was limited to joists and sills accessible in the cellar. Twelve samples from nine timbers were successfully dated, resulting in the measurement of 1,108 rings ranging from 1645 to 1855 (Tables 12.10 and 12.11). An 1856 construction date was identified from three cellar timbers, all white oak and containing wane (THH2A-THH4B). A lone, tulip poplar kitchen ceiling joist contained wane, and had

Figure 12.13
Image of the Justi and Mary Thiebaud House located at the Switzerland County Agricultural Museum near Vevay, Indiana. The main portion of the house was constructed in 1856.

Source: Authors.

Table 12.11 Samples collected from the Thiebaud House (Craig Township, Switzerland County, Indiana). See Table 12.1 for a description of the table contents; a full description of the data presented is provided in Chapter 7.

Sample	*First*	*Last*	*Outer ring*	r	*Timber type*	*Provenience*
THH01A	1719	1836		0.43	White oak	Beam; cellar
THH01B	1774	1837		0.63	White oak	
THH02A	1778	1856	w	0.62	White oak	S sill
THH03A	1763	1856	w	0.41	White oak	E sill
THH04A	1747	1856	w	0.62	White oak	N sill
THH04B	1743	1856	w	0.59	White oak	
THH05A					Tulip	Loose; likely brace from cellar
THH05B					Tulip	
THH06A	1723	1805		0.36	White oak	Tenon from timber joined to THH05
THH07A	1782	1840		0.83	Tulip	Ceiling joist; kitchen addition
THH07B	1804	1850	w	0.83	Tulip	
THH08A	1644	1783			Tulip	Ceiling joist; kitchen addition
THH09A					Elm	S sill; kitchen addition
THH10A	1717	1831		0.59	White oak	Likely replacement from modern porch repair
THH11A	1707	1827		0.67	White oak	Likely replacement from modern porch repair
THH12A					Cottonwood	Scab; breezeway

a harvest date of 1850 (THH07B). However, with only one sample from the kitchen, we are apprehensive to assign a construction date to this structure.

The corncrib is a small timber frame structure sitting on stone piers about two feet above the ground. It measures 10 × 20 feet, and has a metal covered gable roof. The interior is divided into three grain storage bins. The building appears to be constructed from recycled timbers. Only one timber displayed wane, a red elm sleeper with an 1868 death date. As we have shown with the barn, when a new building was constructed or an existing building was expanded, it was common to recycle timbers from other buildings. Since we cannot establish death dates for the other timbers in the corncrib, the structure does not give us other physical clues to its construction, and it is not included in archival sources, we can only conclude with limited certainty that it was constructed in 1868 or sometime after (also the year that the barn was expanded, suggesting substantial changes to the farmstead).

12.5.2 Thiebaud Farmstead summary

Our analysis of the Thiebaud Farmstead has added significant depth to regional tree-ring chronologies. Analysis generated 125 samples from 97 timbers, resulting in the measurement of 12,095 rings ranging from 1564 to 1879. Seven timber types were successfully dated (Table 12.10).

Like the Robert Glenn Farmstead in Chapter 10, very little material culture from the farmstead's initial settlement is intact. Tree-ring analysis of three of the five remaining 19th-century structures (a smokehouse and carriage house are also found at the farmstead) show they were built after Justi had inherited the farmstead from his parents in the 1840s (Knox 1948b), and the 1850 construction date of the press barn suggests Justi was one of the earliest participants in the hay economy.

12.6 Detecting recycled timbers in new press barns

Several of our press barn studies illustrate the historic practice of recycling timbers from old structures in new buildings and in building modifications. They demonstrate the importance of collecting samples from throughout a building or landscape. Limited samples, especially from buildings with recycled timbers, can result in false construction dates.

The region containing Mormon beater hay press barns was first settled in the early decades of the 19th century. For many of the farmsteads we have analyzed, the barns appear to be from a second wave of building construction that took place mid-century or later. Many press barns contain timbers with death dates several decades prior to their construction, indicating that recycling old buildings into new construction was a frugal and common practice. We hypothesize that the recycled timbers are from buildings erected during the region's initial settlement (Roberts 1996; Grissino-Mayer and van de Gevel 2007; Baas et al. 2017).

12.6.1 Posey Township Hay Press Barn

The Posey Township Hay Press Barn (Switzerland County, Indiana) is a double-aisle barn that measures 64 × 66 feet (Figure 12.14). Through tree-ring analysis we conclude that the barn was constructed in 1874. Mixed in with the timbers having 1873 or 1874 death dates (POS03, 05, 10, 12, and 13) are several timbers with 1844 death dates (POS01, 02, 04, 06, and 07; Table 12.12). Additionally, note that the outermost ring (although not cutting dates) in POS08 and 09 suggest a construction date later than 1844. Our initial thought was that the barn's press bents and a hay press were inserted into an 1844 barn, but the random patterns of timbers with different harvest dates supports the barn being built in 1874 using both new and recycled materials.

Our analysis of the Posey Township Hay Press Barn generated 15 crossdated samples from 12 timbers, resulting in 1,166 measured rings ranging from 1725 to 1873. Four timber types were sampled (Table 12.13). A unique component to the double-aisle form are the pair of three-story spaces for curing hay and the presence of aisle guardrails to keep wagons from falling into these spaces.

12.6.2 Anderson Hay Barn

The Anderson Hay Barn, located in Craig Township, Indiana, is also built with recycled timbers (Figure 12.15). The barn measures 59 × 60 feet and has all the character-defining features of a beater press barn except for the bents and bay that house the press (Figure 12.16, ramps, aisles, gable roof, entry on the eave's sides, etc.). Using landownership records we identified that William Johnston purchased and settled the property in 1823, and, using tree-ring analysis we determined that he constructed a log house on the property in 1833 (see section 6.4). The property was deeded to James Anderson in 1865 who likely constructed the barn in 1870. Given the form

Figure 12.14
Plan drawing with sample locations of the double-aisle Posey Township Hay Press Barn located in Switzerland County, Indiana. The barn was constructed in 1874 using new and recycled material.

Source: Authors.

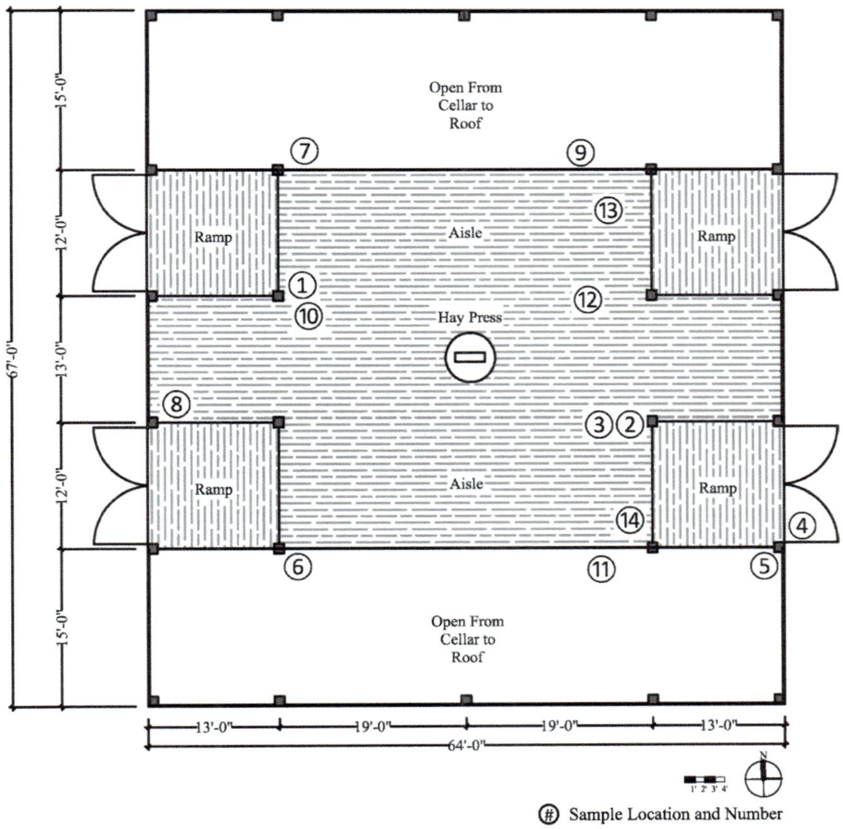

Table 12.12 Samples collected from the Posey Township Hay Press Barn (Switzerland County, Indiana). See Table 12.1 for a description of the table contents; a full description of the data presented is provided in Chapter 7. Sample POS01A was dated but not measured.

Sample	*First*	*Last*	*Outer ring*	r	*Timber type*	*Provenience*
POS01A	1832	1844	w		White oak	NW press bent post
POS01B	1724	1838		0.59	White oak	
POS02A	1750	1844	w	0.76	Tulip	SE press bent post
POS02B	1756	1844	w	0.52	Tulip	
POS03A	1797	1874	w	0.68	White oak	S press bent lower chord
POS04A	1766	1844	w	0.64	Tulip	S hay bent post
POS05A	1765	1861		0.66	White oak	S hay bent angle
POS05B	1863	1874	w	0.60	White oak	
POS06A	1746	1844	w	0.62	Tulip	S hay bent post
POS07A	1740	1844	w	0.46	Tulip	N hay bent post

(Continued)

Table 12.12 Continued

Sample	First	Last	Outer ring	r	Timber type	Provenience
POS08A	1829	1863	w8		Beech	W wall post; S press bent
POS09A	1736	1873			Hickory	"Guardrail"
POS10A	1815	1874	w		Beech	N press bent angle
POS11A					White oak	Tenon remnant from "guard-rail"
POS12A	1769	1873	w		Hickory	N press bent bottom chord
POS13A	1778	1874	w	0.61	White oak	Floor joist
POS14A					Beech	Floor joist

Table 12.13 Summary table of the timber type chronologies prepared from the Posey Township Hay Press Barn (Switzerland County, Indiana). See Table 12.2 for a description of table contents; a full description of the data presented is provided in Chapter 7.

Timber type	Samples	Timbers	Undated	First	Last	Years	r	Mean	SD
Beech	2	2	1	1817	1873	89		1.75	0.77
Hickory	2	2	0	1737	1872	238		0.90	0.47
Tulip	5	4	0	1741	1843	457	0.65	1.74	0.87
White Oak	6	4	1	1725	1873	382	0.63	1.39	0.46
Total	15	12	2	1725	1873	1166			

and layout of the barn, we had hypothesized that the barn predated Hewitt's 1843 patent, because of the absence of the bay and press. However, dendrochronology concluded that it was built in 1870 toward the end of the region's hay culture (late 1870s) rather than its beginnings in the 1840s. It also suggests that the beater press barn form supported other types of hay presses. To date, this is the only example of the press barn form constructed without a beater press we have found. The hay barn, like nearly all the press barns we have documented, was modified for curing tobacco. To accomplish this, Anderson (or a later property owner) used timber cribbing from a recycled structure built in 1858 to support a mow (samples RCH18–23; Table 12.14).

We also identified construction and modification dates for the Johnston House, a two-room log house constructed in 1833, and modified in 1851 and 1872. From the cellar, we were able to sample floor joists from the original log structure and the house's various alterations. Like the barn, the house contains timbers that were recycled into the residence (Baas et al. 2017). Our analysis of the Anderson Barn and Johnston House generated 40 series from 32 timbers, resulting in 3,456 measured rings ranging from 1690 to 1873. Five timber types were sampled (Table 12.15).

Figure 12.15
Photo of the north wall of the single-aisle Anderson Hay Barn. The barn was constructed in 1870. It was modified with recycled material when the barn was converted into a tobacco warehouse.
Source: Authors.

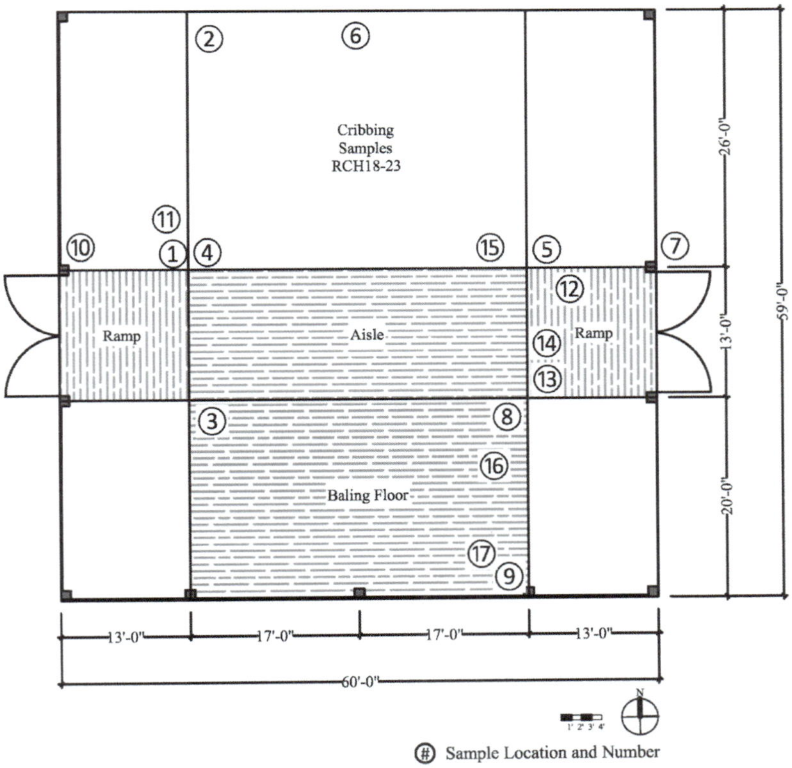

Figure 12.16
Plan drawing of the single-aisle Anderson Hay Barn with sample locations. Note the absence of two bents and a bay for a Mormon beater press, which suggest the use of an alternative press design that did not require the barn structure for operation.
Source: Authors.

Table 12.14 Samples collected from the Anderson Barn and Johnston House (Craig Township, Switzerland County, Indiana). See Table 12.1 for a description of the table contents; a full description of the data presented is provided in Chapter 7.

Barn						
Sample	*First*	*Last*	*Outer ring*	r	*Timber type*	*Provenience*
RCH01A					Hard elm	Post; aisle
RCH02A	1788	1865		0.60	Ash	Nailer; N elevation
RCH03A	1834	1870	w/e	0.66	Ash	Post; aisle
RCH03B	1771	1829		0.64		
RCH04A	1788	1839		0.56	Ash	Beam; aisle
RCH05A	1767	1870			White oak	Post; aisle
RCH06A	1763	1870	w		White oak	Post; N elevation
RCH06B	1773	1866		0.78		
RCH07A	1820	1869	w	0.63	Walnut	Post; E elevation
RCH07B	1823	1855		0.68		
RCH08A					Hard maple	Post; aisle
RCH09A	1843	1870	w/e	0.68	Ash	SE post
RCH09B	1835	1868		0.46		
RCH10A	1800	1874	w/e	0.61	Ash	Post; W elevation
RCH11A					White oak	W sill
RCH12A	1779	1865		0.60	Ash	Joist; N ramp; E addition
RCH13A	1785	1870	w	0.65	Ash	E sill
RCH14A	1831	1863	sap	0.71	Walnut	Joist; ground floor; E addition
RCH14B	1789	1870	w	0.54		
RCH15A	1834	1870	w	0.68	Walnut	N sill
RCH15B	1689	1851	sap	0.60		
RCH16A	1705	1870	w	0.30	Ash	Joist; ground floor
RCH17A	1807	1869		0.29	Walnut	Joist; ground floor
RCH18A	1763	1858	w	0.68	Ash	Cribbing; open bay
RCH19A	1742	1858	w	0.63	Ash	Cribbing; open bay
RCH20A	1740	1858	w	0.65	Ash	Cribbing; open bay
RCH21A	1743	1858	w	0.36	Ash	Cribbing; open bay
RCH22A	1750	1858	w	0.56	Ash	Cribbing; open bay
RCH23A	1753	1858	w	0.49	Ash	Cribbing; open bay
House						
Sample	*First*	*Last*	*Outer ring*	r	*Timber type*	*Provenience*
RAH01A	1707	1832	w	0.53	Ash	Floor joist; ground floor; timber house
RAH02A	1714	1832	w		White oak	Floor joist; ground floor; timber house
RAH03A	1691	1833	w	0.55	Ash	Floor joist; ground floor; timber house

House						
Sample	*First*	*Last*	*Outer ring*	r	*Timber type*	*Provenience*
RAH04A	1727	1833	w	0.52	Ash	Floor joist; ground floor; timber house
RAH05A	1771	1851	b	0.42	Ash	Floor joist; ground floor; SW addition
RAH06A	1747	1830	w		Walnut	Post; most likely W wall of log house based on shape
RAH07A	1728	1872	w	0.42	Ash	Floor joist; ground floor; W addition
RAH08A	1786	1872	b	0.62	Ash	Floor joist; ground floor; W addition
RAH09A	1771	1838		0.49	Ash	Floor joist; ground floor; SE addition
RAH09B	1720	1838	w	0.39	Ash	
RAH10A	1724	1850		0.41	Ash	Floor joist; ground floor; SE addition
RAH10B	1822	1851	w	0.38	Ash	
RAH11A	1772	1851	w	0.60	Ash	Floor joist; ground floor; SW addition
RAH12A	1706	1872	b	0.42	Ash	Floor joist; ground floor; W addition

Table 12.15 Summary table of the timber type chronologies prepared from the Anderson Barn and Johnston House (Craig Township, Switzerland County, Indiana). See Table 12.2 for a description of table contents; a full description of the data presented is provided in Chapter 7.

Barn									
Timber type	*Samples*	*Timbers*	*Undated*	*First*	*Last*	*Years*	r	*Mean*	*SD*
Ash	16	14	0	1707	1873	1323	0.55	1.16	0.37
Hard elm	0	0	1						
Hard maple	0	0	1						
Walnut	7	4	0	1690	1869	446	0.56	0.91	0.44
White oak	3	2	1	1766	1869	242		1.05	0.37
Total	26	20	3	1690	1873	2011			
House									
Timber type	*Samples*	*Timbers*	*Undated*	*First*	*Last*	*Years*	r	*Mean*	*SD*
Ash	12	10	0	1691	1872	1247	0.48	1.04	0.44
Walnut	1	1	0	1748	1829	82		0.60	0.28
White oak	1	1	0	1716	1831	116		0.88	0.41
Total	14	12	0	1691	1872	1445			

12.7 Using dendrochronology to establish a vernacular precedent for beater hay press barns

Dendrochronology, with field documentation, is one effective method that can be useful for determining if regional farmers adopted and transformed an existing vernacular barn type to accommodate the beater press and its operations, or if the barn form was a new invention along with the press. The 1850 construction of the Thiebaud Hay Press Barn is the oldest press barn dated using tree rings, with the majority of the other barns constructed in the 1860s and 70s (Table 12.7). The Thiebaud Hay Press Barn is significant in that it displays the character-defining features of a press barn a mere seven years following the 1843 patent of Hewitt's machine, and since the form came about so quickly it suggests to us the adaption of an existing vernacular barn form (see section 12.5). We hypothesized that the vernacular English barn form was the type adopted by 19th-century barn builders and sought to use dendrochronology to establish its presence in the region prior to the 1840s. Establishing construction dates for English barns predating Hewitt's invention is challenging when the window of time between the region's settlement and the press' patent is a brief 40 years. This search is further hampered by the loss of so many regional barns.

The initial scholarly identification of English barns was made by Fred Kniffen (1965) and further developed by Henry Glassie (1968, 1974). The barn's simple form is derived from the assembly of four parallel bents that create three individual bays. The central bay is a wagon aisle, or driveway. It is flanked by bays for storing grain or hay, or for stabling animals (Figure 2.2). Like press barns, English barns have gable roofs and doors on the eaves' sides that provide access to the aisle. The aisle contains a wood floor where wheat was winnowed by opening the doors to create a wind tunnel effect through the structure. The flanking bays contain mows (mid-level floors) for grain and hay storage.

Geographer Robert W. Bastian describes how the English barn is a "ubiquitous structure in rural Indiana," and is the "prevailing barn" along the Ohio River. In the southern part of the state they account for more than half of all 19th-century vernacular barns (Bastian 1977). The barn's versatility in meeting a farmer's agricultural needs was key to its appeal. The form could be expanded both longitudinally and laterally. To expand the barn longitudinally, additional bays were added to the gable ends of the standard three-bay barn form to increase it to four, five, six, etc., bays (Glassie 1974). Barns were also expanded laterally by adding shed additions to the eave sides of the structure. In regions of dairy and beef cattle production, the three-bay form was elevated and placed onto a stone foundation to create spaces for hay storage above and stabling in the cellar below. This variation of the English barn is commonly referred to as an English raised barn, English bank barn, or multi-level three-bay barn (Bastian 1977; Noble and Seymour 1982; Noble and Cleek 1997). The barns were adapted to landscapes with rolling topography, and were typically built with the cellar partially constructed into a hillside.

The Bakes Road English Barn is located in Jefferson Township, Switzerland County, Indiana (Figure 12.17). It was built on the banks of Indian Creek, a tributary of the Ohio River. The four-bay barn is 30 × 54 feet in size and was later enlarged to 60 × 54 feet when shed additions were constructed on the eave sides (Figure 12.18). The two center bays contain a threshing floor with a swing beam, and the two bays flanking the aisle have a mow. The truss-like design of the bents, which are composed of top and bottom chords connected by a center post and angle bracing, is common in North American English barns. This structural design stiffened the bottom chord swing beam, and bolstered the weight capacity of the mow.

We successfully crossdated 24 series from 18 timbers, resulting in 2,045 measured rings ranging from 1681 to 1823. Five timber types were sampled (Tables 12.16 and 12.17). Based on dendrochronological evidence, we concluded that the Bakes Road English Barn was built late in the 1824 growing season and prior to initiation of the 1825 growing season (since a consistent 1824 outer ring death date was found in the timbers).

The analysis of the Bakes Road English Barn is just one test of our hypothesis that the English barn form is a precedent for the Mormon beater hay press barn. In the absence of archival documents, tree-ring analysis was able to establish an 1824 building date. This date confirms the presence of the vernacular English barn form in what was becoming a hay-producing region prior to Samuel Hewitt's 1843 patent, and it supports our hypothesis that regional farmers and builders adapted the barn type for housing the beater press. We are cautious, however, in using this study as proof of a direct connection between the two barn forms, but simply as scientific evidence that the English barn form existed in the region prior to Hewitt's invention. We do see commonalities in this English barn and the press barns we have documented (e.g., aisles flanked by multiple bays, raised barns with ramps, entry on the eave's sides, gable roof, etc.). The ability, or flexibility, to design and build barns with bent spacings that define and accommodate spaces for specific agricultural activities, whether traditional or new, supports our hypothesis that the bents of English barns were easily repositioned to create interior spaces for storing vast amounts of hay, and to house and operate a beater press.

Figure 12.17
Image of the Bakes Road English Barn located near Mount Sterling, Indiana. The barn was constructed in 1824. Tree-ring analysis was able to establish the regional presence of the vernacular barn form prior to the 1843 invention of the Mormon beater hay press. This image shows the aisle and barn bent. The top of the bent is constructed with a structural chord to span the posts and to carry the weight of hay stored in the mow.

Source: Authors.

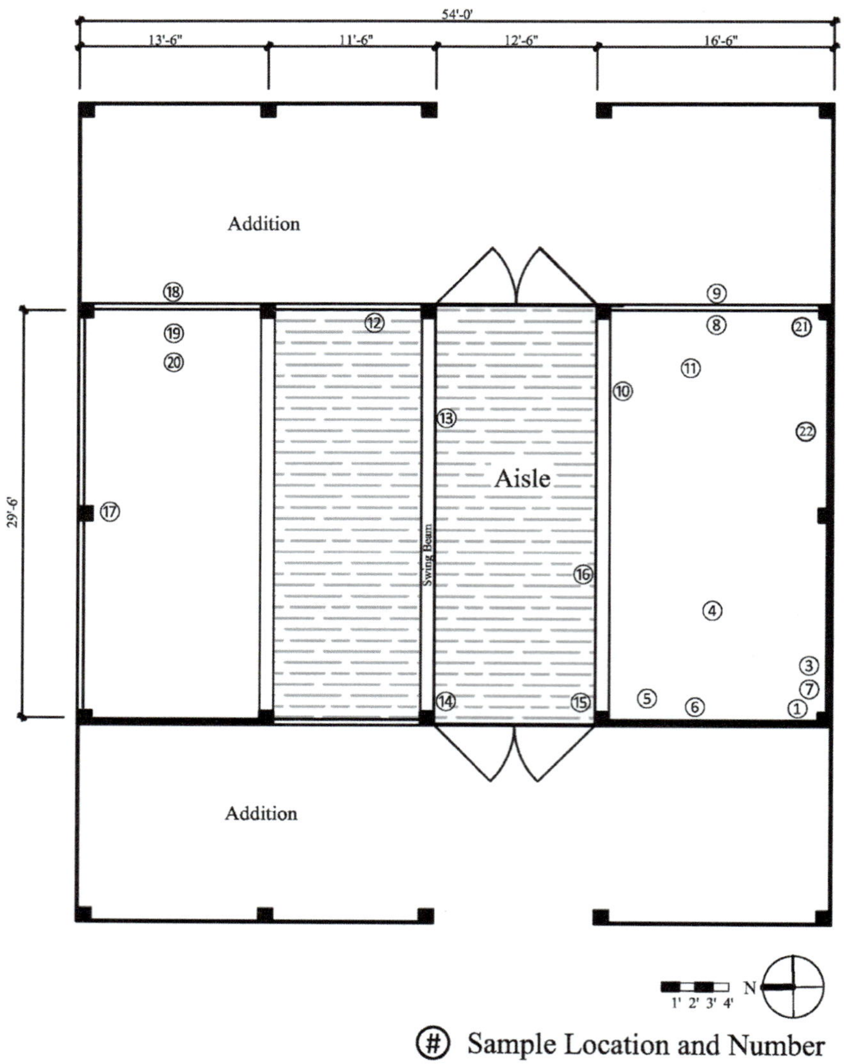

Figure 12.18 Plan drawing of the Bakes Road English Barn with sample locations.

Source: Authors.

Table 12.16 Samples collected from the Bakes Road English Barn (Jefferson Township, Switzerland County, Indiana). See Table 12.1 for a description of the table contents; a full description of the data presented is provided in Chapter 7.

Sample	*First*	*Last*	*Outer ring*	r	*Timber type*	*Provenience*
BKS01A	1694	1824	w	0.66	White oak	SW post
BKS02A	1714	1818		0.49	Walnut	Plate; W elevation
BKS02B	1787	1824	w	0.79		
BKS03A	1733	1823		0.67	White oak	Beam/nailer; mow; S elevation
BKS03B	1712	1743		0.34		
BKS03C	1743	1824	w	0.73		

Sample	*First*	*Last*	*Outer ring*	r	*Timber type*	*Provenience*
BKS04A	1808	1824	w	0.33	Beech	Beam; mow
BKS04B	1735	1824	w	0.58		
BKS05A	1740	1824	w	0.60	Beech	Rafter
BKS06A	1793	1824	w	0.65	Beech	Rafter
BKS06B	1741	1824	w	0.62		
BKS07A	1781	1824	w	0.45	Beech	Rafter/nailer; S elevation
BKX08A					Soft maple	Nailer; S bay; E elevation
BKS09A	1695	1823	w	0.63	Walnut	Plate; E elevation
BKS10A	1749	1824	w	0.57	Beech	Knee brace S aisle bent
BKS11A	1727	1824	w	0.41	Beech	Floor joist; mow
BKS12A	1707	1824	b	0.79	Walnut	Post; aisle
BKS12B	1739	1824	b	0.79		
BKS13A	1723	1824	w	0.52	Beech	Beam; aisle
BKS14A	1704	1824	w	0.78	Walnut	Post; aisle
BKS15A	1697	1824	w	0.52	Walnut	Post; aisle
BKS16A	1702	1791			Ash	Beam; aisle
BKS17A	1719	1820			Ash	Post; N elevation
BKS18A					Beech	Plate; N bay; E elevation
BKS19A					Ash	Nailer; N bay; E elevation
BKS20A					Beech	Nailer; N bay; E elevation
BKS21A	1680	1824	w	0.63	White oak	SE post
BKS22A	1745	1824	w	0.42	Beech	Nailer/mow floor support; S elevation

Table 12.17 Summary table of the timber type chronologies prepared from the Bakes Road English Barn (Jefferson Township, Switzerland County, Indiana). See Table 12.2 for a description of table contents; a full description of the data presented is provided in Chapter 7.

Timber type	*Samples*	*Timbers*	*Undated*	*First*	*Last*	*Years*	r	*Mean*	*SD*
Ash	2	2	1	1703	1819	188		1.58	0.50
Beech	10	8	2	1729	1823	680	0.53	1.29	0.57
Soft maple	0	0	1						
Walnut	7	5	0	1697	1823	708	0.67	1.21	0.66
White oak	5	3	0	1681	1823	469	0.65	1.13	0.40
Total	24	18	4	1681	1823	2045			

12.8 Conclusion

In Chapter 12 we demonstrated how we use dendrochronology to interpret a regional vernacular barn type. In Chapter 13 we show how tree-ring analysis identifies species bias by early barn builders and carpenters. The chapter includes discussions of the abundance of conifer timbers represented in the flooring and siding of the Thiebaud barn and the siding of another hay press barn, the Braytown North press barn (Switzerland County, Indiana).

12.9 Literature cited

Baas, C. 2004. *Thiebaud Farmstead National Register of Historic Places Registration Form.* National Park Service, Washington, DC.

Baas, C., and D.L. Rubino. 2012. Using tree-ring growth patterns to date the construction of a Beater Hay Press Barn in Allensville, Switzerland County, Indiana. *Indiana Archaeology* 7: 35–57.

Baas, C., and D.L. Rubino. 2013. The most successful press in this or any other country: The material culture of 19th-century Beater Hay presses in the Mid-Ohio Valley. *Material Culture* 45: 1–20.

Baas, C., and D.L. Rubino. 2014. Pressing hay in the commonwealth: Using tree-ring growth patterns to date the construction of two Kentucky Beater Hay Press Barns. *Journal of Kentucky Archaeology* 3: 2–31.

Baas, C., D.L. Rubino, and P. Jiang. 2017. Using tree-ring growth patterns to identify construction dates of two nineteenth century timber structures in Craig Township, Switzerland County, Indiana. *Indiana Archaeology* 12: 12–41.

Bastian, R.W. 1977. Indiana folk architecture: A lower Midwestern index. *Pioneer America* 9: 115–136.

Beach, C.D. 1987. *Turn to the River: U.P. Schenk, Swiss Immigrant Entrepreneur, Shipping Magnate, and Family Man*. Kappa, Kappa, Kappa, Inc., Vevay, IN. 43 p.

Brown, I. 1858. *Sixth Report of the State Board of Agriculture*. Joseph J. Bingham, Indianapolis, IN. 771 p.

Cox, E.T. 1872. *Third and Fourth Annual Reports of the Geological Survey of Indiana*. R. J. Bright, Indianapolis, IN. 488 p.

Dufour, P. 1925. *The Swiss Settlement of Switzerland County, Indiana*. Indiana Historical Commission, Indianapolis, IN. 446 p.

Glassie, H. 1968. *Pattern in the Material Folk Culture of the Eastern United States*. University of Pennsylvania Press, Philadelphia, PA. 316 p.

Glassie, H. 1974. The variation of concepts within tradition: Barn building in Otsego County, New York. *Geoscience and Man* 5: 177–235.

Grissino-Mayer, H.D., and Saskia L. van de Gevel, 2007. Tell-tale trees: Historical dendroarchaeology of log structures at Rocky Mount, Piney Flats, Tennessee. *Historical Archaeology* 41: 32–49.

Groover, M.D. 2004. Household succession as a catalyst of landscape change. *Historical Archaeology* 38: 25–43.

Harris, J.K. 1851. Hay presses. *Ohio Cultivator* 7: 259–260.

Kniffen, F.B. 1965. Folk housing: Key to diffusion. *Annals of the Association of American Geographers* 55: 549–577.

Knox, J.L. 1948a. *Some Interesting Pioneer Homesteads in and Around Vevay, Indiana*. Self-published second edition (originally published in 1927).

Knox, J.L. 1948b. The Thiebaud genealogy. *Indiana Magazine of History* 44: 326–334.

McShane, C., and J.A. Tarr. 2007. *The Horse in the City: Living Machines in the Nineteenth Century*. The Johns Hopkins University Press, Baltimore, MD.

Noble, A.G. 1984. *Wood, Brick, and Stone: The North American Settlement Landscape, Volume 1: Houses*. University of Massachusetts Press, Amherst, MA. 160 p.

Noble, A.G., and R.K. Cleek. 1997. *The Old Barn Book*. Rutgers University Press, New Brunswick, NJ. 222 p.

Noble, A.G., and G.A. Seymour. 1982. Distribution of barn types in Northeastern United States. *Geographical Review* 72: 155–170.

Roberts, W.E. 1993. An early hay press and barn on the Ohio river. *Material Culture* 25: 29–36.

Roberts, W.E. 1996. *The Log Buildings of Southern Indiana*. Trickster Press, Bloomington, IN. 228 p.

Strezewski, M. 2004. *An Archaeological Survey of the Thiebaud Property, Switzerland County, Indiana*. Indiana University-Purdue University at Fort Wayne Archaeological Survey, Fort Wayne, IN.

Chapter 13

Unique applications of tree-ring data

13.1 Introduction

There is great satisfaction in successfully crossdating the tree rings in a building's timbers. A set of successfully crossdated timbers allows for the dating of a building or landscape – one can determine when it was built and when/if it was modified. This satisfaction comes not from just successfully executing a set of steps and making pencil marks on pieces of wood. Tree-ring analysis, coupled with historical archives and documents, allows for a better understanding of the past; it allows one to better understand the fleeting, remaining pieces of a region's cultural fabric. In the previous chapters focusing on case studies ranging from communal groups to specialized agriculture, tree rings, and the annual resolution they permit, helped to make sense of how a region has evolved over time and how humans interact with their environment and ultimately impact a landscape.

In short, tree rings provide more than dates. Tree rings can be used to discover and better understand the past and can unlock more information than simply how large or small of a ring one can expect a tulip poplar to deposit in 1656. There are more than growth patterns in the tree-ring record found in a building's timbers. Information about the building, its builder(s), and the culture of the time can be explored (Baillie 1995). Dean (1997) states that applications of tree-ring science in an archaeological context can provide layers of information about the past. For example, a building can be precisely dated using tree rings; a date provides an exact year of construction and modification of a landscape. Other knowledge such as behavioral and environmental information can be obtained from tree-ring analyses. Behavioral information such as how trees were used as a natural resource can be determined. Environmental information such as how tree species composition and availability of certain tree species changed over time can be gleaned from tree-ring analysis too (Dean 1997; Billamboz 2003).

13.2 Using tree rings to make sense of timber selection

When a barn, house, church, etc., was erected many decisions had to be made: Where on the property should the building be erected? Which direction should the front façade face? What should be the size and shape of the building? (Hutslar 1986). Two questions we find extremely compelling are "what type of timber should be used?" and "how should different timber types be used throughout a structure?" By answering these questions, we may gain a better understanding of how builders approached the construction of a structure and how they affected the landscapes they modified and subsequently inhabited. We have had the good fortune to sample over 200 buildings throughout our study area and have been given the opportunity to gain some insight into potential answers. Using construction (and modification) dates and timber species analysis, we have tried to formulate answers to what types of timber and how different timber types were used. In order to better understand how timbers were chosen for construction, one must consider the many factors that would influence timber type selection.

13.2.1 Why use a specific timber type?

It is well established that different species of trees vary in wood characteristics that would make them more or less suitable for construction (Table 13.1). For instance, there is much variation in wood properties, such as rot resistance, tendency to warp, workability, shrinkage, density, elasticity, and chemical composition (Panshin and de Zeeuw 1980; Burns and Honkala 1990; United States Department of Agriculture 2007). Some types of timber, such as white oak are dense and rot resistant (in comparison to other local timber types) and were regularly selected for particular timbers in a building where these properties would be beneficial (Table 13.1; see Chapter 2 for a listing and description of the different structural timbers in historic buildings). For example, the Daniel L. Harmon House (Chapter 14) was constructed from tulip poplar logs, but the sill was made of white oak. White oak was considered "tough," "strong," and "durable" (Flint 1904; USDA 2007). White oak, consequently, is a commonly used timber in many historic buildings and often used exclusively or nearly exclusively in a number of buildings we have analyzed. All accessible timbers in the Stone Barn (Kent, Indiana), which was built in 1856, are white oak. This two-story barn was built with a timber frame interior and a stone façade. Additionally, the buildings of the Fort Wayne military garrison were built entirely of white oak (Woehrmann 1971; Chapter 10). White oak was also used routinely in the construction of hay press barns (Chapter 12) and in the construction of the actual hay press, which would be subjected to excessive mechanical force during the pressing of hay bales (Baas and Rubino 2013).

Different timber types could have also been selected for unique applications based on their working characteristics. For example, species such as white oak, tulip poplar, and chestnut, which are easily split, were used for shingles (Table 13.1). Conversely, hickory, a very hard wood, is not commonly found in historically constructed buildings of our study region. It is a dense, flexible wood that is hard to work and was likely more appropriately used in other applications such as tool handles (Flint 1904; Wigginton 1972).

13.2.2 Are there biases in timber selection?

Given the unique characteristics of different timber types (durability, strength, decay resistance, etc.; Table 13.1), it is reasonable to assume that builders in the region would be biased when felling trees for construction. Barclay and Rayburn (2014) and de Graauw (2017) cite a bias for particular timber types in

Table 13.1 Timber types used in construction of historic buildings in eastern North America. Only timber types used in historic construction are listed (we did not list species used historically for making wooden objects (e.g., furniture, gun stocks, fence posts) or current uses of the various timber types). See Table 3.1 for a listing of the species that are included in the various timber type classifications.

Timber type	*Application*	*Properties*
Ash (white and blue)	General construction timber[3,7]	Fine, straight grain[6] Easily split, pliant, readily smoothed, not as resistant to exposure as European ash[9]
Black locust	Barn timbers[2] Foundation blocks, sleepers, and pegs[6]	Difficult to work, resists rot and termites[6] Strong, heavy, not prone to warp, tolerates exposure to weather[9]
Black walnut	Cabinets[2] Flooring, doors, walls, puncheons, general construction timber, interior finishing[*3] Interior trim[7] Shingles[10]	Hard, easy to split and work, aesthetically pleasing[6] Lighter, probably weaker than English walnut[9]
Butternut	General construction timber[7]	
Cherry	Cabinets[2] Flooring, general construction timber[3] Interior trim[7]	
Chestnut	Shingles[2] Puncheon floors, doors, inside wall boards, lathing, wall logs, sleepers, sills, plates, rafters, joists[6] Interior trim[7]	Soft and light, easy to split and work, long-lasting when exposed[6]
Fir	General construction timber[7]	
Hard pine	General construction timber[7]	
Hemlock	General construction timber[7]	
Hickory	General construction timber, pegs[3]	Heavy, flexible, durable, difficult to work[6] Tough, elastic[9]
Maple	General construction timber[3] Interior trim[7]	(Sugar maple) Finer grained[9]
Oak (unspecified)	Flooring, puncheons, general construction timber[3] Flooring, structural timbers[6] Pegs, structural timbers[7] Clapboard[8] Shingles[11]	Hard, but worked easier than hickory[6] Great strength[7]

Timber type	Application	Properties
Pine	Cabin logs, foundation blocks, paneling, flooring, rafters, joists, doors, door frames, windows frames[6]	Grows straight, durable, works easily with hand tools[6]
Spruce	General construction timber[7]	
Tulip poplar	Interior millwork, siding, fascia, soffits, corbels, windows, doors, trim[1] Flooring, shakes, shingles, doors, walls, general construction timber, interior finishing*[3] Structural timbers[4] Cabin logs, rafters, joists, weatherboard, and paneling (if the tree is old enough to have green heartwood present)[6] General construction timber[7]	Termite resistant[1] Splits well, rot resistant[3] Long and straight trunks suited for timber construction, light weight, insect and decay resistant, machines well with hand tools[4] Long, clear logs, light, strong, soft, easily worked[6] Well suited to board and scantling production[9]
White oak	Barn timbers and flooring[2] Shakes and shingles[3]	Tough, strong, durable[2] Splits well, rot resistant[3] Strong and hard[5] Tough, dense, flexible, splits easily[9]
White pine	Wide variety of purposes other than structural timbers[7] Shingles[9]	Cuts and works easily[7]

*interior finishing = baseboards, chair rails, peg rails, door and window frames, newel posts, handrails, and mantels. See Baas and Rubino (2013) for a description of wood types used in constructing hay presses (Chapter 12). Cherry is *Prunus serotina* Ehrh.; Fir is *Abies* Miller; Hemlock is *Tsuga canadensis* L. (Carrière), Spruce is *Picea* A. Dietr.

Sources: 1 = Cassens et al. (2009); 2 = USDA (2007); 3 = Hutslar (1986); 4 = Hill and Shelly (1985); 5 = Lippincot (1973); 6 = Wigginton (1972); 7 = Stotz (1966); 8 = Latta (1938); 9 = Flint (1904); 10 = McDonough (1883); 11 = Baily (1856).

northeastern New York and West Virginia, respectively. If bias was, in fact, a major factor influencing timber selection, from where did this bias come and how do we identify and measure it?

Experience would lead builders to favor certain timber types over others for a particular construction member (sill or flooring) or for using it at all in a building. Settlers and builders would bring knowledge of wood characteristics (as they did with construction techniques; Chapter 2) to our study region from Europe, and then from eastern and southern North America. These biases would likely be followed unless local knowledge was superior; old traditions could be rejected or hybridized. Ultimately "what worked" would diffuse across the continent as settlement moved westward (Kniffen 1965; Stanton 1985). Wigginton (1972), in his collection of oral histories from the Appalachians, states that a working knowledge of the different timber type characteristics and suitable uses was essential. He further states that one's "survival too often depended on it." Conversely, it has been suggested that early builders learned by trial and error and that their experience informed decisions about which timber type to use in future applications (USDA 2007).

An example of a clear understanding of which timber to use in particular construction situations is evident in the stipulations set forth by Monroe County, Illinois, government officials when they drew up a contract for the construction of a jail in 1818, the first public building in the county. The contract stipulated that the jail was to be constructed of dovetailed, "hughen" oak timber except for two inside tiers, which were to be cottonwood. The corners of the building were to be boxed with oak planks, and the door and gable ends were to be made of oak planks. Black walnut or oak would be used for the shingles (McDonough 1883). Similarly, a contract was drafted on April 5, 1851, to build a courthouse in the county. According to the specifications of the contract, all joists in the first and second floor had to be white or post oak or white or yellow pine; the ceiling joists, roof timbers, shingles, frontispieces, side-light doors, and door shutters were to be made of white pine; stairs and floors were to be yellow pine. The courthouse was completed April 1, 1853 (McDonough 1883). The stipulated use of these timber types relates to their characteristics; oak is a strong wood and withstands exposure to the elements, and white pine, given its straight grain and workability, would serve well for structural timbers and decorative accents.

Further examples of purposefully selecting particular timber types for construction is seen in Harmonist structures in Indiana (Chapter 11). Blair (1964) states that the Harmonist builders preferred tulip poplar. Given its wood properties and working characteristics (Table 13.1) and its versatility, it is easy to see why builders would have preferred it when constructing buildings. Interestingly, the Harmonists, who were German, would not have worked with the species while in Europe – tulip poplars are only native to North America (*Liriodendron tulipifera*) and Asia (*L. chinense* (Hemsl.) Sarg.). Appreciation of tulip poplar timber properties was acquired once the Harmonists reached North America, perhaps when they originally settled in western Pennsylvania (Harmony, Pennsylvania).

Hutslar (1986) states that oak, black walnut, cherry (*Prunus serotina* Ehrh.), maple, hickory, ash, and tulip poplar were commonly used. He further states that white oak, black walnut, and tulip poplar were the most commonly used in 19th-century Ohio structures. In western Pennsylvania, oak was used for structural timbers due to its strength while white pine was used in most other applications; hemlock, butternut, tulip poplar, yellow pine, ash, fir, and spruce were also commonly used (Stotz 1966). However, Hutslar (1986) also states that any timber species would serve as suitable timber for a building if it was of adequate size. He states that wood uses were interchangeable, but builders did have a preference and selection was based on "strength, durability, and ease of handling" Hutslar (1986).

13.2.3 What species of trees are available for construction?

In the region where the case studies in this text were performed, the forest composition (species and relative proportion of species) changes with several factors including soil type (composition, chemistry, how well it drains), glacial history, and time since last disturbance (windstorm, tornado, logging). Forests throughout the area are often classified into particular forest types based on the dominant or most common species found in a community (Braun 1950; Gordon 1969). For example, a large portion of central Indiana falls within the beech-maple forest region; forests of this large area are dominated by American beech and sugar maple. The southeastern corner of the state falls in the western mesophytic forest region. This region's forest contains a mix of species including several different oaks, tulip poplar, maples, American beech, basswood, ash, and buckeye (Braun 1950). Although a particular region may fall into a particular forest type, there is fine-scale variation in forest composition. For instance, along streams, where moisture may be more abundant, different species of trees (or different numbers of these trees) may be found than in adjacent stands that have drier soil. Consequently, timber availability would be influenced by topography (the shape and layout of the land), with species such as tulip poplar, sycamore, and black walnut more prevalent in moister soils and oaks and hickories dominating drier sites (Potzger and Friesner 1934)

If one assumes that historic structures were built with trees that were felled from the forest found on-site (Hutslar 1986), the tree species growing in a particular area will affect timber selection and, ultimately, the timber types found in a building. Simply stated, the composition of the timber types found in a building should mirror that of the local forest. Furthermore, one could assume that the more abundant a species was at a location, the more that species would be found in a historic structure (de Graauw 2017). Consequently, one would expect that with changes in topography and geographic region, changes in forest composition would lead to differences in the number and types of trees used in construction.

13.2.4 Methods for identifying timber bias

The composition (types, abundance, and size of different tree species) of the original forests encountered by European settlers can be deduced through several methods. For example, letters and journal entries by travelers can be used to describe these forests (Flint 1904). A more systematic approach can be performed by studying the remnant patches of old-growth forests that were spared from clearing and land conversion following European colonization and settlement (Lindsey et al. 1969; Homoya et al. 1985). Also, General Land Office (GLO) survey records can be used to describe vegetation found in a region prior to settlement and clearing (Potzger et al. 1956; Chapter 6). Prior to selling newly acquired tracts of land (immense chunks obtained by the government via purchase or treaty), the GLO of the US government dispatched surveyors to divide land into parcels of a predetermined size. To delineate each parcel, surveyors would mark and record trees adjacent to a marker at each of the parcel's corners. Records of these trees can be obtained and studied to identify vegetation types throughout large tracts of land and represent "our most tangible account of Indiana's primeval forests" (Potzger et al. 1956).

To gain a better understanding of how different timber types were used for construction, we identified all of the original timbers (not just those used for tree-ring analysis) in several structures and compared our findings with the GLO survey records for the area surrounding each of the buildings. The Kuns House at Delphi Canal Park (Figure 8.4; Chapter 8) is a single-pen log structure that was built with horizontal logs in 1866. We identified the timber type of all of the timbers used to construct the walls of the building; we did not include the roof or sills since we were uncertain if they were part of the original fabric of the building (Table 13.2). Red oak was used predominantly in the structure. However, only two of the witness trees found in the area were oak. The most common witness tree, American beech, was only used for three of the timbers in the building.

Table 13.2 Comparison of timber types used to construct the Kuns House (moved to Delphi, Indiana) and GLO witness tree records from the construction site and adjacent area. Using wood anatomy only permits identification to timber type, not tree species; please see Table 3.1 for a listing of what species fall under the various timber types (e.g., sugar maple is a type of hard maple).

Timber type	*Timbers*	*GLO trees*	*Tree species*
Red oak	26	16	American beech
White elm	7	8	Hickory
American beech	3	5	Ash
		4	White oak
		2	Sugar maple
		2	Elm
		1	Sassafras
		1	Tulip poplar
		1	Black oak
		1	S. oak

We performed a similar analysis for the Braytown North Hay Press Barn (Switzerland County, Indiana). The barn was constructed from a mix of native hardwood species (Table 13.3). Tree-ring analysis of the barn suggests an 1858 construction and that recycled timbers were used in construction and/or renovation. We identified all accessible timbers (100) in the barn (we were unable to reach timbers in the roof structure such as rafters and roof sheathing) and found American (white) elm, ash, and hard maple to be the most commonly used timber types. Comparison of the timber types used in construction with the types of trees noted by the original GLO survey of the area surrounding the barn show little similarity (Table 13.4).

Table 13.3 Samples collected from the Braytown North Hay Press Barn (Switzerland County, Indiana). "First" and "last" refer to the first and last rings present in the sample. If no date is provided, the sample could not be reliably crossdated. "Outer ring" refers to the outermost ring of the sample; see Table 7.1 for a complete list of codes used for outer-ring condition. "r" is the correlation coefficient obtained by correlating the ring widths of this sample with all other samples of the same timber type from the building (series intercorrelation; Chapter 5). "Timber type" is the type of timber; see Table 3.1 for a list of possible tree species represented by the various timber types. "Provenience" is the timber's function in a structure; see Chapter 2 for a description of the various proveniences. Note: Timber type and provenience are only given once if more than one sample was obtained from an individual timber. An r-value is not reported if less than three timbers of a particular timber type were obtained from a structure or if the sample did not overlap temporally with the other samples of a particular timber type. A full description of the data presented in the table is provided in Chapter 7.

Sample	*First*	*Last*	*Outer ring*	r	*Timber type*	*Provenience*
JUR01A	1772	1853	b	0.27	Ash	Floor joist; ground floor; recycled
JUR02A	1699	1830	w	0.36	Walnut	Floor joist; ground floor
JUR03A	1837	1858	w	0.31	White elm	Floor joist; ground floor
JUR03B	1757	1835		0.57		

Sample	*First*	*Last*	*Outer ring*	r	*Timber type*	*Provenience*
JUR04A	1819	1857			Honey locust	Floor joist; ground floor
JUR05A	1694	1830		0.46	Walnut	Floor joist; ground floor
JUR05B	1789	1837	w	0.33		
JUR06A					Hard maple	Floor joist; ground floor
JUR07A	1748	1858	w	0.43	Walnut	Floor joist; ground floor
JUR07B	1695	1736		0.44		
JUR08A	1826	1858	w	0.50	White elm	W sill
JUR08B	1726	1827		0.38		
JUR09A	1799	1843	w	0.75	Ash	Post; N elevation
JUR09B	1786	1843	w	0.56		
JUR10A	1724	1855	w	0.44	White oak	Joist; ramp
JUR11A	1728	1858	w	0.50	Ash	Joist; ramp
JUR12A	1695	1856	w	0.45	White oak	Joist; ramp
JUR13A	1662	1858	w	0.46	White elm	Post; N press bent
JUR14A	1783	1858	w	0.73	White elm	Post; N elevation
JUR14B	1722	1847		0.57		
JUR15A					Soft maple	Beam; S elevation
JUR16A	1794	1858	w		Tulip	Diagonal brace; W elevation
JUR16B	1800	1858	b			
JUR16C	1751	1787				
JUR17A	1497	1631			White elm	Top chord; S press bent
JUR18A	1724	1855		0.52	Ash	Beam; S elevation
JUR18B	1845	1858	w	0.28		
JUR19A	1729	1858	w	0.43	Ash	Ramp flooring
JUR20A	1745	1855		0.54	White oak	Ramp flooring
JUR21A					White oak	Loose; cellar
JUR22A	1725	1858	w	0.51	Ash	E Sill
JUR23A	1796	1858	w	0.38	White oak	Beam; S press bent
JUR25A	1607	1796		0.54	White pine	Siding; S elevation
JUR26A	1668	1764		0.72	White pine	Siding; S elevation
JUR26B	1676	1764		0.73		
JUR27A	1535	1647		0.55	White pine	Siding; S elevation
JUR27B	1540	1664		0.56		
JUR28A	1608	1773		0.74	White pine	Siding; S elevation
JUR28B	1608	1729		0.69		
JUR29A	1617	1727		0.53	White pine	Siding; S elevation
JUR30A	1623	1702		0.68	White pine	Siding; S elevation
JUR31A					White pine	Siding; S elevation
JUR32A	1568	1775		0.44	White pine	Siding; S elevation
JUR32B	1570	1744		0.55		
JUR33A	1733	1814		0.61	White pine	Siding; S elevation

(Continued)

Table 13.3 Continued

Sample	*First*	*Last*	*Outer ring*	r	*Timber type*	*Provenience*
JUR33B	1741	1805		0.74		
JUR34A	1562	1643		0.57	White pine	Siding; S elevation
JUR35A	1738	1817		0.49	White pine	Siding; S elevation
JUR36A	1676	1749		0.67	White pine	Siding; S elevation
JUR37A	1613	1783		0.67	White pine	Siding; S elevation
JUR37B	1612	1705		0.66		

Table 13.4 Comparison of timber types used to construct the Braytown North Hay Press Barn and GLO witness tree records from the construction site and adjacent area. Using wood anatomy only permits identification to timber type, not tree species; please see Table 3.1 for a listing of what species fall under the various timber types. Surveyors noted American beech and black locust while documenting the forest near the Braytown North Hay Press Barn, but they did not find it at the survey points. Hackberry is Celtis occidentalis *Willd.*

Timber type	*Timbers*	*GLO Trees*	*Tree species*
American elm	28	4	Sugar maple
Ash	20	3	Elm
Hard maple	20	3	Ironwood
Tulip	14	2	Hickory
White oak	9	2	White oak
Black walnut	4	1	Buckeye
Soft maple	4	1	Ash
Honey locust	1	1	Basswood
		1	Hackberry

Therefore, there is not much agreement between the timber types used in both structures and local witness tree distribution. There are trees not found in the buildings, and timber types found in the buildings that are not represented in the forest survey. Also, there is not a pattern in the number of trees found in the GLO survey and the number of timbers in a building; the most common tree is not the most commonly found timber type. These data suggest that there was bias in builders' selection of timber. A multitude of factors could also explain this discrepancy. For instance, surveyors were identifying trees closest to some defined point in space; they were not marking trees that were of the specific size needed for construction. For example, ironwood (*Carpinus caroliniana* Walter) is a small tree/large shrub that would likely not attain a size large enough to be used as a log for construction. Also, perhaps there were buildings present (but subsequently raised) at the sites. These structures may have been built with the number and types of timbers that more closely match the survey data.

13.2.5 Does species availability change over time?

As settlement increased in an area, one can assume that there would be increased pressure for the harvesting of preferred timber types. Hence, a shift in timber type composition would occur with older and newer buildings being built with a unique subset of timber types. Tree-ring analysis of historic buildings would offer an excellent way to observe if such an evolution in timber use has actually occurred. de Graauw (2017) found a shift in timber type from white oak to a mix of white oak, tulip poplar, white pine, and chestnut over time (1750–1899) in the central Appalachians. Barclay and Rayburn (2014) found a shift from oak and ash to hemlock and spruce in New York over the 19th century. A similar shift in timber types found in buildings over a much longer time span (2000 to 500 BC) was found by Billamboz (2003) in southwest Germany.

Given that different timber types vary in their suitability for construction, that builders were biased in their timber selection, and that different types of trees were available for harvesting both spatially and temporally, how can tree-ring analysis help to make sense of timber selection? We have tried to systematically analyze the use of particular timber types over time by examining the timbers we have dated from historic buildings of various ages. We hypothesized that preferred timber types such as white oak would be most common in older (mid to early 19th century in our study area) buildings and diminish in use over time as supplies become limited. Likewise, some of the timber types that would be less ideal for construction (rot resistance, workability, etc.) would increase in use over time and be absent from earlier construction.

To see if our hypothesis was true, we analyzed the use of American beech timber in local buildings. American beech is a very dense, heavy wood. However, its inability to withstand exposure to moisture and to resist rot (Elliott 1912) would likely make it unacceptable for many building applications. American beech trees were often not harvested when forests were cleared (Record 1908; Chapman 1942). Our experience with coring large, live trees and sampling recently dead American beech trees (Chapter 4) suggests that it often experiences heart rot at the stem's base (the inside of the stems is often hollow for a considerable distance from the ground). Consequently, long logs suitable for construction may not be reliably obtained from American beech trees. Interestingly, none of the sources we used to construct Table 13.1 mention American beech as a useful tree for construction. This evidence supports that American beech would have been a non-preferred timber and that we could expect an increase in American beech usage with time – it should be absent in older buildings and more prevalent in newer structures (late 19th century).

To analyze the use of beech over time, we looked at several buildings we have dated in the region. The Tunnel Mill House and Barn are located in Charlestown, Indiana. The two-story brick house's first-floor joists are white oak, and analysis of these timbers provided an 1818 construction date. The horse barn on the property dated to 1896. It was made with many American beech timbers. These findings suggest a shift in timber type use over time on this property. The Cravenhurst Barn in Jefferson County, Indiana, was built in 1910. It is constructed almost exclusively with American beech timbers while other barns in the county that date to the mid- to late-19th century are constructed with white oak. These findings support our hypothesis that timber type use shifted over time. However, the Keigwin House in Jeffersonville, Indiana, was built in 1819. This two-story timber frame house was built completely of American beech. The Little Log Cabin in North Vernon, Indiana, was built in 1817 with American beech and tulip poplar. Many timbers in the *Musée de Venoge* (constructed in 1828; Chapter 9) were also American beech. One would expect to find white oak, the preferred species for construction, to be used extensively in these early (for our study region) structures, but they are not.

Similarly, a temporal pattern in the use of tulip poplar has been difficult to identify. The buildings of New Harmony (Chapter 11) are largely constructed with tulip poplar and were built prior to the Harmonists departure in 1824. However, we have dated several barns from nearby southwest Indiana that date to the 1880s that contain a substantial number of white oak timbers. The Sullivan House (Jefferson County, Indiana; Chapter 8) was built with mostly tulip poplar timbers and was constructed in 1822. Again, many buildings from this county dating from much later were built mostly with white oak, a preferred construction timber type. A scan of the tables listing the ages of different timber types from the case studies in Chapters 8 through 12 further demonstrates the extensive use of tulip poplar throughout the entirety of the 19th century.

13.2.6 Summary of timber type selection

From our analysis of many regional buildings, we have not been able to tease out clear patterns of timber usage over time or location in the mid-Ohio River Valley – we are unable to establish hard and fast timber selection rules used by builders. We have, however, been able to make solid, general observations regarding timber use in the study region. Our work has shown that most buildings were constructed from a subset of timber types even though regional forests contain a multitude of different species. Most regional buildings are built from white oaks, tulip poplar, ashes, and American beech. These species are widely distributed in the region and appear to be preferred by builders throughout the 19th century. The occurrence of certain timber types does show that usage changes with location and suggests that builders would choose from what species were available. For example, buildings erected in valleys are constructed with species such as American (white) elm and black walnut, species often associated with moist soils. Changes with time are also apparent; for example, the use of basswood, an extremely light wood that is not rot resistant, in late 19th-century buildings (Glenn Farmstead; Chapter 10).

Bias, as would be expected, was likely extremely important in regard to timber type selection (de Graauw 2017). Bias would result from experience with using different timber types and a particular wood's characteristics (rot resistance, ease of working, etc.). We come to this conclusion based on the consistent use of certain timber types and the absence or rarity of use of other types in regional buildings. For example, hackberry is a very common tree, and we have not found it used in historic buildings. Sycamore, a widely distributed tree in the region is nearly absent from historical structures; we have found it used as roof sheathing in several structures. Likewise, sassafras has only been encountered in three buildings from the area. Although it is not a dominant species in regional forests, it is quite widespread. Hickory is also common in regional forests, but we encounter it as timber infrequently. We have found black cherry only once in our fieldwork, floor joists from an undateable early 20th-century structure. Given the commonality of these tree species, they were most likely selected against based on their wood properties or were used to build nonstructural wooden objects such as furniture, wagons, farm implements, etc.

Also, it is tantalizing to suggest that historic buildings may be useful in describing pre-European settlement forest composition by examining the types of timber used in construction. In other words, do the timbers found in a building represent a snapshot of the number and types of trees that were found when the land was cleared for settlement? Assuming that historic structures were built (entirely or predominantly) with timber obtained from trees felled from the construction site, the timbers found in a building could provide clues as to the forest composition (types and amounts of various tree species). A systematic analysis of all of the timbers in a building (not just those that were dated), therefore, would help in describing the forest. Hutslar (1986) states that any timber species would serve as suitable timber for a building if it was of adequate size.

From our experience, we strongly recommend against trying to determine forest composition (number and types of trees that make up a stand) using historic building timber types alone in our study region. It is common to find buildings constructed entirely or almost entirely out of a single timber type (for example, the use of tulip poplar in New Harmony (Chapter 11) and in the Sullivan House (Chapter 8) and the use of white oak at Fort Wayne (Chapter 12)). Analysis of old-growth forests and GLO records of the region repeatedly show that a single-species forest did not exist. The presence of a particular timber type in a building suggests its presence in forests, but its commonality or dominance is likely not determinable. Additionally, without a keen understanding of how timbers were incorporated into a building – were they all harvested on-site, were they transported in from a local sawmill, or was a mix of both used – it is difficult or impossible to accurately describe regional forests. For instance, Eleutherian College was built largely from sweet gum. Forests that were dominated with sweet gum would likely not have been found at the site of the college or never existed in the entire region. Timbers would have been sourced, most likely from a sawmill, and the original source(s) of the timber is likely indeterminable. Furthermore, local buildings built later than the college are constructed of more desirable timber types such as oak.

Although it may not be possible to identify clear and specific timber use patterns over space and time or to precisely describe regional pre-European settlement forests of the mid-Ohio River Valley, much information is still gained through tree-ring analysis of historic buildings. Tree-ring analysis allows for dating a building and landscape; it gives a precise date to settlement and subsequent modification of the environment. It also permits a better understanding of how regional forests were harvested and utilized. By using tree-ring analysis, one is afforded the opportunity to examine how timber was utilized over a large geographic region. When coupled with historic documents, tree rings allow for a better understanding of a region's cultural heritage and shed additional light on how humans interact with the environment (Čufar 2007).

13.3 Determining timber origin: dendroprovenancing

As the case studies in the previous chapters have shown, dating of buildings is possible if appropriate reference chronologies are available (timber type, time period covered, geographic region). When European settlers arrived in the region in which we perform our work, they encountered dense forests filled with ample timber for constructing homes, barns, smokehouses, etc. (Latta 1938; Lippincot 1973). Timber resources were so abundant it was often wasted. When tracts of land needed to be cleared for various purposes, such as agriculture, large numbers of tree would be killed and allowed to decay or were burned (Flint 1904; Hutslar 1986). When landscapes were transformed by settlers and buildings were erected, trees would be harvested from on-site for construction or obtained from local sawmills.

Studies of old-growth forests and GLO records for the region have been used to characterize regional forests in regard to what species of trees were present when European settlers arrived. Studies have shown that the region was dominated by hardwoods (angiosperms; Chapter 3), including oaks, American beech, maples, ashes, tulip poplar, hickories, and basswood. Finding timbers in historic buildings made of these trees would be expected. However, we have found many buildings in the region that contain a large number of softwood (gymnosperm) timbers. Softwoods (excluding the ubiquitous eastern red cedar, *Juniperus virginiana* L., which we have never found in a historically constructed building) have an extremely spotty distribution throughout the landscape and often grow in small, isolated stands if they are present at all in the mid-Ohio River Valley (Coulter 1914; Daubenmire 1931; Friesner and Potzger 1932; Jackson 2004). Given the absence and paucity of these species throughout the landscape, it is hard to imagine enough trees could be harvested for timber production. Therefore, when encountering softwoods, such as white pine and hemlock, one must wonder about the origin of the timber.

As was stated in the introduction to this chapter, tree-ring analysis of historically constructed buildings provides information beyond tree-ring widths. Analysis permits precise dating of a structure or landscape, but it can also provide information regarding where timbers originated. Dendroprovenancing, determining the geographic origin of wood using tree-ring analysis, has been performed to successfully identify the origin of wood used for ships, buildings, musical instruments, and panels used for paintings (Bridge 2012). In Europe, dendroprovenancing has been used to better understand historic logging activity and timber trade (Haneca et al. 2005, 2009). In North America, dendroprovenancing has been used to determine the origin of timbers used in ship building (Martin-Benito et al. 2014; Creasman et al. 2015) and in timber used for the construction of canals (Taormina and Speer 2016).

So where does the softwood we find in buildings along the Ohio River originate? Timber production records do provide some clues as to the origin of the timber (Steer 1948). New York and Pennsylvania exported vast quantities of white pine timber in the 1800s with both states, at one time, leading all states in the amount of lumber cut (Clapp 1920). In 1879, 52 million board feet of pine (white and yellow) were produced in Indiana; it is "assumed" that the white pine originated in the northern most counties of Indiana (Miller 1965). The Great Lake states also exported vast quantities of white pine timber in the 19th century (Clapp 1920; Buls 1949).

In an attempt to determine where the white pine found in mid-Ohio River Valley buildings originated, we turned to the tree-ring patterns in the timbers. If we could match the growth patterns in the timbers to white pine tree-ring chronologies from other regions, we may be able to provenance or locate the source of the timbers. The following documents our attempt to provenance white pine timbers collected from three buildings found in southern Indiana along the Ohio River, the Sullivan House (Chapter 8), the Thiebaud Hay Press Barn (Chapter 12), and the Braytown North Hay Press Barn.

13.3.1 Determining timbers' geographic source

Initial analysis of the white pine timbers included wood identification using microscopic wood anatomical features (Chapters 3 and 5). We then followed a multistep procedure to produce a floating chronology (a set of samples that are crossdated among themselves but have no calendar dates assigned to any of the tree rings; Chapter 5). Once a floating chronology was created, we could crossdate it against other white pine chronologies with verified dates. To determine a potential source for the timbers, we would look for a strong correlation (Chapter 5) between the floating chronology and the verified chronologies. The stronger the correlation, the more likely the geographic origin of the timber (Bridge 2012).

To create our floating chronologies, we began by following standard sample preparation procedures, including sanding and assigning arbitrary years (not dates) to each of the rings in the samples. Skeleton plots were then created for each sample, and visual crossdating was performed. Each ring in each of the series was then measured, and we used COFECHA to assist in the crossdating process. We then renumbered each ring in the crossdated skeleton plots to create a floating composite chronology by assigning the first ring in the crossdated chronology as ring 1, the next ring 2, etc. Rings were numbered in this manner until we reached the last ring (most recently formed) in any of the samples. We then verified crossdating and ring-width measurements in the floating chronology using COFECHA. See Chapter 5 for a more thorough explanation of each of the steps used to create a floating chronology.

Once we had created a composite floating chronology for the timbers from the building, we compared it to verified chronologies available from the International Tree-Ring Data Bank (National Centers for Environmental Information 2018). The ITRDB is an online repository of tree-ring data from throughout the world (Grissino-Mayer and Fritts 1997; ITRDB 2018; Chapter 4). We compared our floating chronology to 36 white pine chronologies from eastern North America and the Midwestern United States. This step was performed using COFECHA. We used each ITRDB chronology as our dated chronology and the floating chronology as an undated chronology. COFECHA assisted in date assignment of the floating chronology by breaking it into 50-year-long segments lagged by 25 years (see Chapter 5). Each of these segments was compared to the verified chronology, and potential calendar year assignments were suggested for the segments (e.g., segment 1–49 of the floating chronology matches the reference chronology if 1527 is added to each ring in the segment). This was performed for each of the segments in the floating chronology for all 36 white pine chronologies from the ITRDB. We used r-values to identify strong, significant correlation between the floating chronology's segments and the verified ITRDB chronologies. Therefore, by using statistical and visual crossdating we were able to find which chronology(ies) matched the floating chronology the best (Martin-Benito et al. 2014; Creasman et al. 2015; Taormina and Speer 2016).

13.3.2 Sullivan House

The Sullivan House (Madison, Indiana) was constructed in 1822 (Chapter 8). Most of the timbers analyzed from the house were tulip poplar. In the fall of 2008, a windstorm caused extensive damage to the roof and east façade of the building. During repair of the structure, we were able to collect a large number of white pine timbers (Table 13.5). We obtained 28 white pine timbers from the building and successfully crossdated 23 of them (27 individual series were obtained from these 23 timbers). We created a 263-year floating chronology consisting of 2,504 rings ($r = 0.54$; 1.40 ± 0.65 mm; mean ± SD ring width) that spanned from 1556 to 1818. We found the highest correlation (largest r-value) between the Sullivan House white pine chronology and a chronology from northwestern Pennsylvania (Figure 13.1). Since the original structure was built in 1822, we conclude that the white pine timbers were incorporated into the building during its initial construction and were not added during subsequent renovation to the structure.

Table 13.5 White pine samples collected from the Sullivan House (Madison, Indiana). See Table 13.3 for a description of the table contents; a full description of the data presented is provided in Chapter 7.

Sample	*First*	*Last*	*Outer ring*	r	*Provenience*
SUL01A	1625	1716		0.56	Roof sheathing
SUL02A	1707	1818	w	0.46	Studding; wall or ceiling
SUL16A	1611	1679		0.55	Framing timber
SUL31A					Framing timber
SUL41A	1688	1795		0.60	Roof truss?
SUL43A	1661	1747		0.45	Roof truss?
SUL44A	1586	1661		0.51	Roof sheathing
SUL45A	1588	1672		0.58	Roof sheathing
SUL45B	1594	1695		0.64	
SUL46A					Roof sheathing
SUL47A					Roof sheathing
SUL48A	1608	1708		0.54	Roof sheathing
SUL49A	1651	1723		0.62	Roof sheathing
SUL49B	1652	1721		0.63	
SUL50A	1649	1816	w	0.44	Roof sheathing
SUL51A	1601	1699		0.62	Roof sheathing
SUL52A	1674	1760		0.62	Roof sheathing
SUL52B	1673	1750		0.60	
SUL53A	1609	1699		0.54	Roof sheathing
SUL54A	1599	1718		0.52	Roof sheathing
SUL55A	1621	1705		0.62	Roof sheathing
SUL56A	1609	1687		0.33	Roof sheathing
SUL57A					Roof sheathing
SUL58A	1717	1819	w	0.58	Roof sheathing
SUL58B	1718	1819	w	0.55	
SUL59A	1656	1744		0.49	Roof sheathing
SUL60A	1717	1816	w	0.45	Roof sheathing
SUL61A	1664	1760		0.54	Roof sheathing
SUL62A					Roof sheathing
SUL63A	1685	1771		0.58	Roof sheathing
SUL64B	1555	1695		0.51	Baseboard
SUL65A	1605	1674		0.65	Framing timber

13.3.3 Thiebaud Hay Press Barn

The Thiebaud Hay Press Barn was built in 1850 and was subsequently modified using either newly felled trees for timbers or recycled timbers (Chapter 12). The structural timbers of the barn are all hardwood species (e.g., ash, white oak, and American beech) that were likely sourced from the farmstead (Table 12.9). However, the barn's siding and much of the floor are white pine. We created a white pine chronology from floorboards that spanned from 1613 to 1849 (Tables 12.9 and 12.10). This chronology had the highest correlation with a northwestern Pennsylvania chronology as did the Sullivan House white pine chronology (Figure 13.2). We found four floorboards with a bark or wane (harvest date) of

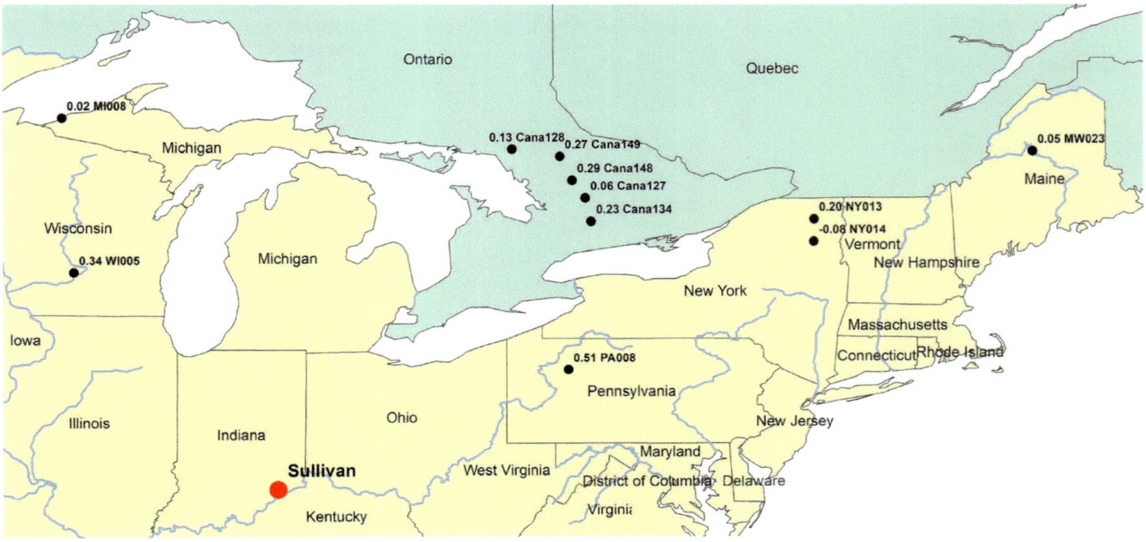

Figure 13.1
Dendroprovenancing white pine timbers from the Sullivan House (Madison, Indiana). We correlated the white pine chronology created from the pine timbers of the Sullivan House to 36 white pine chronologies obtained from the ITRDB. *r*-value and chronology identification are only given for those verified ITRDB chronologies that overlapped with the Sullivan House chronology for at least 100 years. The white pine chronology from the Sullivan House correlated the strongest with chronology PA008.

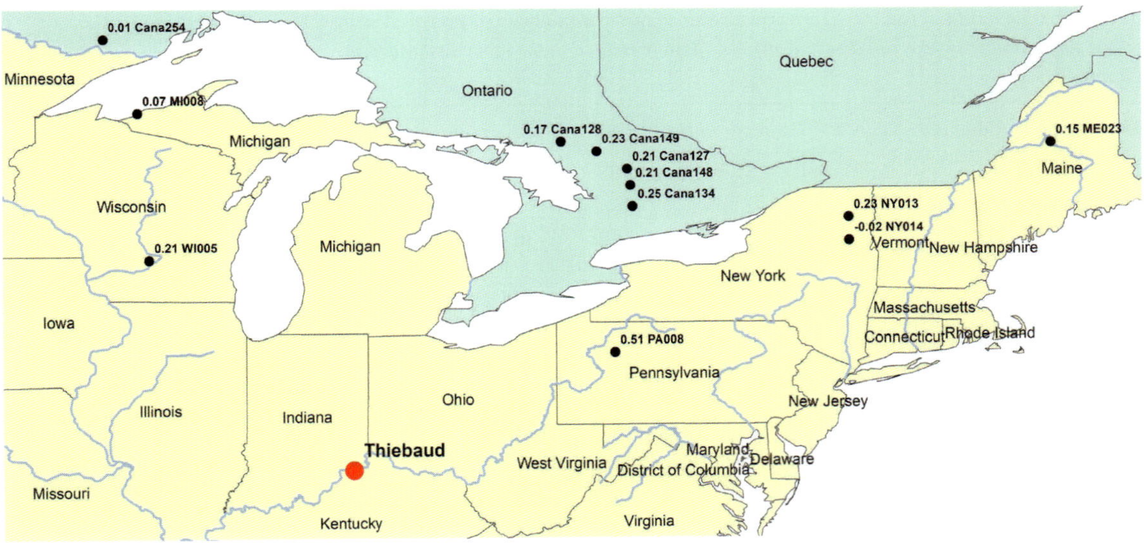

Figure 13.2
Dendroprovenancing white pine timbers from the Thiebaud Hay Press Barn (Switzerland County, Indiana). See Figure 13.1 for a description of figure construction.

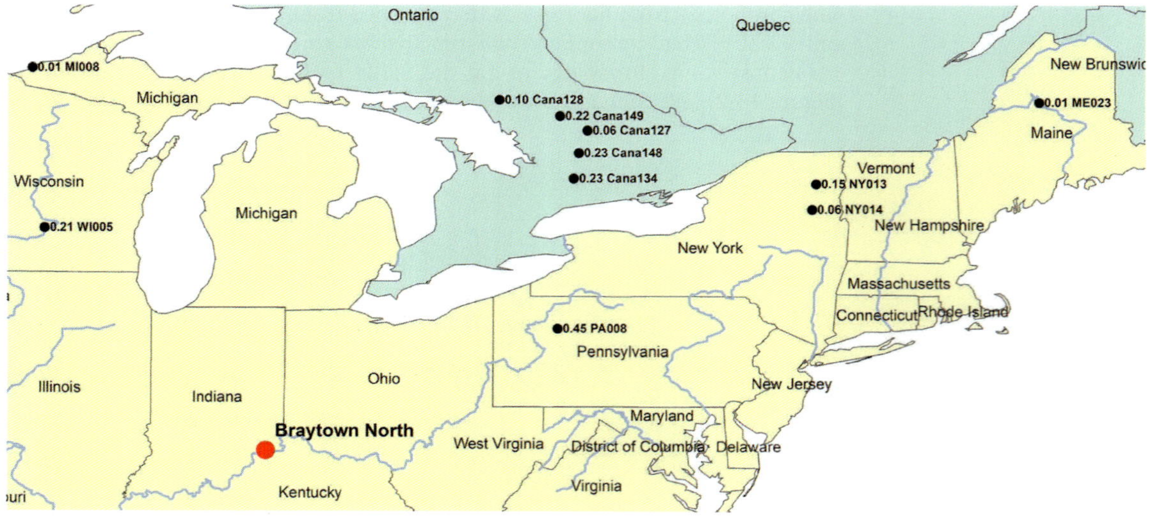

Figure 13.3
Dendroprovenancing white pine timbers from the Braytown North Hay Press Barn (Switzerland County, Indiana). See Figure 13.1 for a description of figure construction.

Table 13.6 Summary table of the timber type chronologies prepared from the Braytown North Hay Press Barn (Switzerland County, Indiana). "Samples" and "timbers" are the number of successfully crossdated samples and timbers from the building. "Undated" is the number of timbers that could not be crossdated reliably. "First" and "last" refer to the first and last dates in the chronology. "Mean" and "SD" are the mean and standard deviation (mm) of the ring widths in the chronology. "r" is the mean series intercorrelation for each timber type (see Chapter 5). r is not provided if less than three timbers were obtained for a particular timber type or if a sample did not overlap temporally with the other samples for a particular timber type. No total r, mean, or SD are provided since each chronology was prepared using only a single timber type. A full description of the data presented in the table is provided in Chapter 7.

Timber type	Samples	Timbers	Undated	First	Last	Years	r	Mean	SD
Ash	8	6	0	1726	1857	707	0.48	1.15	0.50
Hard maple	0	0	1						
Honey locust	1	1	0	1821	1856	36		1.53	0.37
Soft maple	0	0	1						
Tulip poplar	2	1	0	1796	1857	119		0.82	0.46
				1753	1785	33		1.23	0.66
Walnut	5	3	0	1695	1857	437	0.41	1.15	0.60
White elm	7	4	0	1667	1857	614	0.52	0.63	0.29
				1498	1630	133		0.42	0.19
White oak	4	4	1	1696	1857	458	0.46	1.35	0.43
White pine	18	12	1	1536	1816	2083	0.60	1.00	0.43
Total	47	32	4	1498	1857	4620			

1827. Interestingly, floor joists used to construct the west addition to this 1850 barn also had an 1827 felling date (Table 12.9). We hypothesize that the floorboards and structural timbers were recycled from some unknown building that was once present on the farmstead.

13.3.4 Braytown North Hay Press Barn

The Braytown North Hay Press Barn (Switzerland County, Indiana) was constructed of a mix of native hardwood species (Table 13.3). Tree-ring analysis of the barn suggests an 1858 construction of the barn and that recycled timbers were also used in construction and/or renovation. The siding of the barn is white pine. We cut cross sections from 13 pieces of siding from the south elevation to create a chronology spanning from 1536 to 1816 (Table 13.6). Again, this white pine chronology shows the highest correlation with a chronology from northwestern Pennsylvania (Figure 13.3).

13.3.5 White pine in the mid-Ohio River Valley

Dendroprovenancing suggests that the timbers found in the three structures just described originated from northeastern North America, namely northwestern Pennsylvania. Given the ample amount of readily available local timber one must wonder why softwood timbers are found in historically constructed buildings of the mid-Ohio River Valley. As was stated earlier, builders preferred to use certain types of timber in construction based on a wood's mechanical and workability properties. White pine was a top choice for various construction needs (Stotz 1966). Lippincott (1973) states that regionally, "of the soft woods, the absence of white pine was the greatest want." He further states that millions of board feet were supplied from western New York by floating timber down the Allegheny River to Pittsburgh where it could be sent west along the Ohio River (Lippincott 1973). In *Flint's Letters from America – 1818–1820* (1904) he states that "white and yellow pines, similar to those of Canada, are brought from Allegany (*sic*) river, and are now sold here [southeastern Indiana], in boards, at a cent per square foot." He also describes "rafts of timber" descending the Ohio River at Cincinnati, Ohio. In Jeffersonville, Indiana (southeastern Indiana), he describes a shingle maker using pine wood he had brought down the Allegheny River. He further states that shingle making is lucrative in the region since native pine cannot be found (1904). The importance of river transport of timber has been established in Europe using dendroprovenancing (Haneca 2005; Eißing and Dittmar 2011), and it appears it was also very important in 19th-century North America.

Various sources, such as travelers' firsthand accounts and US Department of Agriculture statistics, support that pine was actively exported from Mid-Atlantic states during the 19th century. The demand for white pine was great. Elliott (1912) states that it was, at one time, considered the only valuable softwood and was used in an extremely wide range of applications. By coupling historic sources with tree-ring analysis we are able to better understand how timber was utilized over a wide geographic range and gain insight into 19th-century timber commerce. We can conclude that it was profitable to export timber along the Ohio River over several decades much as it was to the Atlantic Coast (Cox 1982). We have dated several other regional buildings that contained softwood timbers (mostly white pine and hemlock but red and hard pine also) that were built as late as the early 1860s. Dendroprovenancing white pine timber in the mid-Ohio River Valley is an excellent example of how using tree-ring analysis can provide for a better understanding of human behavior and how humans interact with the environment (Dean 1997).

13.4 Conclusion

Tree-ring analysis of timbers offers a unique method to precisely date the construction of historic buildings and landscapes. As can be seen from the preceding discussion on timber selection and provenancing and the case studies of the previous chapters, tree-ring analysis offers scientists, historians, and preservationists so much more than mere dates (Billamboz 2003). Tree-ring analysis provides a clearer understanding of historic events and practices. Analysis gives insights on how humans interact with forests and utilize natural resources (Billamboz 1992). Continued study of timbers from historically constructed buildings will aid in the creation of chronologies that reach farther back in time than living trees. These long chronologies will allow for a better understanding of how previous forests grew, how humans interacted with them, and how forests and climate change over time (Čufar 2007). Trouet et al. (2017) state that tree-ring science is at the "nexus of archeology, climatology, and ecology" and offers a unique opportunity for researchers to understand how humans interacted with their environment in the past.

Current use of tree-ring samples obtained from historically constructed buildings are allowing researchers to better understand the natural world, and exciting new directions in tree-ring science are currently being investigated (Čufar 2007). For example, the tree-ring patterns found in timbers of historic buildings offer a unique insight into the ecology of a region (de Graauw 2017). Forest stand structure and dynamics (types of trees and how they change over time) can be interpreted from dendrochronological studies of timbers (Billamboz 2003). Trouet et al. (2017) were able to study the disturbance history of forests in the Ohio and Hudson River Valleys as far back as the 16th century. The tree-ring patterns in historical timbers can also be useful for understanding ecological phenomena such as masting (years of large seed production; Bleicher (2014)). Dean (1997) states that the tree-ring patterns of timbers may be used in new ways not yet imagined and that future studies will likely allow for an even better understanding of environmental and ecological phenomena.

We began this book with a simple question: "Can you date my building?" Yes, under the right circumstances (suitable samples and reference chronologies), buildings can be precisely dated. Moreover, tree-ring dating offers further information than simply a date. Insight into the builders, the forest from which the timbers were harvested, and the inhabitants/users of the building are made available from the study of the tree-ring patterns in a structure's timbers. In conclusion, tree-ring analysis of historic buildings provides a wealth of information to a wide variety of researchers asking many diverse questions.

13.5 Literature cited

Baas, C., and D. Rubino. 2013. The most successful press in this or any other country: The material culture of 19th-century beater hay presses in the mid-Ohio Valley. *Material Culture* 45: 1–20.

Baillie, M.G.L. 1995. *A Slice through Time: Dendrochronology and Precision Dating*. Routledge, London, UK. 176 p.

Baily, F. 1856. *Journal of a Tour in Unsettled Parts of North America in 1796 and 1797*. Baily Brothers, London, UK. 439 p.

Barclay, D.J., and J.A. Rayburn. 2014. Tree-ring dating of historic buildings in Willsboro, Northeastern New York, and development of regional chronologies for dendroarchaeology. *Tree-Ring Research* 70: 79–90.

Billamboz, A. 1992. Tree-ring analysis from an archaeodendrological perspective. The structural timber from the South West German lake dwellings. Pp. 34–40. In *Tree Rings and Environment*. Lundqua Report 34. (T.S. Barholin, B.E Berglund, D. Eckstein, F.H. Schweingruber, and O. Eggertssson, eds.). Lund University, Lund, Sweden.

Billamboz, A. 2003. Tree rings and wetland occupation in Southwest Germany between 2000 and 500 BC: Dendroarchaeology beyond dating in tribute to F.H. Schweingruber. *Tree-Ring Research* 59: 37–49.

Blair, D. 1964. *Harmonist Construction*. Indiana Historical Society, Indianapolis, IN. 82 p.

Bleicher, N. 2014. Four levels of patterns in tree-rings: An archaeological approach to dendroecology. *Vegetation History and Archaeobotany* 23: 615–627.

Braun, E.L. 1950. *Deciduous Forests of Eastern North America*. The Blackiston Co. Philadelphia, PA. 596 p.

Bridge, M. 2012. Locating the origins of wood resources: A review of dendroprovenancing. *Journal of Archaeological Science* 39: 2828–2834.

Buls, E.J. 1949. The grand portage fur traders route and Indian reservation. *Proceedings of the Indiana Academy of Science* 59: 225–229.

Burns, R.M., and B.H. Honkala (technical coordinators). 1990. *Silvics of North America, Volume 2. Hardwoods*. United States Department of Agriculture, Forest Service. Agriculture Handbook No. 654. Washington, DC. 876 p.

Cassens, D.L., M.O. Hunt, and H.M. Barnes. 2009. *Yellow-Poplar Lumber for Exterior Architectural Applications in New Construction and for Historical Restoration*. Purdue Extension FNR-410-W, Purdue University, West Lafayette, IN. 6 p.

Chapman, A.G. 1942. Forests of the Illinoian till plain of Southeastern Indiana. *Ecology* 23: 189–198.

Clapp, E.H. 1920. *Timber Depletion, Lumber Prices, Lumber Exports, and Concentration of Timber Ownership. Report on Senate Resolution 311*. 2nd ed. United States Department of Agriculture, Forest Service, Government Printing Office, Washington, DC. 73 p.

Coulter, S. 1914. Notes upon the distribution of forest trees in Indiana. *Proceedings of the Indiana Academy of Science* 24: 167–177.

Cox, T.R. 1982. The diary of William Langdon for 1855: A Pennsylvania Lumber Raftman's year. *Journal of Forest History* 26: 124–139.

Creasman, P.P., C. Baisan, and C. Guiterman. 2015. Dendrochronological evaluation of ship timber from Charlestown Navy Yard (Boston, MA). *Dendrochronologia* 33: 8–15.

Čufar, K. 2007. Dendrochronology and past human activity – a review of advances since 2000. *Tree-Ring Research* 63: 47–60.

Daubenmire, R.F. 1931. Factors favoring the persistence of a relic association of eastern hemlock in Indiana. *Butler University Botanical Studies* 2: 29–32.

Dean, J.S. 1997. Dendrochronology. Pp. 31–64. In: *Chronometric Dating in Archaeology*. (R.E. Taylor and M.J. Aitken, eds.). Plenum Press, New York, NY.

de Graauw, K.K. 2017. Historic log structures as ecological archives: A case study from Eastern North America. *Dendrochronologia* 45: 23–34.

Eißing, T., and C. Dittmar. 2011. Timber transport and dendro-provenancing in Thuringia and Bavaria. Pp. 137–149. In: *Tree Rings, Art, Archaeology: Proceedings of an International Conference*. (P. Fraiture, ed.). The Royal Institute for Cultural Heritage, Brussels, Belgium.

Elliott, S.B. 1912. *The Important Timber Trees of the United States: A Manual of Practical Forestry*. Houghton Mifflin Company, Boston, MA. 382 p.

Flint, J. 1904. *Early Western Travels 1748–1846: A Series of Annotated Reprints of Some of the Best and Rarest Contemporary Volumes of Travel, Descriptive of the Aborigines and Social and Economic Conditions in the Middle and Far West, During the Period of Early American Settlement*. Arthur H. Clark Company, Cleveland, OH. 333 p.

Friesner, R.C., and J.E. Potzger. 1932. Studies in forest ecology 1. Factors concerned in hemlock reproduction in Indiana. *Butler University Botanical Studies* 2: 133–144.

Grissino-Mayer, H. D., and H.C. Fritts. 1997. The International Tree-Ring Data Bank: an enhanced global database serving the global scientific community. *The Holocene* 7: 235–238.

Gordon, R.B. 1969. The natural vegetation of Ohio in pioneer days. *Bulletin of the Ohio Biological Survey* 3(2): 113. The Ohio State University, Columbus, OH.

Haneca, K., K. Čufar, and H. Beeckman. 2009. Oaks, tree-rings and wooden cultural heritage: A review of the main characteristics and applications of oak dendrochronology in Europe. *Journal of Archaeological Science* 36: 1–11.

Haneca, K., T. Wazny, J. Van Acker, and H. Beeckman. 2005. Provenancing Baltic timber from art historical objects: Success and limitations. *Journal of Archaeological Science* 32: 261–271.
Hill, D.B., and J.R. Shelly. 1985. *Yellow-Poplar*. FOR-2. Department of Forestry, Department of Forestry, College of Agriculture Cooperative Extension Service, University of Kentucky. 4 p.
Homoya, M.A., D.B Abrell, J.R. Aldrich, and T.W. Post. 1985. The natural regions of Indiana. *Proceedings of the Indiana Academy of Science* 94: 245–268.
Hutslar, D.A. 1986. *The Architecture of Migration: Log Construction in the Ohio Country, 1750–1850*. Ohio University Press, Athens, OH. 558 p.
Jackson, M.T. 2004. *101 Trees of Indiana*. Indiana University Press, Bloomington, IN. 364 p.
Kniffen, F. 1965. Folk housing: Key to diffusion. *Annals of the Association of American Geographers* 55: 549–577.
Latta, W.C. 1938. *Outline History of Indiana Agriculture*. Alpha Lambda Chapter of Epsilon Sigma Phi/Perdue University/Indiana County Agricultural Agents Association, Lafayette, IN. 372 p.
Lindsey, A.A., D.V. Schmelz, and S.A. Nichols. 1969. *Natural Areas in Indiana and Their Preservation*. Indiana Natural Areas Survey, Purdue University, Lafayette, IN. 594 p.
Lippincott, I. 1973. *A History of Manufactures in the Ohio Valley to the Year 1860*. Arno Press, New York, NY.
Martin-Benito, D., N. Pederson, M. McDonald, P. Krusic, J.M. Fernandez, B. Buckley, K.J. Anchukaitis, R. D'Arrigo, L. Andreu-Hayles, and E. Cook. 2014. Dendrochronological dating of the World Trade Center ship, Lower Manhattan, New York City. *Tree-Ring Research* 70: 65–77.
McDonough, J.L. 1883. *Combined History of Randolph, Monroe, and Perry Counties, Illinois*. J.L. McDonough and Co., Philadelphia, PA. 512 p.
Miller, F.T. 1965. A general history of forestry in Indiana. Pp. 1–13. In: *Forestry and Conservation in Indiana: In Commemoration of Fifty Years of Forestry Education at Purdue University and the Purdue University Centennial*. (W.C. Bramble, ed.). Purdue University, West Lafayette, IN.
National Centers for Environmental Information. 2018. *International Tree Ring Data Bank*. National Oceanic and Atmospheric Administration, U.S. Department of Commerce. www.ncdc.noaa.gov/data-access/paleoclimatology-data/datasets/tree-ring. Accessed 11 September 2018.
Panshin, A.J., and C. de Zeeuw. 1980. *Textbook of Wood Technology*, vol. 1. 3rd ed. McGraw-Hill Book Company. New York, NY. 705 p.
Potzger, J.E., and R.C. Friesner. 1934. Some comparisons between virgin forest and adjacent areas of secondary succession. *Butler University Botanical Studies* 3: 85–98.
Potzger, J.E., M.E. Potzger, and J. McCormick. 1956. The forest primeval of Indiana as recorded in the original U.S. land surveys and an evaluation of previous interpretations of Indiana vegetation. *Butler University Botanical Studies* 13: 95–111.
Record, S.J. 1908. *Suggestions to Woodlot Owners in the Ohio Valley Region*. Circular 138. U.S. Department of Agriculture, Forest Service. Washington, DC. 15 p.
Stanton, G. 1985. *Brought, Borrowed, or Bought: Sources and Utilization Patterns of the Material Culture of German Immigrants to Southeastern Indiana, 1833–1860*. PhD Dissertation. Indiana University, Bloomington, IN.
Steer, H.B. 1948. *Lumber Production in the United States 1799–1946*. United States Department of Agriculture Miscellaneous Publication No. 669, Washington, DC. 233 p.
Stotz, C.M. 1966. *The Architectural Heritage of Early Western Pennsylvania*. University of Pittsburgh Press, Pittsburgh, PA. 293 p.
Taormina, R., and J.H. Speer. 2016. Forgotten waterways: Analyzing beams from the Wabash and Erie Canal. *Tree-Ring Research* 72: 78–90.
Trouet, V., M. Domínguez-Delmás, C. Pearson, N. Pederson, and D. Rubino. 2017. "Dendro-archeo-ecology" in North America and Europe: Re-purposing historical materials to study ancient human-environment interactions. Pp. 365–394. In: *Dendroecology: Tree-Ring Analyses Applied to Ecological Studies*. (M. Amoroso, L. Daniels, P. Baker, and J.J. Camarero, eds.). Springer, Cham, Switzerland.
United States Department of Agriculture (USDA). 2007. *The Encyclopedia of Wood*. Skyhorse Publishing Inc., New York, NY. 496 p.
Wigginton, B.E. 1972. *The Foxfire Book*. Anchor Books, Garden City, New York, NY. 384 p.
Woehrmann, P. 1971. *At the Headwaters of the Maumee: A History of the Forts of Fort Wayne*. Indiana Historical Society, Indianapolis, IN. 306 p.

Chapter 14

Active inquiry

Dating a 19th-century log house using historical documents and tree-ring science

14.1 Introduction

The preservation and interpretation of historic structures and landscapes is often dependent on the accurate identification of significant historic dates. Historians typically use fieldwork and archival sources to document sites (e.g., photographs, newspapers, legal documents, maps, atlases, journals, and diaries). However, rarely do archival documents explicitly identify precise construction dates for historic sites, leaving historians to infer dates based on available archival evidence. Dendroarchaeology is the use of tree rings to assign construction dates to historic structures. The results of tree-ring analysis, paired with archival resources, provide managers of historic sites with the most accurate information for making preservation and interpretive decisions.

We find great satisfaction in dating historically constructed buildings. Determining the precise calendar date in which a timber's tree rings are formed and performing the detective work necessary to track down reliable historic documents can be time-consuming and frustrating, but determining the date of a building's or a landscape's construction is, as we repeatedly find, worth the effort. Determining a construction date informs proper interpretation of a building. It also generates chronologies or databases of accurately dated tree rings that can be used to better understand how forests grew in the past and how humans have historically interacted with the environment to create the material culture of a region (Chapter 13).

Our goal for this chapter is to provide the reader with historical documentation and tree-ring samples so that you can independently determine the construction date of the Daniel L. Harmon House, a 19th-century log house from southern Indiana (Figure 14.1). In this chapter we provide the reader with the all of the information (historical and dendrochronological) necessary to date the building. We opened this text with a simple question posited to us by many property managers and owners: "Can you date my building?" In this chapter, we ask you, the reader: "Can you date my building?"

When designing this chapter's exercises, we wanted to provide the reader with an authentic tree-ring dating experience. Therefore, it includes the use of actual historical documents (obituaries, maps, census data, and family histories). Performing dendrochronological analysis of timber samples requires expensive, specialized equipment that few, if any, of our readers have at their disposal. Therefore, we have

Figure 14.1
Photo of the south façade of the Daniel L. Harmon log house. The log structure is covered with clapboards.
Source: Jeff Harmon.

created the tree-ring portion of these exercises so that they can be performed with the simplest of materials: pencils, erasers, transparent tape, and scissors.

"Tree-ring samples" were created by taking the actual samples from the structure and plotting their ring widths so that they can be printed and examined without the use of a microscope (Figure 14.2). In these tree-ring samples, individual rings are delimited by horizontal lines, a tree-ring width is simply the distance between two successive vertical lines. In the actual tree-ring samples obtained from the structure, the mean ring width was a mere 1.13 mm, and many tree rings were less than a half millimeter in width – a scale that unquestionably necessitates the use of a microscope to identify individual growth rings. Therefore, we scaled the size of each ring so that a sample could be printed on a standard piece of paper. While attempting to date the Daniel L. Harmon House, you will not be working with the true ring widths found in the wood, but with scaled ring widths. The

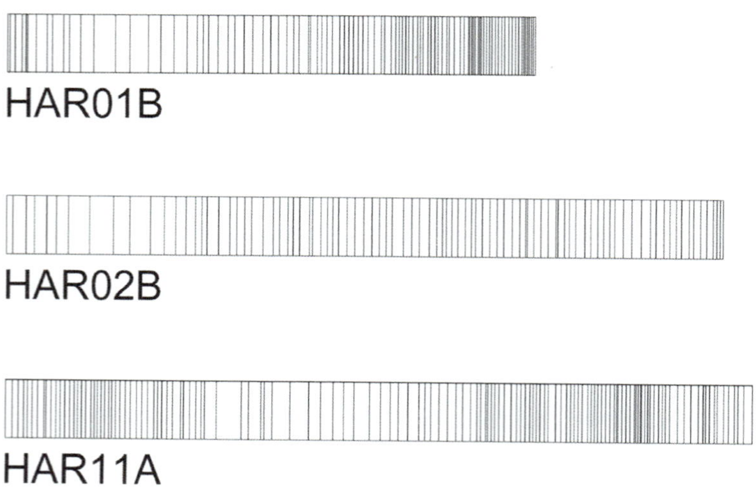

Figure 14.2
Tree-ring samples collected from the Daniel L. Harmon House. These samples were created by using the actual measurements of the tree-ring widths in the timbers sampled from the building.

patterns of ring sizes are maintained in each sample; small rings will still be small compared to neighboring rings, and the patterns of growth will be maintained so that accurate crossdating is possible.

14.2 Historical documentation

As was noted in Chapter 6 (and in the case studies of Chapters 8 through 13), a myriad of primary and secondary sources of historical information can be used to interpret historically constructed buildings and landscapes. Combining these sources with tree-ring analysis of timbers provides a potent combination of information that can be useful for understanding a building's or landscape's history. Combing these diverse pieces of information allows a wide variety of researchers (historians, preservationists, dendrochronologists, ecologists) to ask and answer many unique questions.

Our search for historical documentation concerning the Daniel L. Harmon House resulted in the following:

- The US Population Census (1850 through 1880) reports that Daniel L. Harmon was born about 1828 and lived in Oregon Township, Clark County, Indiana (Figure 14.3).
- Daniel passes in 1881.
- Daniel was married to Martha Daily in 1851, and then to Rachel Carolyn Bower in 1865 (following Martha's passing in 1863).
- Family oral history says that Daniel's father, William, died in November 1854 in the Ohio River at Shawneetown, Illinois, supposedly when robbed and thrown overboard returning home from St. Louis on a steamboat.
- Daniel was drafted into service in 1864 (Company C, 59th Regiment, Indiana Infantry) and was discharged in 1865.
- A plat map published in 1875 shows Harmon's two parcels of land totaling 152 acres (Figure 14.4).
- A 1948 newspaper article reported that heavy winds caused damage to the house, then owned by Milton Harmon, Daniel's son. The article reported the "loss of some of the roof on his house and a cedar tree in the corner of his yard which had been there since 1865" (Charlestown Courier 1948).

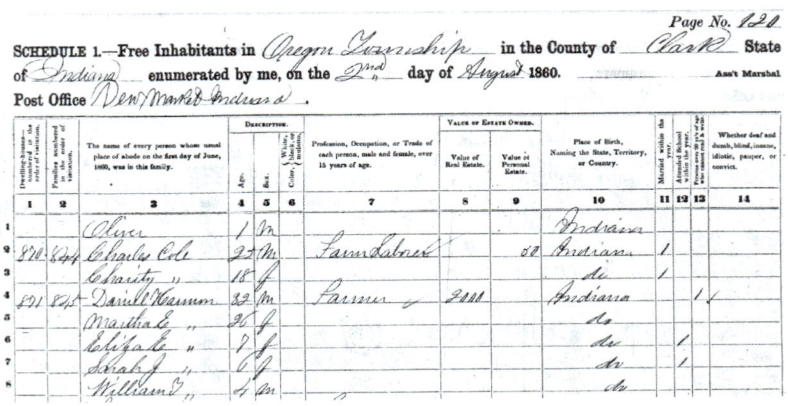

Page No. 120

SCHEDULE 1.—Free Inhabitants in Oregon Township in the County of Clark State of Indiana enumerated by me, on the 2nd day of August 1860. Ass't Marshal

Post Office New Market Indiana

Figure 14.3
US Census data (1860) for the Daniel and Martha Harmon family (Dwelling 871). In 1860, Daniel was 32 years old, is employed as a farmer, and lists Indiana as his place of birth. Also living in the household are his wife Martha, three daughters, and a son.

Source: US Census Bureau.

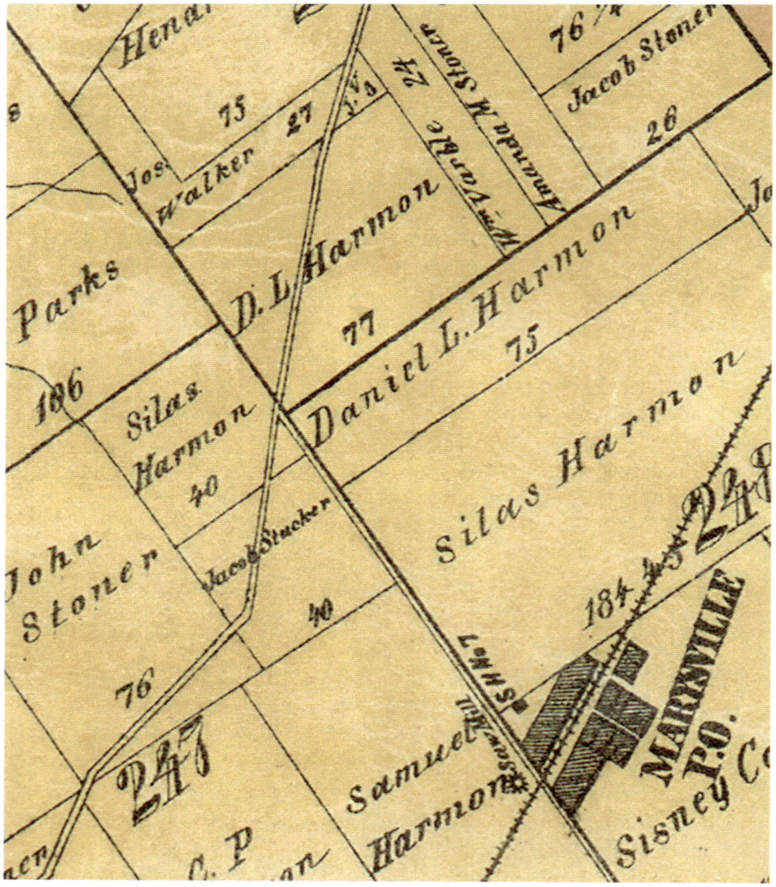

Figure 14.4
Map of Clark County, Indiana, (1875) showing property owned by Daniel L. Harmon (Davis 1875).

Source: Library of Congress, Geography and Map Division.

- The 1958 obituary of Milton Harmon, Daniel's son, included the statement that he was the "only son of Daniel L. and Rachel E. Harmon, pioneer residents of this community, who settled here in 1865 – when the land was all a track of timber and Indian huts" (Charlestown Courier 1958).
- Oral history provided by the previous owner in 2008 states that Daniel Harmon's great-granddaughter described the house being constructed in 1854.

14.3 Tree-ring samples

The Daniel L. Harmon House is a two-room, one and a half story log house located near Marysville, Indiana, in northern Clark County. The house is 21 ft × 30 ft in size, has two rooms on each story, and was constructed using half-dovetail corner joinery, a common construction method in southern Indiana (Figure 14.1; McAlester and McAlester 1988; Roberts 1996). The samples used to date the Daniel L. Harmon House were obtained while the residence was being renovated; we had access to nearly all of the structural timbers of the house (Figure 14.5). Cores were taken from each elevation and from both stories of the building so that

Figure 14.5
Interior photo of the Daniel L. Harmon log house showing sample locations. Left: HAR01B (west wall) and HAR02B (north wall). Right: HAR11A (interior wall separating the two rooms on the lower story).

Source: Authors.

we were likely to date original timbers from the building and not replacements. Sampling was concentrated on wane- and bark-bearing timbers. Cross sections were obtained from timbers that had deteriorated and were going to be replaced in the subsequent renovation. Cross sections were also obtained (using a chain saw) from an internal wall on the first floor, which was going to be shortened in the renovation to accommodate the creation of a larger passageway between the two rooms. The ground level floor joists and staircase were removed, and the timbers were stored on site. We searched through them for wane- and bark-bearing timbers. Using a circular saw, we sectioned timbers with wane or bark. During sampling, we photographed each sample and recorded its provenience. The footprint of the house was measured and sketched (Figure 14.6).

We collected tree-ring samples from 26 timbers (tulip poplar, white oak, and sweet gum) and followed standard procedures for surface preparation (mounting and sanding; Chapter 5). Five of the tulip poplar samples collected from the house are provided to you for subsequent analysis and dating (Table 14.1 and Figure 14.7). All samples have bark present. Therefore, the year of harvest (death date) can be determined for each timber if crossdating is successful.

14.4 Crossdating and date assignment

Determining the harvest date of each of the timbers is a multistep process. If a similar harvest or cutting date is found for all of the timbers in a building, one can assume the construction date of the building since timbers were used green, or unseasoned. First, years (not dates) need to be assigned to the individual tree rings in each sample. Next, skeleton plots must be constructed. Skeleton plots are used to graphically highlight unique growth patterns in each of the timbers. Once skeleton plots have been created, individual samples can be crossdated against each other. Crossdating (Chapters 4 and 5) entails matching growth patterns among individual tree-ring samples. At this stage of analysis, the timbers are considered floating since calendar dates will still not have been assigned to individual tree rings.

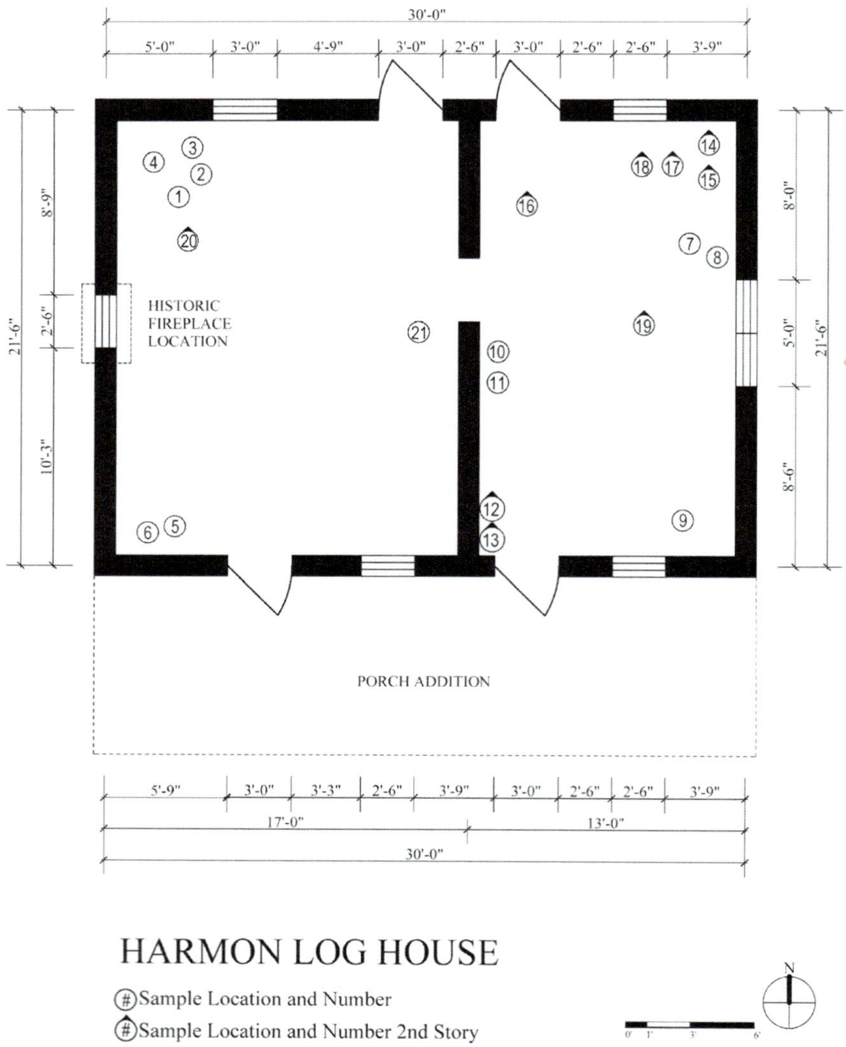

Figure 14.6
Illustration of the plan and sample locations for the Daniel L. Harmon Log House. The sample locations are the numbers within the circles.

Source: Authors.

Table 14.1 Sample location, provenience, and collection method for the Daniel L. Harmon log house tree-ring analysis exercise. All of the samples are tulip poplar. See Figure 14.7.

Sample	*Provenience*	*Collection technique*
HAR01B	Wall log; west elevation	Core
HAR02B	Wall log; north elevation	Core
HAR11A	Internal wall separating pens on ground floor	Chain saw
HAR20A	Framework of staircase	Circular saw
HAR23A	Floor joist first floor	Circular saw

Figure 14.7
The prepared samples (cores and cross sections) from the Daniel L. Harmon log house. The samples have been sanded and are ready for tree-ring analysis.
Source: Authors.

To anchor the samples in time and assign dates to individual tree rings, the undated, floating series must be compared to a regional chronology. A chronology is an accurately dated set of tree rings that has been collected from other buildings or from living trees (Chapter 4). We have provided you with the master regional tulip poplar tree-ring chronology that we have created by sampling many tulip poplar timbers from other historically constructed buildings in the region. Cross-dating the floating series with the master chronology is performed by matching the growth patterns of the skeleton plots of the undated timbers with the skeleton plots of the master chronology. Once a match is obtained, you can transfer calendar dates from the regional chronology to the floating chronology.

The preceding (much abbreviated) explanation of dating timbers of unknown age is provided to remind the reader of how tree-ring samples are ultimately dated. We strongly suggest revisiting Chapter 5 for a full discussion of the various steps needed to accurately date tree-ring samples of unknown age. Additionally, online resources for this text offer an in-depth, step-by-step explanation of how to date the five tree-ring samples we have prepared for you. At the site you will also find downloadable files containing the individual tree-ring samples (Figure 14.2), graph paper for skeleton plotting, and the regional master tulip poplar chronology skeleton plot.

We originally designed the content of this chapter for college and graduate-level students focused on preservation since dendrochronology is typically taught in the sciences (botany, ecology, geography, etc.) and not in courses, labs, and studios in which these students normally enroll. Tree-ring science is a powerful tool that can play an important role in preservation, and we felt it important to make the subject matter available to a wider audience. While this lesson plan is for programs of higher education, we have successfully presented tree-ring analysis at the K–12 level (Rubino and McCarthy 2002; Rubino and Hanson 2009). We believe tree-ring analysis provides an excellent approach to bridging the sciences (life and earth sciences) and the humanities (history). A future goal of our work is to create lesson plans that meet Next Generation Science Standards (NGSS 2018) dimensions for learning science (Concepts, Practices, and Core Ideas).

Instructors wishing to use this exercise will find class handouts along with detailed instructions for incorporating the exercises into their classes at the text's website. We have also prepared a lecture highlighting the process of dating timbers (along with a primer on botany) and a highly detailed script for the lecture. Goals, teaching objectives, and tips for creating assignments are also provided.

14.5 Conclusion: can you date my building?

After reading this text and using online tools, the reader will be able to apply the basics of tree-ring science and interpretation of historic documentation to independently determine a construction date for the Daniel L. Harmon House. The exercises have been classroom tested with graduate and undergraduate students enrolled in a historic preservation program (Baas et al. 2018). Assessment informed us that students found the exercise interesting and that they enjoyed using scientific inquiry to investigate preservation questions.

14.6 Literature cited

Baas, C., D.L. Rubino, and J. Spodek. 2018. *Teaching Dendrochronology to Preservation Students: Linking Historic Sources to Scientific Inquiry*. Paper presentation, Indiana Academy of Science Annual Meeting, Indianapolis.

Davis, G.W. 1875. *Map of Clark County, Indiana*. Library of Congress, Geography and Map Division, Philadelphia, Braden & Burford. www.loc.gov/item/2013593193/. Accessed 25 June 2018.

Charlestown Courier. 1948. *Milton Harmon House*. April 15.

Charlestown Courier. 1958. *In Memoriam*. December 11.

McAlester, V., and L. McAlester. 1988. *A Field Guide to American Houses*. Alfred A. Knopf, New York, NY. 525 p.

NGSS. 2018. *Next Generation Science Standards*. www.nextgenscience.org/. Accessed 23 June 2018.

Roberts, W.E. 1996. *The Log Buildings of Southern Indiana*. Trickster Press, Bloomington, IN. 228 p.

Rubino, D.L., and B.C. McCarthy. 2002. Dendropedagogy: Teaching botany, ecology, and statistical principles through tree-ring studies. *The American Biology Teacher* 64: 689–694.

Rubino, D.L., and D. Hanson. 2009. CSI for trees: Dendrochronology gets upper elementary students thinking like scientists. *Science and Children* 47: 21–25.

United States Census Bureau. 1860. *Oregon Township*, Clark County, IN.

Index

Note: Page numbers in *italics* indicate figures, in **bold** indicate tables.